The Oceans: Key Issues in Marine Affairs

The GeoJournal Library

Volume 78

Managing Editor: Max Barlow, Concordia University,
   Montreal, Canada

Founding Series Editor:
   Wolf Tietze, Helmstedt, Germany

Editorial Board: Paul Claval, France
   Yehuda Gradus, Israel
   Risto Laulajainen, Sweden
   Sam Ock Park, South Korea
   Herman van der Wusten, The Netherlands

The titles published in this series are listed at the end of this volume.

The Oceans: Key Issues in Marine Affairs

edited by

HANCE D. SMITH

*University of Cardiff,
U.K.*

KLUWER ACADEMIC PUBLISHERS
DORDRECHT / BOSTON / LONDON

A C.I.P. Catalogue record for this book is available from the Library of Congress

ISBN 1-4020-2746-X (HB)
ISBN 1-4020-2780-X (e-book)

Published by Kluwer Academic Publishers,
P.O. Box 17, 3300 AA Dordrecht, The Netherlands.

Sold and distributed in North, Central and South America
by Kluwer Academic Publishers,
101 Philip Drive, Norwell, MA 02061, U.S.A.

In all other countries, sold and distributed
by Kluwer Academic Publishers,
P.O. Box 322, 3300 AH Dordrecht, The Netherlands.

Printed on acid-free paper

Funded by European Union (ERBIC 18CT 970152)

All Rights Reserved
© 2004 Kluwer Academic Publishers
No part of this work may be reproduced, stored in a retrieval system, or transmitted in any form or by any means, electronic, mechanical, photocopying, microfilming, recording or otherwise, without written permission from the Publisher, with the exception of any material supplied specifically for the purpose of being entered and executed on a computer system, for exclusive use by the purchaser of the work.

Printed in the Netherlands.

CONTENTS

ACKNOWLEDGEMENTS xiii
AFFILIATIONS xv
INTRODUCTION 1
HANCE D. SMITH 1
   1. Introduction 1
   2. The global setting 1
   3. The uses of the sea 2
   4. Regional geography 3
CHAPTER 1. THE GEOGRAPHY OF THE SEA 5
HANCE D. SMITH 5
   1. Introduction 5
   2. Marine Geography 5
     2.1. EVOLUTION AND STRUCTURE 5
     2.2. PARADIGMS 6
     2.3. SEA, LAND AND THE HUMAN EFFORT 8
   3. From Traditional to Urban Industrial to Post Industrial 8
     3.1. THE TRADITIONAL SOCIETY 8
     3.2. THE URBAN INDUSTRIAL SOCIETY 11
     3.3. THE NATURE OF THE MILLENNIUM 14
   4. Marine Geography, Ocean and Coastal Management: Setting a Course ... 16
   5. Conclusion: A Research and Teaching Agenda 17
   References 18
CHAPTER 2. THE LAW OF THE SEA AT THE MILLENNIUM: 25
E.D. BROWN 25
   1. UNCLOS at the Millennium 25
     1.1. THE STATUS OF UNCLOS AT THE MILLENNIUM 26
     1.2. THE ESTABLISHMENT OF UNCLOS INSTITUTIONS 29
        1.2.1. Seabed Institutions 29
        1.2.2. Commission on the Limits of the Continental Shelf 31
        1.2.3. Meetings of States Parties 33
        1.2.4. United Nations Open-ended Informal Consultative Process on Oceans and the Law of the Sea 34
        1.2.5. Dispute Settlement Institutions 35
   2. Settlement of Disputes under UNCLOS 38
     2.1. THE DISPUTE SETTLEMENT SCHEME IN OUTLINE 38
        2.1.1. General Provisions 38
        2.1.2. Compulsory Procedures Entailing Binding Decisions 39
        2.1.3. Limitations and Exceptions to Applicability of Section 2 39
        2.1.4. Sea-bed Mining Disputes 42
     2.2. AN OVERCOMPLEX SCHEME? 43
        2.2.1. The Limited Scope of Compulsory Procedures Entailing Binding Decisions 43

CONTENTS

    2.2.2. Will the System Achieve the Definitive Settlement of Disputes and the Development of a Uniform and Consistent Jurisprudence? 43
   2.3. A TENTATIVE EVALUATION 46
    2.3.1. Declarations on Choice of Dispute Settlement Mechanisms 46
    2.3.2. State Practice in Resolving Disputes in UNCLOS Era 49
    2.3.3. Tentative Conclusions 61

CHAPTER 3. MARITIME BOUNDARIES 63
G.H. BLAKE 63
  1. Introduction 63
  2. The 1982 UN Convention on the Law of the Sea (UNCLOS) 64
   2.1. BREADTH OF THE TERRITORIAL SEA 64
   2.2. STRAITS USED FOR NAVIGATION 65
   2.3. THE EXCLUSIVE ECONOMIC ZONE (EEZ) 65
   2.4. DEFINITION OF THE CONTINENTAL SHELF (CS) 66
   2.5. THE INTERNATIONAL SEABED 67
  3. Boundary Delimitation 68
   3.1. DELIMITATION PRINCIPLES 68
   3.2. COMMON MARITIME ZONES 69
  4. Unfinished Business 69
   4.1. UNRESOLVED DISPUTES 70
   4.2. EXCESSIVE CLAIMS 73
   4.3. LIMITS OF THE CONTINENTAL SHELF BEYOND 200 NM 73
  5. Conclusion 74
  References 76

CHAPTER 4. GEOGRAPHY AND GEO-STRATEGY OF THE OCEANS 77
A. VIGARIE 77
  1. Theoretical Foundations of the Geo-Strategy of the Oceans 78
   1.1. ALL MODERN COUNTRIES NEED EXTERNAL RELATIONS 78
   1.2. THIS LEADS TO A GEO-POLICY TO SATISFY NEEDS 79
  2. General Methodological Approach of Research into the Geo-Strategy of the Oceans 81
   2.1. NATIONAL MARITIMISATION FACTORS AT THE BEGINNING OF THE 21ST CENTURY 81
   2.2. THE MEANS: THE CAPACITY FOR INTERVENTION IN THE OCEANIC SPHERE 85
  3. Research Perspectives and Directions for the Future 87
   3.1. THE DIVERSITY OF THEMATIC AREAS 87
   3.2. A DYNAMIC VISION OF THE WORLD OCEAN 88
  4. Conclusions 88
  References and Notes 89

CHAPTER 5. MARITIME TRANSPORT 91
J. MARCADON 91
  1. The Recent Evolution of the World Fleet and the Revolution in Sea Transport. ... 91
   1.1. RECENT EVOLUTION ACCORDING TO FLAG. 91
   1.2. SHIP TECHNOLOGY, SIZE AND SPECIALISATION 94

    1.2.1. The supremacy of the diesel engine and the arrival of fast ships..............94
    1.2.2. Size and Specialisation in the last 40 Years 95
    1.2.3. Company Mergers 95
  2. Fleet Developments and Organisation...97
    2.1. BULK FLEETS AND INCREASED SAFETY AWARENESS 97
      2.1.1. Ship Owners and Operators...97
      2.1.2. Fleet Diversity in Liquid Bulk 98
      2.1.3. Environment and Safety Measures. 99
    2.2. FLEETS ON REGULAR ROUTES..100
      2.2.1. The Conventional Freight Sector..100
      2.2.2. The Container Sector 101
    2.3. MARITIME PASSENGER TRANSPORT 102
  3. Conclusion...103
CHAPTER 6. OFFSHORE OIL AND GAS AT THE MILLENNIUM......................105
H. J. PICKERING 105
  1. Introduction 105
  2. Market and Price Trends 105
  3. Contribution of Offshore Production..110
  4. Offshore Exploration and Development...113
  5. Reserves 119
  6. Opportunities for the Millennium...125
    6.1. TECHNOLOGICAL DEVELOPMENTS 126
    6.2. STATE OWNERSHIP TO PRIVATE INVESTMENT 127
    6.3. CORPORATE RESTRUCTURING 130
  7. Challenges for the Millennium 131
    7.1. CIVIL UNREST AND POLITICAL UNCERTAINTY...............................131
    7.2. BORDER DISPUTES 132
    7.3. ENVIRONMENTAL IMPACT MITIGATION..133
  8. Conclusions 136
  References 137
CHAPTER 7. FISHERIES 145
J.L.SUÁREZ, G.GONZÁLEZ, S. FERIA...145
  1. The State of Resources. The Generalisation of Overfishing...............................145
    1.1. THE EVOLUTION OF CATCHES THROUGHOUT HISTORY 145
    1.2. GEOGRAPHICAL DISTRIBUTION OF FISHING POTENTIAL..............148
  2. New Fishing Powers: The Boom in the Market, A Slump for Fishers. 149
    2.1. FISHING IN THE POST-INDUSTRIAL ERA...150
    2.2. SUPPLY AND DEMAND OF FISH PRODUCTS.......................................152
  3. Fishing and the Governance of the Oceans 154
    3.1. THE CRISIS OF THE BIO-ECONOMIC MODEL AND THE
    NATIONALIZATION OF RESOURCES 154
    3.2. THE HIGH SEAS: AN INTERNATIONAL CHALLENGE........................156
    3.3. NEW APPROACHES TO FISHERIES MANAGEMENT:
    DEREGULATION, DECENTRALIZATION AND PRIVATIZATION 157
  References 159

CHAPTER 8. THE USE OF THE SEA FOR RECREATION AND TOURISM 161
M.B. ORAMS 161
1. Introduction 161
2. The Importance of the Sea 162
3. The Importance of Marine Settings for Recreation and Tourism 163
   3.1 THE RAPID GROWTH OF MARINE RECREATION AND TOURISM 164
   3.2 THE DISTRIBUTION OF MARINE TOURISM ACTIVITIES 167
4. Problems and Challenges for the 21st Century 169
   4.1 POTENTIAL SOLUTIONS AND AREAS FOR RESEARCH 171
References 172

CHAPTER 9. MANAGING MARINE WASTE DISPOSAL 175
R.C. BALLINGER 175
1. Introduction 175
2. The Waste Issue 176
   2.1 WASTE CHARACTERISATION 176
   2.2 GEOGRAPHY OF MARINE WASTE 177
3. The Context for Marine Waste Management: General Issues and Trends 178
   3.1. EARLY TRENDS 178
   3.2. AWARENESS AND INSTITUTIONALISATION OF THE
   ENVIRONMENT 178
   3.3. GENERAL WASTE AND POLLUTION CONTROL TRENDS 179
      3.3.1. Waste reduction and minimisation 179
      3.3.2. Waste recycling, recovery and reuse 180
      3.3.3. Waste treatment and disposal 180
      3.3.4 The solution to pollution is dilution 181
      3.3.5 Points of entry to environment 181
   3.4. INSTITUTIONALISATION OF WASTE MANAGEMENT 182
      3.4.1. Global and regional institutions 182
      3.4.2 Sectoral institutionalisation of waste management 182
   3.5. THE PRECAUTIONARY APPROACH AND MARINE WASTE
   DISPOSAL 183
4. Discussion: Current, Emerging and Future Marine Waste Issues 184
   4.1. CURRENT ISSUES 184
      4.1.1. Pollutant sources 184
      4.1.2. Economic issues 187
      4.1.3. Public awareness issues 187
   4.2. EMERGING AND FUTURE ISSUES 188
      4.2.1 Emerging issues 188
      4.2.2 Future issues 189
5. Conclusions: Final Comments 190
References 191

CHAPTER 10. MARINE CONSERVATION AND RESOURCE MANAGEMENT 199
S. PULLEN 199
1. Introduction 199
2. Out of Sight - Out of Mind 201

3. Marine Resource Management - Sustainable Development!..................203
   3.1. FISHERIES 203
   3.2. COASTAL RESOURCES – CONSERVING COASTLINES.....................205
   3.3. OFFSHORE RESOURCES...205
4. Marine Nature Conservation - Protecting Marine Wildlife and Habitats 206
   4.1. PROTECTING BIODIVERSITY 206
   4.2. MARINE SITE PROTECTION 207
5. Marine Pollution – Prevention and Cure!..208
   5.1. LAND BASED MARINE POLLUTION 208
   5.2. SHIPPING – REDUCING THE RISK FROM OFFSHORE ACTIVITIES 209
6. Awareness, Human Resources and Partnerships 210
References 213

CHAPTER 11. THE NORTH ATLANTIC 215
LEWIS M. ALEXANDER..215
1. Geographic Site..216
   1.1. BATHYMETRY 216
   1.2. OCEAN CIRCULATION 217
   1.3. LIVING MARINE RESOURCES..217
   1.4. OFFSHORE OIL AND GAS 218
2. Historical Background..218
3. The 18th and 19th Centuries..220
4. The 20th Century..220
5. Managing the North Atlantic Ecosystem...222
6. Living Marine Resources..222
7. Northeast Atlantic...224
8. Northwest Atlantic 224
9. The High Seas 225
10. Mariculture...226
11. Marine Pollution...226
12. Commercial Shipping...227
13. Other Activities..228
14. Potential Future Trends in the North Atlantic 228
References and Notes 229

CHAPTER 12. THE MEDITERRANEAN AND BLACK SEAS 231
A. VALLEGA..231
1. The Large Mediterranean Marine Region 231
2. Articulated Physical Contexts 232
3. Variability in Ecological Conditions 234
4. The Changing Geopolitical Context..235
5. Sea Use Development 240
6. Regional Co-operation..244
References 250

CHAPTER 13. MULTILATERAL MANAGEMENT OF NORTHEAST ASIAN SEAS: PROBLEMS AND PROGNOSIS 253
MARK J. VALENCIA..253

1. Introduction 253
2. Ongoing and Potential Cooperation: Enhancing the Epistemic Community.......255
   2.1. SAFETY AT SEA 255
      2.1.1. Law and Order at Sea 256
   2.2 FISHERIES 257
      2.2.1. Existing International Regimes and Their Inadequacies........................257
China and Japan 257
   2.3. MARINE ENVIRONMENTAL PROTECTION 264
      2.3.1. Overview 264
      2.3.2. Initiatives for Marine Environmental Protection and Their Deficiencies
     266
      2.3.3 Inadequacies of Existing Regimes and Ways Forward.........................268
   2.4. MARINE SCIENTIFIC RESEARCH 270
   2.5 NAVAL COOPERATION..271
   2.6 THE NEXT STEP: A REGIME REGULATING THE USE OF FORCE AND MILITARY AND INTELLIGENCE GATHERING ACTIVITIES IN THE EEZ
  273
      2.6.1 The Way Forward...274
3. Conclusions 275
References and Notes 278

CHAPTER 14. THE ARCTIC OCEAN...283
Yu.G. MIKHAYLICHENKO 283
1. Introduction 283
2. Resources 283
   2.1. MINERAL RAW MATERIAL RESOURCES 283
      2.1.1. Hydrocarbon raw materials 283
      2.1.2. Minerals..284
   2.2. SPATIAL RESOURCE..284
   2.3. TRANSPORTATION RESOURCE...285
      2.3.1. Present status and close prospects 285
      2.3.2. Potential resource 287
   2.4. BIORESOURCES 288
   2.5. TOURISM 290
3. Environmental Status 290
4. Ethnic Problems 291
5. International Environmental and Scientific Co-operation 292
   5.1. INFRASTRUCTURE...292
   5.2. COOPERATION IN GLOBAL CHANGE STUDIES..................................292
6. Perspective 294
   6.1. ECOSYSTEM RESOURCE...294
   6.2. THE RUSSIAN FACTOR..295
7. Conclusion..295
Main Sources..296

CHAPTER 15. THE SOUTHERN OCEAN 297
DONALD R. ROTHWELL..297

1. The Southern Ocean Area 297
   1.1. THE SOUTHERN OCEAN ENVIRONMENT 298
   1.2. SOUTHERN OCEAN MARINE LIVING RESOURCES...........................300
2. The Southern Ocean and the Antarctic Treaty Regime 300
3. Southern Ocean Maritime Zones...301
   3.1. THE NEED FOR A COASTAL STATE 301
   3.2. TERRITORIAL SEA 301
   3.3. CONTINENTAL SHELF...302
   3.4. EXCLUSIVE ECONOMIC ZONE..302
   3.5. THE SOUTHERN OCEAN DEEP SEA-BED...303
4. Antarctic Environmental Protection: The 1991 Madrid Protocol........................304
5. Southern Ocean Resource Management...304
   5.1. CCAMLR..304
   5.2. ANTARCTIC SEALING...306
   5.3. SOUTHERN OCEAN MINING..307
6. Other International Regimes operating in the Southern Ocean 308
7. Conclusion..309
References..309
INDEX 313

ACKNOWLEDGEMENTS

Thanks are due first of all to the authors, listed under Affiliations for their hard work and patience required to co-ordinate a substantial number of contributions over an extended period.

Secondly, the editor is indebted to Mrs Sylvie Beresford for translation of Chapters 4 and 5; and to Dr Stephen Pettit of the Marine and Coastal Environment Group at Cardiff University for preparation of the camera-ready copy.

Finally, thanks are also due to the International Geographical Union for financial support for publication of this book; and to the Commission on Marine Geography of the IGU for the general support of its members, several of whom have contributed to this volume.

Hance D Smith
Editor

AFFILIATIONS

L. M. ALEXANDER. Department of Marine Affairs at the University of Rhode Island, Kingston, Rhode Island, USA.

R. C. BALLINGER. School of Earth, Ocean and Planetary Sciences, Cardiff University, Cardiff, United Kingdom.

G.H. BLAKE. International Boundaries Research Unit, University of Durham, United Kingdom.

E. D. BROWN. Emeritus Professor of International Law, Cardiff University, Cardiff, United Kingdom.

S. FERIA. Departamento de Geographia Humana, Universidad de Sevilla, Spain.

G. GONZÁLEZ. Departamento de Geographia Humana, Universidad de Sevilla, Spain.

J. MARCADON; University de Nantes, I.G.A.R.U.N., Nantes, France.

YU. G. MIKHAYLICHENKO. Department of Life and Earth Sciences, Ministry of Industry, Science and Technology of the Russian Federation, Moscow, Russia.

M. B. ORAMS. Centre for Tourism Research, Massey University, New Zealand.

H. PICKERING, CEMARE, Portsmouth University, Portsmouth, United Kingdom.

S. PULLEN, Head, Living Seas Programme, WWF-UK, United Kingdom.

D. R. ROTHWELL, Faculty of Law, University of Sydney, Sydney, Australia.

H. D. SMITH. School of Earth, Ocean and Planetary Sciences, Cardiff University, Cardiff, United Kingdom.

J. L. SUÁREZ. Departamento de Geographia Humana, Universidad de Sevilla, Spain.

M. J. VALENCIA, East-West Center, Honolulu, Hawaii, USA.

A. VALLEGA. Dipartimento BOUS, Universita di Genoa, Genoa, Italy.

A. VIGARIE. University de Nantes, Nantes, France.

INTRODUCTION

HANCE D. SMITH

1. Introduction

The millennium has been widely regarded as a time to take stock among other things, of the planet and its relationships with the people who inhabit it. Why this should be so is not always clear, apart from an apparent fixation with counting years. Further, the fact that the counting itself is based on the origins of the Christian religion is, for many, not of primary significance. Even if it is, counting in this way may still not be clear.

And yet there are arguably sound reasons for taking stock at this time. Most obviously since the European 'industrial revolution' pressures on the environment caused by human activities have increased inexorably. At global level, over the past three decades in particular, environmental concerns have undoubtedly become a major political force. In a marine context, the Law of the Sea Convention concluded in 1982 is one manifestation of this. These pressures are in turn linked to technological and economic factors which are inextricably intertwined in any understanding of environmental impacts. Of special note are the implications of this for the exploitation of marine resources, in the periodic 'long waves' of economic expansion and contraction, each lasting for around or just over half a century, coupled with technological advances evident in the fishing and offshore oil industries in particular. Beyond these are cultural, including social and political dimensions, linked in turn to major shifts in perception and thought, perhaps akin in scale to the transition between the European Middle Ages and the modern era some half a millennium ago. All of these factors operate on a variety of time scales. It is in the context of these time scales that the year 2000 should be considered.

2. The global setting

At the global geographical scale the millennium does indeed have significance in various ways. The first main chapter accordingly considers the geography of the sea, summarising a variety of themes, including the current state of sea uses and the intellectual evolution of marine geography. This forms the backdrop to three themes, namely the status of the legal framework governing the oceans; the enclosure by states in the form of maritime boundaries, which stems in part from the provisions of the legal framework; and the economic and political factors which ultimately drive the global marine economy and the associated maritime trade and marine resource development patterns.

The modern law of the sea has developed rapidly over the past half-century or so, in association with the enormous expansion of the global economy. The UNCLOS III negotiations (1973-82) provided a comprehensive convention – in effect an economic treaty. Chapter 2 considers the status of the law of the sea, including participation by states; and the establishment of UNCLOS institutions, notably the International Seabed Authority which will be responsible for the Area beyond state jurisdiction, and the Convention on the Limits of the Continental Shelf which geographically demarcates the area within state jurisdiction relative to the ocean beyond. The arrangements for dispute settlement are also discussed. Since the Convention on the Law of the Sea came into force in 1994, there is now a record of cases to consider.

Chapter 3 on maritime boundaries then takes up the theme of enclosure in detail, considering in turn the several categories of boundaries – some one third of the possible boundaries between states have been settled in the past 50 years. Lists of agreements and disputes are provided, and the need for a shift in approach to management and collaboration is underlined.

The focus then shifts in Chapter 4 to the relationships between geopolitics and international trade, with particular reference to the maritime dimension, as around 80 per cent of international trade by volume goes by sea. Both a theoretical foundation and general methodological approach are outlined, leading to the consideration of regional outcomes and the future.

3. The uses of the sea

The development of sea uses considered here concentrates on five of the eight major sea use groups. Four of these – shipping, offshore hydrocarbons, fisheries and leisure – in their various ways have to do with the development of the world economy and are major industries in that context. Waste disposal is also a major industry in its own right, but largely reflects the economic development of the land, from whence derives around nine-tenths of the waste entering the marine environment. Marine conservation which, although aimed at a measure of control over the interactions between sea uses and the marine environment, may also be regarded as a use, employing thousands of people and often linked through both the natural environment and the maritime heritage to the leisure industries. This book does not deal with the remaining two major use groups: the military uses which are a reflection of the geopolitics already considered in Chapter 4; and marine science, itself an economic activity of considerable and increasing scale.

Chapter 5 takes up the trade theme introduced in Chapter 4, and is focused entirely on the world's merchant shipping fleet, systematically considering its specialisation, the outcome of international commodity trade development over the past half century; ship size; ownership; and environment and safety. Also dealt with are the geographical patterns of shipping routes, ultimately a function of the regional development of the world economy; and short sea and cruise shipping, the latter also being a major component of the maritime leisure group of industries.

By far the most important activity in the field of marine mineral and energy resource development during the past half century has been the exploitation of offshore oil and gas. Chapter 6 begins by contrasting short term price instability with long term demand, before discussing systematically in turn the contribution of offshore activity to the overall global pattern of hydrocarbon exploitation; assessment of reserves; and technology. Corporate restructuring is considered before moving on to other significant factors which influence the development of this industry including political instability, border disputes and environmental considerations.

Chapter 7 begins by reviewing the somewhat parlous state of the world's fish resources, before concentrating on markets for fish products and implications for fleet size and employment. The key role of ocean governance is highlighted, before considering management measures, including enclosure, modelling, deregulation, and the contribution of privatisation.

The marine leisure industries considered in Chapter 8 are in many parts of both the developed and developing worlds the largest and fastest growing maritime sector. This importance is assessed in terms of growth, location and activities, before consideration of overall geographical patterns; problems and challenges.

Chapter 9 on waste begins by characterisation and geography, and treatment. Then institutional aspects are discussed, before looking at the role of the precautionary approach. Major issues are dealt with, including sources, economics, awareness, surveys, and the future.

In Chapter 10, marine conservation is discussed in terms of perception, marine wildlife and marine pollution, often with a regional focus on the United Kingdom. This regional theme permeates much of the patterns of sea uses, and the final section of the book accordingly deals with the regional geography of the oceans in detail.

4. Regional geography

The regional geography of the oceans which emerges from the complex interactions of the physical environment, law, boundaries, development and geopolitics, and sea uses, is as complex as the regional geography of the land. As a first order approximation at the global scale it is useful to distinguish between development regions on the one hand, focused on the global economy and its associated sea uses; and the major environmental regions of the temperate, tropical and polar regions, together with the large marine ecosystems on the other. In the first category this book considers respectively the North Atlantic and Mediterranean/Black Seas as ocean/sea centred approaches; and the NW Pacific as a land focus of ocean development. The remaining three chapters focus on environmental regions: the Mediterranean, the Arctic and Antarctic respectively. The other major set of environmental regions, namely the tropical and sub-tropical oceans are not dealt with here.

Chapter 11 on the North Atlantic adopts in part a classical geographical approach, considering physical geography, resources and historical geography; before moving on to a managerial theme based in part on large marine ecosystems. The next following chapter on the Mediterranean and Black Seas is in part similar in approach, although geopolitical issues naturally receive special mention. Chapter 13 on the NW Pacific highlights the political fragmentation of the region, the need for co-operation and problems which derive from this.

The final two chapters, on the polar regions, focus first on the nature of the natural resources and environments of the respective regions. The special nature of international scientific and environmental co-operation is naturally underlined in both chapters. The contrast between the politics of the Arctic on the one hand, until relatively recently on the geographical front line of the Cold War; and the truly international approach of the Antarctic Treaty System on the other, is notable. Some of the special management measures in both regions are considered, including the role of the native peoples of the Arctic; the important role of the Arctic river basins; and the role of living resources conventions in the Southern Ocean.

CHAPTER 1. THE GEOGRAPHY OF THE SEA

HANCE D. SMITH

1. Introduction

The origins of this chapter lie with the 28th International Geographical Congress in Den Haag in 1996. The Congress was remarkable for its emphasis on the sea, not least because of the location in Den Haag, and the strong and well known maritime tradition of the Dutch hosts. The sea ranked equal with the land and human effort in the Congress title – some 31 Programme items were directly concerned with aspects of marine geography, physical or human. And yet, even although seven-tenths of the Earth's surface consists of oceans and seas, marine geography is a comparatively new specialism within geography.

The purpose of this first chapter is accordingly to review the geography of the sea as an academic subject at the turn of the century and, indeed, the millennium, through consideration of its evolution, structure and paradigms; and to set it in the context of geography as a whole. Key ideas in human geography are then focused on, which seem to be especially valuable in the study of the sea, emphasising large scale social organisation into traditional and urban industrial societies; and the nature of temporal changes occurring at the millennium. Fundamental directions in the study of marine geography are then considered, including the nature of sea uses, time scales and periodicities, and regional geography. Finally, a research agenda is put forward which evaluates in turn traditional studies, the evolution of theory, and the practical application of marine geography.

2. Marine Geography

2.1. EVOLUTION AND STRUCTURE

Despite its recent substantive phase of development, marine geography has considerable antecedents, the terminology first appearing in the early nineteenth century. The subject in both its physical and human dimensions rapidly grew in importance as the century wore on, parallelling the several stages of global economic expansion spanning that period, all of which depended on maritime trade links and naval power. [1] Of special significance was the work of Maury, Hydrographer of the US navy, who gathered together a vast amount of invaluable information on the physical geography of the sea [2] with the practical purpose, among other things, of determining optimum sailing ship routes for ocean voyages. This work was greatly amplified by that of the *Challenger* expedition of 1872-76, the first large-scale systematic attempt at ocean exploration, and which was to become both an inspiration and a model for many subsequent expeditions up to the middle of the twentieth century. Also of continuing relevance in the nineteenth century was the work of hydrographic surveying and production of navigational charts led by the British, and gradually encompassing the efforts of the other major maritime nations; and the establishment of marine science laboratories. [3]

Parallel to the developments in surveying and exploration at sea were academic developments. These included the establishment and work of the national geographic societies; their journals were not infrequently used to publish papers on what might broadly be termed the physical geography of the sea. In the second half of the nineteenth century the International Geographical Congresses were also significant for the reporting of developments in oceanography, navigation and maritime commerce.

Until the middle of the twentieth century, publication in the field of marine geography – seldom, if ever referred to as 'marine geography' – was an integral part of the writing of the times. Emphasis was on first the physical geography of the sea as an integral component of the marine natural sciences, encouraged both by the marine research cruises already noted, as well as by the marine laboratories and the establishment of the International Council for the Exploration of the Sea. [4,5] Also important was work in economic geography – the commercial geography of the late nineteenth and early twentieth centuries. [6] Since the middle of the twentieth century, scholarly work has evolved first along 'orthodox' systematic lines, [7] especially in coastal geomorphology, ports and shipping, fisheries and offshore hydrocarbons studies. The emergence of 'integrated' approaches dates from the 1980s, and has been given impetus principally through the study of marine resource development and ocean and coastal management. [8,9,10,11]

In a wider academic subject context, there continues to be an appreciable quantity of writing of geographical significance contributed by other disciplines. This is especially the case in the sectoral studies in oceanography, [12] marine ecology [13,14] and fisheries. [15] On the social science side there is a substantial literature on the law of the sea [16,17,18,19] and maritime boundaries. [20,21,22,23,24,25] Work in fisheries economics is of regional significance. [26,27] Also, in the wider multidisciplinary context must be noted the work of higher education institutions, mainly in Europe and the English-speaking world elsewhere, which principally relate to ports, shipping and the environment. It is from such a background, for example, that undergraduate education in marine geography has developed. [28]

2.2. PARADIGMS

In common with other academic fields, geography has developed major schools of thought – approaches or paradigms – each of which has evolved to cope with specific aspects of the subject and areas of work.[29,30] Five of these are of special importance in marine geography, namely, the ecological, spatial analysis, regional, behavioural and general systems analysis approaches. In addition, it is important to take account of the relationships between geography and other subjects as, for example, in the recently conceived co-operation between the International Geographical Union (IGU) and the Intergovernmental Oceanographic Commission (IOC).[31] Although all the paradigms have been developed in a land context, these are equally valuable in a marine setting.

The human ecology approach to geography is especially suited to studies of natural resource use, both resources extracted and considering the environment itself as a resource. By far the most important areas of work are devoted to economic sectors. This includes the fisheries,[32,33,34,35] and mineral and energy resources.[36,37] It is also a valuable approach in studies of waste disposal and pollution,[38] leisure and recreation,[39,40] and conservation.

The spatial analysis approach in the marine field is most relevant to studies of shipping, ports and strategic themes. In shipping geography there are several notable studies,[41,42,43] as is the case with ports.[44,45,46,47] There are also studies of strategic interests[48] and naval topics.[49,50] From a spatial analysis point of view, in a world which still depends overwhelmingly on sea transport, there are important links with work on the regional development of the global economy.[51]

The regional geography of the sea is not well developed,[52] but notably is advancing through the promotion of journal special issues.[53,54,55] Studies of marine resource development noted above already also fall into this category to varying degrees. Apart from notable early work on the North Atlantic,[56] and work on the Pacific,[57,58] a significant current strand relates to the natural sciences studies of large marine ecosystems,[59,60] while earlier work on the physical geography of the sea has a strong regional dimension.[61,62,63] Great potential exists in the development of regional studies incorporating physical and human geography which are linked to the development of the global economy and related coastal and ocean management issues.

Behavioural approaches in marine geography are few and far between. However, there are substantial contributions with a geographical component by anthropologists[64] and sociologists.[65,66] The value of these lies especially in environmental management studies[67,68,69,70,71] in the contexts of organisations and decision-making, and individual behaviour.

Beyond the 'big four' paradigms noted above, lies developing work in systems approaches[72] which, although rooted to some extent in spatial analysis and regional approaches, now transcends both to cut across the other four, and extends beyond into informing relationships between geography and other disciplines. Systems analysis takes into account theories of complexity and chaos, which have recently become fashionable. It is particularly useful in dealing with ocean and coastal management.

Finally to be considered is the relationship between geography and other subjects. [73] On the one hand are the natural sciences where, as already pointed out, most geographical work has been done by non-geographers. However, there have been notable contributions in the field of the physical geography of the oceans and coasts already noted. In the social sciences there are two groups of subjects concerned. The first is those concerned with the human dimension *per se*, including psychology, anthropology, sociology, politics and law; and those which have an important focus on the material relationships of humanity, including archaeology, economics, geography and history. [74]

2.3. SEA, LAND AND THE HUMAN EFFORT

It is possible then, to conceive of a geography of the sea which can stand alongside the more conventional geography of the land, and share the same orthodox structure and paradigms, even although there are inevitably more gaps in marine geography than there are in land geography. What then of the 'specialness' of the sea, which give the dominating characterisation of planet Earth as the 'pale blue dot' described by Carl Sagan in his lecture to the 27th International Geographical Congress in Washington D.C. in 1992. [75] At least five different ideas highlight this 'specialness': the nature of the human mind; the strong maritime traditions based on human experience and thinking about the sea; the distinctive maritime communities which have existed down the ages defined initially in terms of occupation; the singular technologies required to master the marine environment; and the unique nature of that environment in a planetary context.

In the context of the 28th IGC – this sub-heading is a paraphrase of the title of that Conference – with which this chapter opened, no experience surpasses that of the Dutch. At a physical level, the reclamation of much of the land area from the sea is well documented. [76] In terms of marine economic activity, the maritime Dutch nation was substantially built upon trade, whaling, fishing and naval activity in the sixteenth and seventeenth centuries. [77] In pursuing these activities Dutch society was effectively transformed into the first maritime industrial nation. The maritime communities which emerged are graphically illustrated in the Panorama Mesdag of Scheveningen in the 1880s, with all the Dutch herring busses hauled up on the sandy beach in front of the dunes. With this example in mind, we now turn to the development of human life and experience in relation to the sea.

3. From Traditional to Urban Industrial to Post Industrial

3.1. THE TRADITIONAL SOCIETY

The Panorama Mesdag at Scheveningen is a potent maritime symbol of the great transition in human society between the traditional, mainly rural form of social organisation, to the urban industrial. [78] The technology of the herring busses hauled up on the beach, with their sails and drift nets, was traditional; the Dutch herring industry which they were part of in terms of its scale e.g. numbers of boats and organisation of herring processing – wet salting and packing in barrels with a strict quality control system – was modern and industrial. At the turn of the millennium, the urban industrial societies are apparently transforming into a post-industrial phase, in which information management is becoming a key feature. [79] This section considers in turn the traditional and urban industrial societies which still predominate in the maritime world, and examines the nature of the millennium which emerges, as it were, from the human dimensions of global change. The basic approach adopted here is that of considering human culture in relation to the environment at three levels, namely, the material, social and mental.

In both prehistoric and historical maritime societies, as demonstrated by archaeological, historical and linguistic evidence; as well as in contemporary traditional societies in which subsistence takes priority over trade, by far the most important activity is the extraction of food from the sea. The role of fisheries is well known; less well known is the antiquity of aquaculture, which has a history of at least four millennia in the East Asian culture realm. [80,81,82] A primary characteristic of traditional fisheries and aquaculture is a very highly developed awareness of the necessary balance between human activities on the one hand, and the ability of the natural environment to support these on the other. This can still be seen today in societies where subsistence fishing remains important, most notably in the South Pacific islands. [83,84]

The second major maritime activity in traditional societies has been and remains trade. There is ample archaeological evidence of this in Europe, as far back as the Mesolithic, [85] while evidence of early voyages in all major culture regions points in the same direction, [86] although the very earliest voyages involved simply migration and settlement, and extend to at least 30,000 years ago in the Pacific. [87] Some exchanges were not commercial; carrying gifts was a significant purpose.

Turning now to the social dimension of traditional societies, this still give form to the major civilisations or culture regions today. Those identified by Broek and Webb [88] are divided into two categories: the major culture realms of East Asia, South Asia, the area covered by the modern Islamic world, and the Occident; and the minor realms of South East Asia, Africa south of the Sahara, and the maritime Pacific. These are all of great antiquity, with roots in all cases extending back to around 4,000 years.

Of the large geographical scale culture realms, only that of the Pacific is entirely maritime; relatively recently it has been demonstrated that the peoples of the Pacific originated in East Asia. [89] However, in all the major civilisations there have been peoples with a strong, even predominant maritime orientation. Thus, for example, in the evolution of Occidental or Western culture, Greeks, Romans and some Celtic peoples had distinctive maritime cultures. [90,91] Later, in the European medieval period, important maritime peoples became established, including the Catalans and the Vikings. Major spheres of maritime trade became established in northern Europe under the

Hansa, [92] and in southern Europe focused on the northern Italian city states of Venice and Genoa in particular. [93] It was the latter which of course in the European early modern period provided much of the maritime capability of the Spanish and Portuguese Empires, the starting point for the first phase of subsequent European maritime expansion, including the discovery of the Caribbean region of the Americas, the rounding of the Cape of Good Hope, and the first circumnavigation of the globe, all achieved between 1492 and 1522. The maritime localities associated with these peoples in turn were to form the roots of the great European maritime empire building nations of the modern era, notably Spain, Portugal, the Netherlands, England and France.

Despite appearances to the contrary, the traditional societies remain a force to be reckoned with. First, simply in terms of geographical extent, these are much more extensive than the truly urban industrial societies, significantly cutting across the conventional distinction between developed and developing worlds. Of special note are the circumpolar Arctic peoples: the Inuit, Saami and those of Siberia; certain of the North American Indian peoples; the desert societies from the Sahara eastwards to the Indus Valley, south west Africa and Australia; the peoples of the world's monsoon and tropical rain forests: and the peoples of the Pacific. Secondly, in the developed countries especially, there is now emerging a re-assertion of the rights of traditional societies to the resources of land and sea. This is most evident in Canada, in the campaigns of the 'first nations', both Inuit and Indian; in the case of the Saami of northern Scandinavia; and regarding the position of the Torres Strait islanders of Australia. [94] Also significant are the efforts of the Pacific islanders to secure their traditional resources, both in the reef lagoons and in the open ocean. [95] Thirdly, there is an increasing public awareness in the industrial world of the values of traditional societies, especially of their sensitivity in maintaining a balance between the natural world on the one hand, and human activity on the other. A notable manifestation of this is the growth of the wilderness ethos in North America and beyond. [96]

A final point to note in the context of traditional societies, is the strength of the maritime traditions, and the durability of maritime communities with which these are associated. In parts of the world where traditional maritime communities are still extant, this is evident in active traditions which have been transmitted across many generations. However, in the industrial world, roots of maritime traditions reside in their traditional society antecedents, generally in fisheries, shipping and ports. For example, in the former sphere of influence of the medieval Vikings in Europe, maritime traditions remain evident in such areas as place names [97] the design of traditional boats, [98] the legends of the sagas, [99] and historical knowledge of Viking times. The age-old activities of fishing, maritime trade, and warfare, together with acquiring knowledge of the sea, became the starting points for the profound transition from traditional ways to the complexities of the urban industrial societies which were to follow.

3.2. THE URBAN INDUSTRIAL SOCIETY

At the heart of the industrialisation of the world ocean was a process of economic specialisation, graphically described by Adam Smith at the outset of the 'industrial revolution' in the 1770s.[100] Although the traditional maritime communities were very strong in those regions affected by industrialisation, the emergence of increasingly distinct economic sectors in turn transformed the social structure of communities along economic lines. In the long term also, as the role of the state strengthened, a distinction between private, public and voluntary sectors grew for each major grouping of sea uses. These economic and social transformations were reflected in expansion and specialisation of coastal settlements. It became possible to conceive of urban, rural and wilderness coasts and seas, as well as lands, distinguishable by degrees of intensity of sea uses. Again, in the long term, on a global scale, a large proportion of the population became concentrated within 50-100 km. Of the coast. This growing population placed constantly increasing pressure on marine resources of food, minerals and energy; while the sea became the ultimate sink for all forms of waste production of the industrial age.

Broadly speaking, it is possible to consider sea use groups in two categories: the 'old' group of four already mentioned; and the 'new' group of mineral and energy resources, leisure, waste disposal and conservation, which became really important only in the second half of the twentieth century. The processes of industrialisation have highlighted the fundamental nature of these use groups.[101]: two – shipping and strategy – are primarily concerned with the spatial organisation of sea uses. Fisheries and aquaculture, mineral and energy resources, are about the material resources of the marine environment. The remainder: leisure, science and conservation concern non-material uses.

The use of ships and boats of all kinds incorporates the activities of all the other groups, although the primary focus is upon commercial shipping, which has been the cement holding together the trade of all maritime communities and their parent civilisations down the ages. The history of commercial shipping,[102] with associated navigation and port development constitutes a primary record of the development of the global economy, especially in the past half-millennium of increasingly dominant European influence, through first the commercial then industrial revolutions. The evolving strategic use of the sea also parallels the commercial shipping development and overall regional development of the world economy. This has been the case not only for naval warfare itself, with its associated technological advances over the 'European' half-millennium, but also for the geostrategic evolution of the global system,[103] evidenced, for example, in the evolving spheres of geographical influence of major maritime powers,[104,105] and in the development of maritime boundaries already noted above.

The fisheries and aquaculture as major providers of food for human populations share with shipping and naval activity a very long period of evolution. More than any other sea use group, the exploitation of biological resources reflects an enormous range of regional cultural diversity, both in pre-industrial and industrial eras. Although perhaps at least 90 per cent of all fish production is now commercial rather than traditional subsistence, these commercial fisheries and aquaculture activities grew out of traditional roots. It is particularly notable that the expansion of the fisheries during the 'industrial revolution' itself and the succeeding 'long waves' which lasted into the 1870s-1880s, primarily involved a scaling up of traditional fisheries in terms of numbers of boats and fishermen, and a retention of traditional, mainly static gears. The process of concentration of fishing operations and fishing ports in both distant water and inshore fisheries did not really accelerate until the coming of steam and the internal combustion engine in the 1890s-1920s, parallelled by the roles of the railways and the telegraph in linking major fishing ports to their principal inland markets in Europe and North America. Nonetheless the fisheries in particular have above all been associated with the continuation of truly maritime communities.

In contrast to the antiquity of biological resource exploitation, mineral and energy use is very recent. Indeed on a large scale it dates only from the 1940s, with the development of the offshore oil industry in the Gulf of Mexico. Rapid expansion elsewhere is a product of the 1960s and 1970s. [106] The principal reasons for grouping mineral and energy resources as a single fundamental use category is that the resource which is by far the most important – hydrocarbons – is both mineral and energy; and the exploitation of all the major categories – hydrocarbons, aggregates, placers and renewables – are the preserve of large-scale private sector corporate investment and management. Above all, mineral and energy resource exploitation, especially oil and gas, represent a field of rapid and sophisticated technological advance. [107, 108] It has been most influential both in maritime boundary delimitation on the world's continental shelves, and in the provision of the 'common heritage of mankind' concept and its practical implementation in the negotiations leading to the Third Law of the Sea Convention in 1982. All this is despite the fact that deep sea mining remains a remote possibility under existing economic conditions. Much more immediate is the use of renewable energy from the marine environment – wind, waves and tides, now on the very of large scale development, and surely to be a hallmark of the first long wave of the 21st century. [109]

The large scale use of the marine environment for waste disposal is a product of nineteenth century industrialisation and urbanisation, especially after the 1870s in Western Europe and North America. It is a function of both economic expansion and population growth. As such, large scale development of public sector works and private sector industrial waste belongs principally to the long wave of the second half of the twentieth century. The management of both land pollution sources, together with pollution from ships, also belongs to this period. It is probably true to say that pollution levels for all these major sources continued to increase until the 1970s, but that increasingly effective management regimes have evolved from the 1980s onwards. Despite the element of truth in the sea being regarded as the ultimate sink, in reality much marine pollution is very localised. [110]

Next to fisheries and aquaculture, the use of the sea for leisure purposes is most widespread and diverse. [111,112,113] As with large scale waste disposal, its roots lie in the second half of the nineteenth century, in the development of the seaside holiday in Western Europe especially. Likewise its large scale development has been a product of economic expansion and population growth in the long wave of the second half of the twentieth century. The range of activities is vast, both water and coastal land based. Organisation is also diverse, from the corporate giants of the cruise and hotel industries, to millions of small businesses providing activities and services. Tourism and recreation industries also transcend the development divisions, being equally important at regional level in both developed and developing countries.

Marine scientific research also has its roots in the nineteenth century. [114] Above all it has been concerned with the practicalities of using the sea, for example, in the work of Maury on shipping routes, and the Admiralty in hydrographic surveying and the production of charts; as well as exploration, as with the *Challenger* expedition. Marine science tends not to be thought of as a use *per se*. And yet since the 1950s it has been a large industry, involving many hundreds of laboratories and research vessels; and monitoring systems which will shortly culminate in the fully fledged Global Ocean Observing System (GOOS). [115] Despite the apparent importance of 'pure' research in the natural sciences of the sea, in reality the driving force in public, private and voluntary sectors is applied.

The final major use – also not thought of as an economic activity – is conservation. Conservation in its widest sense includes not only the natural, but also the human heritage. As such it involves the full range of natural and social sciences, although by far the greatest investment and effort is in the natural sciences. Much of the innovative thinking, political influence and practical implementation is the work of the voluntary sector organisations at local, national and international levels. It is being increasingly strongly underpinned by global agreements, at least in principle, ranging from the Law of the Sea Convention through the Biodiversity Convention to the Framework Convention on Climate Change. Perhaps more than any other use group, it has promoted truly integrated approaches to ocean management, based on realisation of the pivotal importance of the inter-related natural and human heritages.

Right from the beginning the social structure associated with the emergence of the maritime components of the urban industrial society had three elements, as the 'industrial revolution' phase of the late eighteenth/early nineteenth centuries, namely, private, public and voluntary sectors. No doubt this was due in part to the leading role played by Britain throughout the nineteenth century. While the lead was taken by the private sector in the development of shipping and fisheries, in the best Victorian tradition there was a parallel development of regulation of these industries, coupled with the build up and maintenance 'public sector' naval power in the maritime nations generally. Meanwhile, the work of voluntary effort was marked in marine safety, with the establishment of the Royal National Lifeboat Institution in the 1820s, the early marine laboratories, and ultimately the marine conservation movement of the latter half of the twentieth century.

The coastal settlement patterns which have emerged from the economic and social developments outlined above are dominated by urbanisation, which has rapidly spread from the conventional developed world centred on the North Atlantic and North Pacific, to the developing world where most of the coastal megacities are emerging. [116] The approximate ratio of world population of developed and developing regions respectively is currently one third/two thirds for a total global population of 6 billion. This population is likely to increase substantially – perhaps as much as double, and gradually stabilise in the second half of the present century. It is probable that more than half of this population will be concentrated in a coastal land zone between 100 and 200 km. wide. Geographically localised 'urban' sea areas will require spatial planning akin to adjacent urban land as well as extensive coastal engineering. 'Rural' sea areas on continental shelves where marine resource extraction and navigation are primarily concentrated will require intensive multiple use management systems, as on land. However, the marine wilderness beyond, which will cover most of the coasts and world oceans and seas, will remain relatively little used, although vulnerable to human impacts, especially in the form of pollution and ecosystem modification.

It is this urban/rural/wilderness seascape approach which provides a useful first approximation for dealing with environmental impacts of the human population upon the marine environment, and the management response overall. Already these impacts – resource depletion and pollution especially – weight heavily upon the oceans and seas, and are being increasingly well defined in scientific terms. A considerable organisational response to management needs has emerged over the past century especially. It is with these thoughts in mind that we now turn to an assessment of the nature of the millennium.

3.3. THE NATURE OF THE MILLENNIUM

In assessment of the nature of the millennium it is useful to distinguish on the one hand between an older world – that of the traditional society discussed above and, on the other, a newer world of industrialisation. Beyond this a number of trends are important, relating to environment, science and technology, economic factors, social factors and management.

In the urban industrial world, the traditional society has largely become heritage, a process which accelerated in the long wave between the 1870s and 1930s, under the influence of the enormous maritime technological and economic changes of the period. However, in the developing world, and in the isolated, lightly populated fastnesses of the developed world as well, the traditional society is still very much with us, although under great strain as its environment, economy, society and culture are being transformed by the continuing spread of industrialisation. The threat is being countered by the growth of 'first nation' political activities in some key countries in the West, such as Australia, Canada and Colombia, but the end of the old ways of life is foreseeable.

The overarching theme of the 'newer' world of the past 500 years might well be termed the 'European experience', at least in a geographical sense. It of course began with the maritime expansion of Europe from the late fifteenth century onwards.[117] Although the second half of this period has been dominated by progressive, high profile industrialisation, at a deeper level have been the processes of economic specialisation emerging from the decline of the European Late Middle Ages. The earlier manifestations of this were primarily commercial and maritime,[118] rather than industrial *per se*, in the sense of manufacturing. Indeed, the earliest significant maritime industrialisation was that of the Dutch in the course of the seventeenth century. In a European historical context, the half-millennium is termed 'modern', although 'modern' is also applied more narrowly and in a European cultural context, to the long wave between the 1870s and 1930s, which indeed witnessed the peak of industrialisation in Europe itself, and the apogee of European cultural, political and military influence worldwide. In the next following long wave of the second half of the twentieth century, the United States/Canada and Japan/South Korea regions pulled decisively ahead.

As noted in the introduction to this section, distinguishing thematic trends have been apparent throughout, which can cast some light on both the historical and geographical nature of the millennium. The development of European science and technology has been crucial since the end of the Middle Ages in explaining many aspects of maritime experience, including navigation, ship design, fishing gear and oceanography itself. American inspired offshore engineering has also been developed in a European geographical location at the very end of the period. However, it is the workings of technological and economic factors in tandem which more than anything else defines the onward progress of industrialisation itself, including especially the half-century or so stages or 'long waves' already alluded to above, which have been especially notable since the 1770s.[119] Closely allied to this process of technological and economic specialisation has been the rise of the urban industrial society, first alongside, then rapidly replacing the traditional settled agricultural and nomadic tribal societies. This has been associated with first, rapid population increase, then population stabilisation in parallel with stabilisation of the urban industrial world. This demographic transition has now been passed through by approximately one third of the world population, with the rest likely to stabilise within the next century, giving a world population substantially higher than at present (see above). The environmental impacts of economic specialisation and population increase have been enormous, if still uncertain in some respects, especially when allied to environmental influences which operate at least partly independently of human activity, notably climate change and its knock-on effects.

It is in the evolution of systems of thought to cope with these impacts, and the profound transition from a traditional to a post-industrial technological, economic and social world, which is arguably the only truly millennial perspective of the present time, comparable with the dawn of the Middle Ages in Europe a thousand years ago, and the subsequent transition from medieval to modern times. A key part of this system of thought revolves around marine geography and related sciences, and ocean and coastal management concepts, aimed at pulling together and understanding both the interactions of these trends and the human response to them. It is to this marine geography and related environmental management process that we now turn.

4. Marine Geography, Ocean and Coastal Management: Setting a Course

The geography of the sea is inevitably not as well developed as the geography of the land. On the physical side there remains much to be done on surveying and charting the sea bed, and gathering information on the sea itself. On the human side also, there is much research needed on the uses of the sea and their development and management. There are three especially useful starting points for evaluating progress. The first is an awareness of the fundamental nature of sea uses and the implications which arise therefrom. The second consideration relates to the temporal dimension of marine geography —both overall time scales and stages of development of sea uses. Thirdly is the marine regional geographies which are associated with the first two themes.

The fundamental nature of sea uses points towards the need to develop a model or models of humanity which can explain the interactions between people and the marine environment. There are already within the social sciences in general, and geography in particular, schools of thought which can contribute to such a model. Of special note in geography are the ecological, spatial, regional and behavioural approaches which focus respectively on environmental, economic, geographical and social/psychological dimensions of geographical inquiry.

The classification of use groups outlined above is based on human needs for communication and defence; for food, material and energy resources; for disposal of waste; and for support of predominantly mental processes – leisure, education and research, and conservation. This classification is also applicable to the land. Such a fundamental approach also makes it possible to make a clear intellectual link between theory and practice: the nature of marine geography on the one hand, and its practical application on the other.

In consideration of the applied nature of the subject, the dominant contemporary theme which has emerged over the past 30 years is that of sustainability. Sustainability may involve maintaining a balance among environment, economy and society, but it remains in essence a late twentieth century political concept. Much of it is not new. The underlying idea of this balance can be detected in Adam Smith's perceptions in *The Wealth of Nations*, for example; while the landowners of Britain at the time of the industrial revolution spoke of 'improvement' – moving this balance in the direction of development.[120]

Consideration of time scales and stages involves looking for regularities in key relationships between human activities and the marine environment. The initial classification of these relationships may be based on the fundamental use categories discussed above. There are several time scales of importance. First is that of the major cultures or civilisations – something in the order of 4,000 years. [121] Second, the time scales of regional rural and urban societies extends over many centuries – the *longue durée* of Braudel. [122] Third are the time scales of individual lifetimes. Also important is the distinction between the traditional, pre-industrial societies on the one hand, and the urban industrial societies of the past half-millennium on the other.

The traditional pre-industrial societies are rapidly becoming a thing of the past. Nonetheless, as already noted, these still occupy most of the world's land surface, in both nomadic and settled agricultural forms. Before the onset of industrialisation, the traditional societies represented long term stability between environment and human activities on both land and sea. Their roots extend far beyond the past 4,000 years into prehistory.

By contrast, the global urban industrial society of today is characterised by staged development patterns. Initially focused on Europe, it has expanded to its present global geographical extent over the past 500 years. A variety of development time scales have been identified: 150, 50 to 60, 25 and 10 years; and the business cycle of 40 months, relating to a range of technological and economic factors and their interactions. Of these, 150 years is significant for European early modern development, prior to the eighteenth century, [123] the 50-60 year 'long wave' already noted several times above has dominated major long term development patterns since the 'industrial revolution' between the 1770s and 1830s.

The third, regional geography theme is composed of both systematic and spatial elements. The systematic elements can be classified in terms of the human mind, culture, society, economy and environment. Maritime development regions – around 30 [124] are all at different stages of development. Of special interest in this regional palimpsest is the presence of truly maritime communities dependent upon the sea. These have been characteristic of both traditional and urban industrial societies. Indeed industrialisation has focused and strengthened the interdependence between human activity and the sea. As the industrial age draws to a close, however, communities of people in specific locations are being replaced by communities of interest, geographically dispersed.

5. Conclusion: A Research and Teaching Agenda

There are three concluding considerations. The first concerns what may be termed traditional research and teaching in marine geography. Here, because the modern development of the subject is limited relative to conventional, land-based geography, there remains much valuable work to be done. This effort falls into two main areas, namely, the economic sectors – the uses of the sea; and the full range of geographical studies integrated physical and human aspects of the subject within the 'traditional' paradigms of ecological/environmental, spatial/economic, regional and behavioural approaches.

The second consideration relates to the future development of theory, where three points arise. The first concerns basic theoretical understanding, where there is a need for marine geography to develop as part of geography as a whole, as well as continuing along traditional lines as enunciated above. The second is to develop a paradigm which integrates the development and management of the world ocean. Sustainability could be an element of such a paradigm which should, however, be wider and deeper. Thirdly there is a need to develop further a practical management paradigm which applies the subject.

A suggested applied paradigm for sea use management is to distinguish between physical interactions between human activities on the one hand, and environmental impacts of those activities on the other; and the specifically human dimension of management, and use these to distinguish in turn between technical and general management dimensions. The technical management dimension includes information management and assessment, and professional practice. The general management dimension deals with technical management co-ordination, organisation-based decision-making, policy and strategic planning of organisations involved in marine environmental management.

A good illustration of the enormous regional diversity involved is in fisheries management. Fisheries management may be viewed in terms of major natural regions: polar, tropical and temperate seas, together with large marine ecosystems; and regional development areas: North America, Western Europe, Eastern Europe, and Japan/South Korea may be regarded as 'first order' regions; while the developing countries contain a substantial number of 'second order' regions of more limited geographical extent. Much of the technical management infrastructure is necessarily closely related to the environmental regional pattern just noted, whereas the general management structure is primarily related to the development regions. The long term failure of fisheries management in the past century is a graphic illustration of the distance which still has to be covered to effectively apply marine geography.

References

1. Smith, H.D. (2000) The industrialisation of the world ocean. *Ocean & Coastal Management* **43** (1) 11-28.

2. Maury, M.F. (1858) *The physical geography of the sea*, 6th Edition. Harper & Bros., New York.

3. Schlee, S. (1973) *A history of oceanography: the edge of an unfamiliar world.* Hale, London.

4. Went, A.E.J. (1972) Seventy years a-growing: a history of the International Council for the Exploration of the Sea. *Rapp. Proc.-V. Reun. Cons. Explor. Mer* **165**.

5. Thomasson, E.M. (ed) (1981) *Study of the sea*. Fishing News Books, Farnham.

6. Chisholm, G.G. (1889, 1966) *Handbook of commercial geography*. 1st and 18th Editions. London.

7. Haggett, P. (1972) *Geography: a modern synthesis*. Harper & Row, New York; (2001) *Geography: a global synthesis*. Prentice-Hall, New York.

8. Couper, A.D. (1978) Marine resources and environment. *Progress in Human Geography* **2** (2) 296-308.

9. Couper, A.D. (ed.)(1983) *The Times Atlas of the Oceans*. Times Books, London; (ed.) (1989) *The Times Atlas and Encyclopaedia of the Sea*. Times Books, London.

10. Vallega, A. (1999) *Fundamentals of integrated coastal management*. Kluwer, Dordrecht.

11. Vallega, A. (2001) *Sustainable ocean governance: a geographical perspective*. Routledge, London.

12. Tchernia, P. (1980) *Descriptive regional oceanography*. Pergamon, Oxford.

13. Alexander, L.M., Sherman, K., Gold, B.D. (eds.) (1990) *Large marine ecosystems: patterns, processes and yields*. American Association for the Advancement of Science, Washington D.C.

14. Longhurst, A.R. (1999) *The ecological geography of the sea*. Longman, London.

15. Hall, S.J. (1999) *The effects of fishing on marine ecosystems and communities*. Blackwell Science, London.

16. O'Connell, D.P. (1982) *The international law of the sea*. The Clarendon Press, Oxford.

17. Brown, E.D. (1984) *Sea-bed energy and mineral resources and the law of the sea*. Graham & Trotman, London; (2001) *Sea-bed energy and minerals: the international legal regime Vol. 2: Sea-bed mining*. Martinus Nijhoff, the Hague/Boston/London.

18. Churchill, R.R., Lowe, A.V. (1987) *The law of the sea*. Manchester University Press, Manchester.

19. Somers, E. (1990) *Inleiding tot het internationaal zeerecht*. Kluwer, Dordrecht.

20. Shalowitz, A.. (1963) *Shore and sea boundaries*. U.S. Department of Commerce, Washington D.C.

21. Blake, G. (ed.) (1987) *Maritime boundaries and ocean resources*. Croom Helm, Beckenham.

22. Prescott, J.R.V. (1985) *The maritime political boundaries of the world*. Methuen, London.

23. Charney, J.I., Alexander, L.M. (eds.) (1993, 1998) *International maritime boundaries*. Martinus Nijhoff, The Hague/New York/London.

24. Various Authors (various dates) *Limits in the seas (series)*. U.S. State Department, Washington D.C.

25. Various Authors (various dates) *Publications of the International Boundaries Research Unit, University of Durham*. IBRU and other publishers.

26. Lawson, R. (1984) *The economics of fishery development*. Frances Pinter, London.

27. Cunningham, S., Dunn, M.R., Whitmarsh, D. (1985) *Fisheries economics: an introduction*. Mansell, London.

28. Smith, H.D. (1986) The geography of the sea. *Geography* **71** (4) 320-324; (1987) Marine geography: applications in coastal and sea use management. *Cambria* **14** (2) 105-114.

29. Holt-Jensen, A. (1980) *Geography – its history and concepts: a student's guide*. Harper & Row, London.

30. Johnston, R.J. (1979) *Geography and geographers: Anglo-American human geography since 1945*. Edward Arnold, London.

31. Vallega, A., Augustinius, Pieter G.E.F., Smith, H.D. (eds.) (1998) *Geography, oceans and coasts: towards sustainable development.* Franco Angeli, Milano.

32. Bartz, F. (1964) *Die Grossen Fischereraume der Welt*. F. Steiner, Wiesbaden.

33. Coull, J.R. (1972) *The fisheries of Europe*. Bell, London.

34. Coull, J.R. (1993) *World fisheries resources*. Routledge, London.

35. Tomas, P. (1987) *La pesca*. Editorial Synthesis, Palma.

36. Chapman, K. (1976) *North Sea oil & gas*. David & Charles, Newton Abbot.

37. Earney, F.C.F. (1980) *Petroleum and hard minerals from the sea*. Edward Arnold, London; (1989) *Marine mineral resources*. Routledge, London.

38. Ballinger, R.C. (2003) Marine waste disposal. Chapter 9 this volume.

39. Orams, M. (1999) *Marine tourism: development, impacts and management*. Routledge, London; (2003) Marine leisure. Chapter 8, this volume.

40. Hall, C.M. (2001) Trends in ocean and coastal tourism: the end of the last frontier? *Ocean & Coastal Management* **44** 601-618.

41. Couper, A.D. (1971) *The geography of sea transport*. Hutchison, London.

42. Vigarie, A. (1979) *Ports de commerce et vie littorale*. Hachette, Paris.

43. Vallega, A. (1980) *Per una geografia del mare: trasporti marittimi e rivoluzioni economiche*. Mursia, Milano.

44. Bird, J.H. (1963) *The major seaports of the United Kingdom*. Hutchinson, London; (1971) *Seaports and seaport terminals*. Hutchinson, London.

45. Hoyle, B.S., Pinder, D.A. (eds.) (1981) *Cityport industrialisation and regional development: spatial analysis and planning strategies.* Wiley, New York.

46. Hoyle, B.S., Hilling, D. (eds.) (1984) *Seaport systems and spatial change: technology, industry and development strategies.* Pergamon, Oxford.

47. Hoyle, B.S., Pinder, D.A., Husain M.S. (eds.) (1988) *Revitalising the waterfront: international dimensions of dockland redevelopment*. Belhaven, London.

48. Vigarie, A. (1990) Economie maritime et geostrategie des oceans. Paradigme, Caen.

49. Gorchkov, S.G. (1979) *The seapower of the state*. Pergamon, Oxford.

50. Smith, H.D., Pinder, D.A. (eds.) Geo-strategy, naval power and naval port systems. *Marine Policy* **21** (4) 289-408; Pinder, David and Smith, Hance (eds.) (1999) Heritage resources and naval port regeneration. *Ocean & Coastal Management* **42** (10-11) 857-984.

51. Smith, H.D. (1994) The development and management of the world ocean. *Ocean & Coastal Management* **24**(1) 3-16.

52. Morgan, J.R. (1991) Marine regions and the law of the sea. *Ocean & Coastal Management* **15** (4) 261-272.

53. *GeoJournal*.

54. *Ocean & Coastal Management*.

55. *Marine Policy*.

56. Alexander, L.M. (1963) *The offshore geography of Northwestern Europe*. Association of American Geographers, Chicago.

57. Buchholz, H.J. (1987) *Law of the Sea zones in the Pacific Ocean*. Institute of Asian Affairs, Hamburg and Institute of Southeast Asian Studies, Singapore.

58. Couper, A.D. (ed.) (1989) *Development and social change in the Pacific Islands*. Routledge, London.

59. Alexander, L.M. (1993) Large marine ecosystems: a new focus for resources management. *Marine Policy* **17** (3) 186-198.

60. Alexander, L.M., et.al. op.cit. 13 above.

61. Guilcher, A. (1957) *Coastal and submarine morphology*. Methuen, London.

62. King, C.A.M. (1974) *Introduction to marine geology and geomorphology*. Edward Arnold, London; (1975) *Introduction to physical and biological oceanography*. Edward Arnold, London.

63. Gierloff-Emden, M.G. (1980) *Geographie des meeres, oceane und kusten*. Walter de Gruyter, Berlin.

64. Andersen, R., Wadel, C. (eds.) (1971) *North Atlantic fishermen: anthropological essays in modern fishing*. Newfoundland Social and Economic Papers No. 5. Institute of Social and Economic Research, University of Newfoundland. St. John's.

65. Maiolo, J.R., Orbach, M.K. (eds.) (1982) *Modernisation and marine fisheries policy*. Ann Arbor Science, Ann Arbor.

66. Thompson, P. with Lummis, T. (1983) *Living the fishing*. Routledge & Kegan Paul. London.

67. Ratcliffe, T. (1991) Responsibility for watersports development and management. *Ocean & Coastal Management* **18** (2-4) 259-268.

68. Pickering, H.J. (1994) The role of corporations in the management of the marine environment. *Marine Pollution Bulletin* **28** (10) 629-637.

69. Nicholls, C. (1992) *The environmental management of the marine aggregates dredging industry in England and Wales*. PhD Thesis, Unpublished, University of Wales.

70. Cole-King, A. (1995) Marine conservation: a new policy era. *Marine Policy* **17** (3) 171-185.

71. Owen, J. (1999) The environmental management of oil tanker routes in UK waters. *Marine Policy* **23** (4-5) 289-306.

72. Vallega, A. (1992) *Sea management: a theoretical approach.* Elsevier Applied Science, Barking.

73. Haggett, P. (2001) op.cit. 7 above.

74. Smith, H.D. (1998) IGU Oceans Project integrated ocean management: the role of the social sciences and geography, in: Vallega, A., Augustinius, P.G.E.F., Smith, H.D. (eds.) *Geography, oceans and coasts: towards sustainable development.* Franco Angeli, Milano, pp. 125-138.

75. Sagan, C. (1993) Is there intelligent life on Earth? *27th International Geographical Congress Proceedings, Closing Ceremony,* Washington D.C., pp. 87-98.

76. Lambert, A. (1971) *The making of the Dutch landscape: an historical geography of the Netherlands.* Academic Press, London.

77. Aymard, M. (ed.) (1982) *Dutch capitalism, world capitalism.* Cambridge University Press, Cambridge.

78. Broek, J.O.M., Webb, J.W. (1968,1973) *A geography of mankind.* Prentice-Hall, New Jersey.

79. Castells, M. (1996-98) *The information age: economy, society and culture.* 3 vols. Blackwell, Oxford.

80. Bartz, F. (1964) op.cit. 32 above.

81. Von Brandt, A. (1984) *Fish catching methods of the world. 3rd Edition.* Fishing News Books, Farnham.

82. Coull, J.R. (1993) op.cit. 34 above.

83. Johannes, R.E. (1981) *Words of the lagoon: fishing and marine lore in the Palau district of Micronesia.* University of California Press, Berkeley.

84. Couper, A.D. (ed.) (1989) op.cit. 58 above.

85. Cunliffe, B (2001) *Facing the ocean: the Atlantic and its peoples, 8000BC-AD1500.* Oxford University Press, Oxford.

86. Heyerdahl, T. (1980) *Early man and the ocean: the beginning of seaborne civilisations.* George, Allen & Unwin, London.

87. Sykes, B. (2001) *The seven daughters of Eve.* Bantam Press, London.

88. Broek, J.O.M. (1968) op.cit. 78 above.

89. Sykes, B. (2001) op.cit.87 above.

90. Meijer, F. (1986) *A history of seafaring in the classical world.* Croom Helm, Beckenham.

91. Moffat, A. (2001) *The sea kingdoms: the story of Celtic Britain and Ireland.* Harper Collins, London.

92. Dollinger, P. (1970) *The Hansa.* Macmillan, London.

93. Horden, P., Purcell, N. (1987) *The Mediterranean world: man and environment in Antiquity and the Middle Ages.* Blackwell, Oxford.

94. Aboriginal and Torres Strait Islander Commission (1995) *Recognition, rights and reform: report to the Government on social rights measures.* Commonwealth of Australia, Canberra.

95. Couper, A.D. (ed.) (1989) op.cit. 58 above.

96. Martin, V., Inglis, M. (eds.) (1984) *Wilderness: the way ahead*. Findhorn Press/Lorian Press. Findhorn.

97. Small, A. (1968) The historical geography of the Norse Viking colonisation of the Scottish Highlands. *Norsk Geogr. Tidss.* **22** 1-16.

98. McGrail, S. (1981) *Rafts, boats and ships: from prehistoric times to the medieval era.* HMSO, London.

99. Jones, G. (1986) *The Norse Atlantic saga*. Oxford University Press, Oxford.

100. Smith, A. (ed. E Cannan) (1950) *The wealth of nations. 6th edition.* London.

101. Smith, H.D. (1985) The administration and management of the sea. *Area* **17**(2) 109-115.

102. Gardiner, R. (ed.) (1992-95) *Conway's history of the ship. 12 vols.* Conway Maritime Press. London.

103. Vigarie, A. (1990) op.cit. 48 above.

104. Mahan, A.T. (1890/1987) *The influence of sea power upon history, 1660-1783.* Dover Edition, New York.

105. Gorchkov, S.G. (1979) op.cit. 49 above.

106. Earney, F.C.F. (1980, 1989) op.cit. 37 above.

107. Pinder, D.A. (2001) Offshore oil and gas: global resource knowledge and technological change. *Ocean & Coastal Management* **44** (2-3) 579-600.

108. Pickering, H.J. (2003) Offshore oil and gas. Chapter 6 this volume.

109. Charlier, R.H., Justus, J.R. (1993) *Ocean energies: environmental, economic and technological aspects of alternative power sources.* Elsevier, Amsterdam.

110. Clark, R.B. (1986) *Marine pollution*. Oxford University Press, Oxford; Ballinger R.C. (2003) Marine waste disposal. Chapter 19 this volume.

111. Hall, C.M. (2001) Trends in ocean and coastal tourism: the end of the last frontier? *Ocean & Coastal Management* **44** (2-3) 601-618.

112. Orams, M. (1999) *Marine tourism: development, impacts and management.* Routledge, London.

113. Orams, M. (2003) Marine leisure. Chapter 9 this volume.

114. Schlee, S. (1973) op.cit. 3 above.

115. Global Ocean Observing System (various dates) *Reports of meetings of experts and equivalent bodies.* GOOS Reports. Intergovernmental Oceanographic Commission, Paris.

116. Barbiere, J., Li, H (eds.) (2001) Third Millennium Special Issue: Megacities. *Ocean & Coastal Management* **44** (5-6) 283-450.

117. Smith, H.D. (1992) The British Isles and the Age of Exploration – a maritime perspective. *GeoJournal* **26**(4) 483-487.

118. Padfield, P. (1999) *Maritime supremacy and the opening of the Western mind.* John Murray, London.

119. Smith, H.D. (2000) *op cit*. 9 above.

120. Smout, T.C. (1964) Scottish landowners and economic growth, 1650-1850. *Scott. J. Pol.Econ.* **11** 214-234.

121. Galtung, J., Rudeng, E., Heistad, T. (1979) On the last 2,500 years in Western history, and some remarks on the coming 500, in: Burke, P. (ed.) *The New Cambridge Modern History, 13: Companion Volume.* Cambridge University Press, Cambridge, pp.318-361.

122. Wallerstein, I. (1974) *The modern world system: capitalist agriculture and the origins of the European world-economy in the 16th century.* Academic Press, New York.

123. Ibid.

124. Smith, H.D. (1994) op.cit. 51 above.

CHAPTER 2. THE LAW OF THE SEA AT THE MILLENNIUM:

The status of UNCLOS and its dispute settlement system

E.D. BROWN

This chapter is based on a paper presented at a conference on *The Oceans at the Millennium,* held at the National Maritime Museum in Greenwich in April 2000, but has been updated to 6 February 2002.

The chapter falls into two main parts. Part 1, on *UNCLOS at the Millennium*, presents a snapshot of the status of the United Nations Convention on the Law of the Sea, 1982 (UNCLOS) as at 6 February 2002. One of the more interesting innovations in UNCLOS was the introduction in Part XV of a complex system for the *Settlement of Disputes* and Part 2 reports on experience so far in operating this system. The two parts, taken together, give a fair indication of the status of the new legal framework of the oceans some seven years after the entry into force of the Convention on 16 November 1994.

1. UNCLOS at the Millennium

This section is concerned with the current status of UNCLOS in two senses. First, it investigates the progress made so far in achieving the aim of universal participation in the Convention and considers in this context the significance of the declarations and statements made by States Parties when signing or ratifying the Convention. Secondly, a progress report is presented on the establishment of the various institutions called for by the Convention.

1.1. THE STATUS OF UNCLOS AT THE MILLENNIUM

On 6 February 2002, 137 States and the European Community had become parties to UNCLOS. [1] Ratifications or accessions are still trickling in, with 3 States havingbecome parties in 2001 (Bangladesh, Madagascar and Yugoslavia) and 1 so far in 2002 (Hungary on 5 February 2002). However, at 138, the number of parties still falls far short of the number of States members of the United Nations, and the Secretary-General noted in his 1999 Report to the UN General Assembly that, five years after the entry into force of UNCLOS on 16 November 1994, only 117 (77.4 per cent) of 151 coastal States and 15 (35.7 per cent) of 42 land-locked States had become parties.[2] Two years later, in his 2001 Report, the Secretary-General provided a roll-call of the absent States. They included 30 coastal States (5 in Africa; 12 in the Asian and Pacific region; 7 in Europe and North America and 6 in the Latin American and Caribbean region) and 27 landlocked States. [3] Notable coastal-State absentees include Canada, Israel, Turkey, Venezuela and the United States. Of those 5 States, Israel, Turkey and Venezuela did not even sign either UNCLOS or the 1994 Agreement relating to the Implementation of Part XI of the Convention; [4] Canada signed both instruments but has not ratified them; and the United States signed only the 1994 Agreement.

Clearly, universality of participation is desirable, both in order to demonstrate unanimous agreement on the substantive rules embodied in UNCLOS and to tie States Parties to the new dispute settlement provisions of Part XV of the Convention. However, failure to become a party does not necessarily signify disagreement with the provisions of the Convention and, as time goes on, there is likely to be more and more evidence that most, at least, of the provisions of the Convention are being recognised as having become part of international customary law. Indeed, most of the States which have failed so far to become parties are supportive of most of the provisions of UNCLOS but have reservations in relation to particular issues. The most important absentee is of course the United States, but there is ample evidence that the Administration accepts the substantive rules of UNCLOS and, following *signature* of the 1994 Agreement on 29 July 1994, even the revised regime of seabed mining. [5]

---

[1] *Status of the United Nations Convention on the Law of the Sea, of the Agreement relating to the implementation of Part XI of the Convention and of the Agreement for the implementation of the provisions of the Convention relating to the conservation and management of straddling fish stocks and highly migratory fish stocks, Table recapitulating the status of the Convention and of the related Agreements, as at 8 January 2002* (www.un.org/Depts/los/convention, accessed 7 February 2002); updated from *Chronological lists of ratifications of, accessions and successions to the Convention and related Agreements*, updated to 6 February 2002, *ibid.,* accessed 7 February 2002.
[2] *Oceans and the law of the sea. Report of the Secretary-General*, A/54/429, 30 September 1999, para. 10.
[3] *Oceans and the law of the sea. Report of the Secretary-General*, A/56/58, 9 March 2001, para. 17, figures revised by reference to A/56/58/Add. 1, 5 October 2001.
[4] Agreement of 28 July 1994 relating to the Implementation of Part XI of the United Nations Convention on the Law of the Sea of 10 December 1982 (Implementation Agreement), *Law of the Sea Bulletin* (LOS Bull.), Special Issue IV, 1994, p.10; Misc. No. 44(1994), Cm 2705; E.D. Brown, *The International Law of the Sea,* Vol. II, 1994, p.366.
[5] Although the U.S. Administration has so far been unable to persuade Congress to support its wish to accede

Given the organic link between UNCLOS and the Implementation Agreement, [6] it is also necessary to be aware of the status of the later instrument. Any ratification of or accession to UNCLOS made after 28 July 1994 represents consent to be bound by the Implementation Agreement too [7] and States may only establish their consent to be bound by the Agreement if they have previously established, or establish concurrently, their consent to be bound by UNCLOS. [8] On the other hand, States or entities which became parties to UNCLOS prior to 28 July 1994 are under no obligation to become parties to the Agreement. The complexity of this linkage was further aggravated by the provision made in the Agreement for provisional application of the Agreement – which terminated on its entry into force on 28 July 1996 – and membership of the Seabed Authority on a provisional basis, which terminated on 16 November 1998. [9] On 6 February 2002, the position was that 104 Parties to UNCLOS (103 States and the European Union) were bound by the Implementation Agreement. [10] To put it another way, there were still 34 States Parties to UNCLOS which had failed to ratify or accede to the Implementation Agreement. It is notable too that 6 of the States which enjoyed provisional membership of the Authority down to 16 November 1998 (Belarus, Canada, Qatar, Switzerland, the United Arab Emirates and the United States) failed to become parties to the two instruments thereafter and thus lost membership status.

Finally, mention must be made of the position of the Agreement on Fish Stocks of 4 August 1995. [11] Although this Agreement too is intended to implement part of UNCLOS, it is not linked to UNCLOS in the same way as the 1994 Implementation Agreement. It entered into force on 11 December 2001, 30 days after the deposit of the thirtieth instrument of ratification or accession. [12] As at 6 February 2002, 30 States had so expressed their consent to be bound, including Australia, Brazil, Canada, Iceland, Norway, the Russian Federation and the United States but not including the European Union [13] (representing the fishery interests of the United Kingdom and its European Union partners [14]).

---

to UNCLOS, its positive attitude to the Convention is reflected in, *inter alia*, the following documents:- United States: Presidential Statement on United States Oceans Policy, 10 March 1983, 9 *International Legal Materials* (ILM) (1970), p. 807; and *Protest from the United States of America* [against Iranian legislation alleged to be contrary to UNCLOS], 11 January 1994, LOS Bull., No. 25, June 1994, pp. 101-103. Although the U.S. had not signed or acceded to UNCLOS, it did sign the 1994 Implementation Agreement on 29 July 1994.

[6] Under Art. 2(1) of the Implementation Agreement, the provisions of the Agreement and Part XI are to be interpreted and applied together as a single instrument.

[7] Implementation Agreement, Art. 4(1).

[8] Art. 4(2).

[9] See further E.D. Brown, "The 1994 Agreement on the Implementation of Part XI of the UN Convention on the Law of the Sea: breakthrough to universality?", 19 *Marine Policy* (No. 1, 1995), pp. 5-20.

[10] Table and List cited in note 1 above.

[11] Agreement for the Implementation of the Provisions of the United Nations Convention on the Law of the Sea of 10 December 1982 relating to the Conservation and Management of Straddling Stocks and Highly Migratory Fish Stocks, A/CONF.164/37, LOS Bull. No. 29, 1995, p. 25.

[12] Under Art. 40.

[13] Table and List cited in note 1 above.

[14] Member States have transferred competence to the European Community with regard to the conservation

In evaluating the current status of the UNCLOS regime, it is also necessary to be aware of the declarations and statements which States Parties have made under Article 310 of the Convention. [15] This article allows States, when signing, ratifying or acceding to the Convention, to make declarations or statements with a view, *inter alia*, to the harmonisation of their laws with the provisions of the Convention. However, since reservations are prohibited under Article 309, declarations or statements must not purport to exclude or modify the legal effect of the provisions of the Convention in their application to the States concerned.

So far, declarations or statements have been made by 35 States upon signature and declarations have been made by 50 States and the European Community upon ratification, accession or formal confirmation of the Convention.[16] A significant number of these declarations and statements are quite clearly not in conformity with Articles 309-310 and have been objected to by other States Parties. Indeed, the level of concern has been such that the General Assembly, in resolutions adopted every year since 1997, has called upon States to ensure that their declarations or statements are in conformity with UNCLOS and to withdraw those which are not. [17] So far, the appeals have fallen on deaf ears and no withdrawals have been made. [18] This must be a matter of some concern since the declarations or statements refer to a wide range of important issues. As is noted in the Secretary-General's 1999 Report:

> Declarations and statements generally considered not to be in conformity with articles 309 (prohibiting reservations) and 310 include: (a) those which relate to baselines not drawn in conformity with UNCLOS; (b) those which purport to require notification or permission before warships or other ships exercise the right of innocent passage; (c) those which are not in conformity with the provisions of UNCLOS relating to: (i) straits used for international navigation, including the right of transit passage; (ii) archipelagic States' waters, including archipelagic baselines and archipelagic sea-lane passage; (iii) the exclusive economic zone or the continental shelf; and (iv) delimitation; and (d) those

---

and management of sea fishing resources. See further Declaration concerning the competence of the European Union with regard to matters governed by the Agreement, LOS. Bull. No. 32, 1996, pp. 26-28. Although, therefore, the UK, as an EU member State, is not bound by the Agreement, it did ratify it on 10 December 2001 on behalf of Pitcairn, Henderson, Ducie and Oeno Islands, Falkland Islands, South Georgia and South Sandwich Islands, Bermuda, Turks and Caicos Islands, British Indian Ocean Territory, British Virgin Islands and Anguilla.

[15] All declarations and statements made before 31 December 1996 have been analysed and published in *The Law of the Sea: Declarations and statements with respect to the United Nations Convention on the Law of the Sea and to the Agreement relating to the Implementation of Part XI of the United Nations Convention on the Law of the Sea* (UN Publication, Sales No. E.97.V.3), hereafter cited as *Declarations and Statements (1997)*. Subsequent instruments are published in the *Law of the Sea Bulletin* and can also be found on the UN website at www.un.org/Depts/los or www.un.org/Depts/Treaty.

[16] *Ibid.* (www.un.org/Depts/los, accessed 7 February 2002).

[17] The latest appeal is in A/56/12, 28 November 2001, operative para. 3.

[18] *Loc. cit.* in note 3 (A/56/58/Add.1), para. 13.

which purport to subordinate the interpretation or application of UNCLOS to national laws and regulations, including constitutional provisions. [19]

1.2. THE ESTABLISHMENT OF UNCLOS INSTITUTIONS

Prior to UNCLOS, there were of course a large number of institutions on both the global and regional levels which were concerned with law of the sea matters and most of them are still continuing their work in the UNCLOS era. However, this section is concerned with the additional new institutions for which provision is made in UNCLOS. They are considered below under five heads:
- Seabed Institutions;
- Commission on the Limits of the Continental Shelf;
- Meetings of States Parties;
- United Nations Open-ended Informal Consultative Process on Oceans and the Law of the Sea
- Dispute Settlement Institutions.

1.2.1. *Seabed Institutions*
Provision was made in Part XI of UNCLOS for the establishment of the International Seabed Authority (the "Authority") as the body charged with the organisation and control of "activities in the Area" [20], that is, activities of exploration for, and exploitation of, the resources of the seabed Area lying seaward of the outer limit of the continental shelf. [21] In discharging this function, the Authority acts in accordance with Part XI of UNCLOS, as substantially amended [22] by the 1994 Implementation Agreement. The Authority commenced its functions upon the entry into force of UNCLOS on 16 November 1994. Prior to that date, an enormous amount of preparatory work had been undertaken by the Preparatory Commission (PREPCOM) set up under Resolution I of the UN Conference on the Law of the Sea (UNCLOS III) and indeed PREPCOM was only wound up at the conclusion of the first session of the Authority's Assembly. [23]

The Authority consists of three "principal organs" – the Assembly, the Council and the Secretariat; three "subsidiary organs" – the Economic Planning Commission, the Legal and Technical Commission, and the Finance Committee; and the "Enterprise". In addition, reference has also to be made to the International Tribunal for the Law of the Sea and, especially, its Sea-Bed Disputes Chamber. The establishment and entry into operation of these various organs reflect the development of the UNCLOS regime of

---

[19] *Loc. cit.* in note 2, at para 16.
[20] UNCLOS, Arts. 156(1) and 157(1), confirmed by Implementation Agreement, Annex, Section 1(1).
[21] UNCLOS, Art. 1(1)(3).
[22] Despite the inclusion of the euphemistic term "Implementation" in its title, the Agreement amends UNCLOS quite substantially.
[23] See further, E.D. Brown, " 'Neither necessary nor prudent at this stage'. The regime of seabed mining and its impact on the universality of the UN Convention on the Law of the Sea", 17 *Marine Policy* (No. 2, 1993), pp. 81-107; and *loc. cit.* in note 9, at p. 17.

seabed mining as a whole; both have gone through a difficult evolutionary process in the context of changing ideologies and market conditions. [24]

The Assembly. The Assembly is the plenary organ of the Authority and currently consists of the 138 members of the Authority. Although the Assembly is in a sense the supreme organ of the Authority and is endowed with a wide range of powers and functions, in reality many of its powers can be exercised only on the basis of recommendations from the Council or within the confines of formulae established by the Convention.

The Council. The 36-member Council, the executive organ of the Authority, is elected on the basis of rules laid down in the 1994 Implementation Agreement, which rendered inapplicable the original UNCLOS rules. Five groups of States are identified and, in addition, the 36 members are divided into four "chambers" for the purposes of the rules on decision making in the Council The negotiations required to reach agreement on the composition of the various groups proved to be difficult and protracted but, eventually, an extremely complex system of rotation and duration of terms was adopted. The arrangements agreed upon included not only firm nominations for the first, 1996 elections, but also understandings on later elections. Taken with the Council's voting rules, these arrangements placed the industrialised States in a significantly more advantageous position than they would have had under the original UNCLOS rules. [25]

The Legal and Technical Commission. UNCLOS makes provision for two 15-member Commissions. [26] However, the functions of the Economic Planning Commission are being performed by the (now 24-member) Legal and Technical Commission until such time as the Council decides otherwise, or until approval of the first plan of work for exploitation of seabed mineral resources. [27] The first elections took place in 1996 and fresh elections for a further 5-year term were held in July 2001. [28] The functions of the Commission include such important matters as the review of seabed mining work plans, the supervision of activities in the Area, and the assessment of the environmental impact of activities in the Area.

The Finance Committee. Specific provision for a 15-member Finance Committee is made in the 1994 Implementation Agreement, giving effect to a more general requirement in UNCLOS, [29] and elections for 5-year terms were held in 1996 and 2001.

---

[24] See further Brown, *loc. cit.* in note 23 (1993), at pp. 97-107.
[25] For a full analysis, see E.D. Brown, *Sea-Bed Energy and Minerals: The International Legal Regime, Vol. 2: Sea-Bed Mining,* Martinus Nijhoff, Publishers, 2001, Chap. 8: "The UN Convention regime of sea-bed mining: VI. The institutional aspects of the regime".
[26] UNCLOS, Art. 163.
[27] Implementation Agreement, Annex, Section 1(4).
[28] At the first election in 1996, the number of members was increased from 15 to 22 under UNCLOS, Art. 163(2). Prior to the 2001 election, the number was further increased to 24 (ISBA/7/C/6, 5 July 2001).
[29] Implementation Agreement, Annex, Section 9(1), giving effect to UNCLOS, Art. 162(2)(y).

Decisions by the Assembly and the Council on a wide range of financial and budgetary matters must take into account recommendations of the Finance Committee. [30]

The Enterprise. The Enterprise was originally conceived as the organ of the Authority which would carry out mining activities in the Area directly, as well as transporting, processing and marketing minerals recovered from the Area. [31] It was to have operational autonomy and operate on commercial principles. [32] However, the 1994 Implementation Agreement altered the role of the Enterprise quite significantly. It confirmed that the Enterprise would conduct its initial mining operations by way of joint ventures. [33] Moreover, it adopted the general principle applicable to all of the organs of the Authority, that their setting up and functioning should be based on an evolutionary approach, taking into account their functional needs, in order that they might discharge their responsibilities effectively at various stages of the development of sea-bed mining. [34] Accordingly, in the beginning, the functions of the Enterprise are to be performed by the Secretariat of the Authority. [35] The question of the functioning of the Enterprise independently of the Secretariat will be taken up by the Council either upon the approval of a plan of work for exploitation for an entity other than the Enterprise, or upon receipt by the Council of an application for a joint venture operation with the Enterprise. If such joint ventures "accord with sound commercial principles", the Council must then issue a directive providing for such independent functioning of the Enterprise. [36]

The Tribunal and the Sea-Bed Disputes Chamber. The Sea-Bed Disputes Chamber is established in accordance with Part XI, Section 5 of UNCLOS and Article 14 of the Statute of the International Tribunal for the Law of the Sea. It deals with disputes arising out of the exploration for and exploitation of the resources of the Area. Its competence extends over disputes involving States, the Authority, companies and private individuals. It is also empowered to give advisory opinions at the request of the Council or the Assembly of the Authority. [37] The Chamber consists of 11 Judges selected by Members of the Tribunal from their own number to serve for renewable terms of 3 years. The first Chamber was constituted in February 1997 [38] and reconstituted in October 1999. [39] So far, no disputes have been referred to it.

1.2.2. Commission on the Limits of the Continental Shelf

The problems associated with the determination of the outer limit of the continental shelf prior to UNCLOS are well known and the definition of the continental shelf in

---

[30] See Implementation Agreement, Annex, Section 9(7).
[31] UNCLOS, Art. 170 and Statute of Enterprise (UNCLOS, Annex IV), Art. 1(1).
[32] Statute of Enterprise, Art. 2(2) and 1(3).
[33] Implementation Agreement, Annex, Section 2(2).
[34] *Ibid.*, Section 1(3).
[35] *Ibid.*, Section 2(1).
[36] *Ibid.*, Section 2(2).
[37] UNCLOS, Art. 191; see too Art. 159(10) and Annex VI, Art. 40(2).
[38] ITLOS/Press 5, 3 March 1997.
[39] ITLOS/Press 31, 4 October 1999.

Article 76 of the Convention has introduced a much needed scientific precision to the process. [40] An important role in that process is played by the Commission on the Limits of the Continental Shelf set up under Annex II to the Convention. [41]

Where a coastal State intends to establish the outer limits of its continental shelf beyond 200 nm, it has to submit particulars of such limits to the Commission, together with supporting scientific and technical data "as soon as possible but in any case within 10 years of the entry into force of [the]…Convention for that State". [42] The Commission has then to make recommendations in accordance with Article 76. [43] The crucial role of the Commission is reflected in Article 76(8), which provides that, "The limits of the shelf established by a coastal State on the basis of these recommendations shall be final and binding".

A second useful function of the Commission is to provide scientific and technical advice during the preparation of the data to be submitted to it, if so requested by the coastal State concerned. [44]

The 21 members of the Commission were first elected for 5-year terms in 1977 [45] and they have made good progress in developing the system by adopting *Rules of Procedure*, [46] *Scientific and Technical Guidelines*, [47] and a paper on the Commission's *Modus Operandi*. [48]

Even a glance at the *Guidelines* prepared by the Commission will confirm that the process of determining the outer limit of the continental shelf in accordance with Article 76 of UNCLOS is a task of considerable technical and scientific complexity. [49] Fortunately, the Commission is aware of the scarcity of trained manpower in this area and of the need to provide training to make good the shortfall. In addition to agreeing upon an action plan for training and addressing letters to the President of the UN General Assembly, the Intergovernmental Oceanographic Commission (IOC) and the International Hydrographic Organisation (IHO), [50] the Commission also drafted an outline for a proposed 5-day training course for practitioners [51] and, in May 2000,

---

[40] See further, E.D. Brown, *Sea-Bed Energy and Minerals: The International Legal Regime, Vol. I: The Continental Shelf,* 1992, Chap. 4; and E.D. Brown, *The International Law of the Sea,* Vol. I, 1994, Chap. 10.
[41] See further, *op. cit.* in note 40 (1992), pp. 28-32 and *op. cit.* in note 40 (1994), pp. 144-146.
[42] UNCLOS, Annex II, Art. 4.
[43] UNCLOS, Art. 76(8) and UNCLOS, Annex II, Art. 3(1)(a) and Art. 6.
[44] UNCLOS, Annex II, Art. 3(1)(b).
[45] Under UNCLOS Art. 76(8) and UNCLOS Annex II, Art. 2.
[46] Originally adopted as CLCS/3/Rev.2, 4 September 1998; revised at seventh and eighth sessions and reissued as CLCS/3/Rev.3, 6 February 2001.
[47] *Scientific and Technical Guidelines of the Commission on the Limits of the Continental Shelf* (CLCS), 13 May 1999. Annexes to the Guidelines (CLCS/11/Add.1) include flowcharts providing a simplified outline of the Guideline procedures.
[48] *Modus Operandi of the Commission*, CLCS/L.3, 12 September 1997.
[49] *Op. cit.* in note 47. See also *Modus Operandi of the Commission*, CLCS/L.3, 12 September 1997.
[50] See Statement by the Chairman of the Commission on the Limits of the Continental Shelf on the Progress of Work in the Commission, CLCS/18, 3 September 1999, paras. 13-16.
[51] CLCS/24, 1 September 2000 and CLCS/24/Corr. 1, 9 February 2001. See further Statement by the

organised an Open Meeting to explain how it considered that the *Guidelines* should be applied in practice. In order to assist developing States to prepare for the submission of information under Article 76 and to defray the cost of their participation in Commission meetings, the UN General Assembly has established two trust funds. [52]

Given the political sensitivity of all boundary questions, there can be no guarantee that coastal States will carry out their obligations under Article 76 of UNCLOS and Article 4 of Annex II. However, it can at least be said that they now have a relatively precise definition in Article 76 and a ready source of expert assistance to help them in discharging their obligations.

1.2.3. *Meetings of States Parties*

Under Article 319(2)(e) of UNCLOS, the UN Secretary-General is required to "convene necessary meetings of States Parties in accordance with this Convention". On this unpromising foundation, Meetings of States Parties (SPLOS) have become yet another Law of the Sea institution, complete with its own Rules of Procedure. [53] In accordance with these Rules, meetings may be convened by the Secretary-General when he considers this necessary or upon the request of a State Party supported by a majority of States Parties. [54] Rule 4 lists the meetings which have to be held regularly in accordance with UNCLOS:

- Triennial meetings to elect members of the International Tribunal. [55]
- Quinquennial meetings to elect members of the Commission on the Limits of the Continental Shelf.[56]
- Meetings as necessary to deal with remuneration of members of the Tribunal and expenses of the Tribunal and other matters concerning the organisation of the Tribunal. [57]
- Meetings on the dates of elections to fill vacancies in the membership of the Tribunal. [58]
- Meetings on the dates of elections to fill vacancies in the membership of the Commission on the Limits of the Continental Shelf.[59]

Unsurprisingly, perhaps, some delegations have expressed ambitions to extend the role of the SPLOS meetings beyond the mundane administrative tasks referred to above. [60]

Chairman of the Commission on the Limits of the Continental Shelf on the Progress of Work in the Commission, CLCS/25, 1 September 2000, at para. 8.

[52] A/55/7, 30 October 2000. The terms of reference of the two trust funds are in Annexes I and II to this resolution.
[53] Rules of Procedure for Meetings of States Parties, SPLOS/2/Rev.3, 26 July 1995; and Addemdum (SPLOS/2/Rev. 3/Add.1, 12 June 1997).
[54] Rule 3.
[55] Rule 4(1), citing Art. 4 of Statute of Tribunal (Annex VI to UNCLOS).
[56] Rule 4(2), citing Art. 2 of Annex II to UNCLOS.
[57] Rule 4(3), referring to Arts. 18 and 19 of the Statute of the Tribunal.
[58] Rule 4(4)(a), referring to Art. 6(1) of the Statute of the Tribunal.
[59] Rule 4(4)(b).

They have proposed that reports on activities should be presented to SPLOS meetings by the Commission on the Limits of the Continental Shelf and the Seabed Authority. [61] Other delegations called for further consideration of the role of SPLOS meetings in dealing with matters relating to UNCLOS, "taking into account the recent recommendations on the establishment of an appropriate forum for dealing with ocean issues made by the seventh session of the Commission on Sustainable Development". [62] However, several delegations opposed these proposals, stressing that the functions of SPLOS meetings were laid down in UNCLOS and that they had no competence to assume other functions. [63] They pointed out too that the Commission and the Authority were autonomous bodies, and not required to report to SPLOS meetings. In their view, the only appropriate forum for the discussion of issues of a general nature related to UNCLOS was the UN General Assembly. [64] Given this diversity of opinion, it was agreed that the question should be debated further at the tenth meeting of SPLOS (22-26 May 2000). [65] The further debate at the tenth and eleventh meetings was inconclusive and it was decided to retain the question as an agenda item for the next meeting in 2002. [66]

1.2.4. United Nations Open-ended Informal Consultative Process on Oceans and the Law of the Sea

As has been seen, those who advocated an expanded role for SPLOS referred to a recommendation made by the Commission on Sustainable Development. [67] The objective of the recommendation was to enhance the effectiveness of the UN General Assembly's annual debate on oceans and the law of the sea. To this end, it was recommended that the General Assembly should establish "an open-ended informal consultative process, or other processes which it may decide, under the aegis of the General Assembly, with the sole function of facilitating the effective and constructive consideration of matters within the General Assembly's existing mandate". [68] The General Assembly accepted this recommendation and decided to establish what has later been known as the United Nations Open-ended Informal Consultative Process on Oceans and the Law of the Sea (UNICPO). [69]

This new institution or consultative process has already shown in its first two meetings that it is likely to become a very useful forum in which UN member States will have the

---

[60] Report of the Ninth Meeting of States Parties, SPLOS/48, 15 June 1999, paras. 49-53; *Oceans and the law of the sea: Report of Secretary-General*, A/54/429, 30 September 1999, para. 65.
[61] SPLOS/48, para. 49.
[62] SPLOS/48, para. 51.
[63] SPLOS/48, para. 52.
[64] *Ibid.*
[65] *Ibid.*, para. 53.
[66] See Report of the tenth Meeting of States Parties, SPLOS/60, 22 June 2000, paras. 73-78 and Report of the eleventh Meeting of States Parties, SPLOS/73, 14 June 2001, paras. 85-92.
[67] For the CSD recommendation, see Commission on Sustainable Development, *Report of the seventh session (1 May and 27 July 1998, and 19-30 April 1999)*, Economic and Social Council Official Records, 1999, Supplement No. 9, 1999, Decision 7/1. Oceans and seas, at paras. 38-45.
[68] *Ibid.*, paras. 38(d) and 39.
[69] Resolution 54/33, 24 November 1999.

opportunity to consider ocean affairs and law of the sea issues in much greater depth than has previously been possible in the annual General Assembly debate on the Secretary-General's *Reports on oceans and the law of the sea.* [70]

1.2.5. Dispute Settlement Institutions

Part XV of UNCLOS embodies a complex new system for the settlement of disputes. In order to bring it fully into operation, several new institutions or mechanisms had to be established: the International Tribunal for the Law of the Sea (ITLOS) and its various chambers, including the Sea-Bed Disputes Chamber; a list of arbitrators from which arbitral tribunals would be constituted to hear particular cases under Annex VII; lists of experts from which special arbitral tribunals would be constituted for particular cases under Annex VIII; and a list of conciliators from which conciliation commissions could be constituted in particular cases under Annex V. [71]

The International Tribunal on the Law of the Sea. ITLOS has its seat in Hamburg and consists of 21 independent members, [72] no two of whom may have the same nationality. [73] It must include not less than three members from each of the five United Nations regional groups [74] and the representation of the principal legal systems must be assured, as must be equitable geographical distribution. [75] Each State Party may nominate not more than two persons [76] and election is by secret ballot at a meeting of States Parties on the basis of a list of nominees. [77] Those nominees are declared elected who obtain the largest number of votes and a two-thirds majority of the States Parties present and voting, provided that such majority includes a majority of the States Parties. [78]

In accordance with Annex VI, Article 4(3), the election of judges was supposed to have taken place within six months of the entry into force of the UN Convention, that is, by 16 May 1995. However, at a meeting of States Parties to the UN Convention on 22 November 1994, it was decided to defer the first election of judges until 1 August 1996, though the nomination of candidates was to commence on 16 May 1995. [79] This allowed time for more States to become parties to the UN Convention and thus qualify to take part in the election. A State in the process of becoming a party to the UN Convention was permitted to nominate candidates provisionally but they were to be included in the final list of candidates only if ratification or accession took place before

---

[70] See further the Reports of the first meeting of UNICPO (30 May – 2 June 2000), A/55/274, 31 July 2000 and of the second meeting (7-11 May 2001), A/56/121, 22 June 2001.
[71] As at 6 February 2002, only 18 States had nominated arbitrators under Annex VII and 10 States had nominated conciliators under Annex V. For current lists, see http://www.un.org/Depts/los. For the purposes of Annex VIII, FAO, UNEP, IOC and IMO maintain lists of experts in their respective areas of competence.
[72] Statute of ITLOS (UNCLOS, Annex VI), Art. 2(1).
[73] *Ibid.*, Art. 3(1).
[74] *Ibid.*, Art. 3(2).
[75] *Ibid.*, Art. 2(2).
[76] *Ibid.*, Art. 4(1).
[77] *Ibid.*, Art. 4(4).
[78] *Ibid.*
[79] *Report of the Meeting of States Parties* (on 21 and 22 November 1994), SPLOS/3, 28 February 1995, p.7, para. 16 (reproduced in part in LOS Bull., No. 30, 1996, p.83).

1 July 1996. [80] Thirty-three candidates were nominated for the twenty-one places available. [81] Although three candidates were nominated by States other than their own (Anderson by France, [82] Kolodkin by Georgia and Rosenne by Austria), it is notable that no United States jurist was nominated.

Judges are elected for 9 years and may be re-elected. However, a system of rotation operates under which the term of office of one third of the 21 judges expires every three years. [83] The first elections duly took place on 1 August 1996 and on 24 May 1999, at the first triennial election of seven judges, 6 of the original judges were re-elected, together with 1 new judge. [84]

Sea-Bed Disputes Chamber. As was seen above, this Chamber of ITLOS was first constituted in 1997 and reconstituted in 1999. [85]

Special Chambers. In addition to the Sea-Bed Disputes Chamber, ITLOS is also empowered to form special chambers of various kinds:

Summary Procedure Chamber. First, it is required annually to form a special chamber for the speedy dispatch of business by summary procedure. [86] This chamber consists of five members of the Tribunal, with two alternates. [87] National members retain their right of participation in cases involving their home States [88] and, as in the full Tribunal, an *ad hoc* member may be substituted for a regular member if one of the parties to the case does not have a member of its nationality in the chamber. [89] Such an *ad hoc* member will be drawn from the membership of the Tribunal if it includes a member of the required nationality. Failing this, or if such member is unable to be present, a non-member of the tribunal may be specially chosen by the parties. [90] It follows that a judgment may be given by a five-member summary procedure chamber on which only three regular members of the Tribunal are sitting. [91]

---

[80] *Ibid.*
[81] *Election of the Members of the International Tribunal for the Law of the Sea. List of Candidates submitted by Governments*, SPLOS/10, 2 July 1996.
[82] France had apparently decided against nominating a French national. The United Kingdom was unable to make a nomination because of the failure to accede to the UN Convention before 1 July 1996, due to concern over the implications for UK fishery zone limits in the Rockall area which accession might have. See further the explanation given by Baroness Chalker (Minister of State, Foreign and Commonwealth Office) in the House of Lords on 20 June 1996 (*House of Lords Weekly Hansard*, No. 1671, 17-20 June 1996, Cols. 455-457).
[83] Annex VI, Art. 5(1).
[84] UNCLOS, *Report of the Ninth Meeting of States Parties*, SPLOS/48, 15 June 1999, paras. 31-34. Five judges were elected at the first round and the remaining two (Judge Kolodkin of the Russian Federation and Judge Engo of Cameroon) at the third round.
[85] See above, section 1.2.1.
[86] Annex VI, Art. 15(3).
[87] *Ibid.*
[88] Annex VI, Art. 17(1) and (4).
[89] Annex VI, Art. 17(2)-(6).
[90] Annex VI, Art. 17(4).
[91] The other two being *ad hoc* judges not regular members of the Tribunal.

THE LAW OF THE SEA AT THE MILLENNIUM 37

The Chamber was first constituted in October 1996 and has been reconstituted annually. So far, no cases have been referred to the Chamber.

Particular category chambers. Secondly, the Tribunal may form such special chambers, composed of three of more of its members, as it considers necessary for dealing with particular categories of disputes. [92] Given the fact that such a chamber might consist of only three members and that two of them might in certain circumstances be *ad hoc* judges, [93] it is possible that a judgment might be given by a chamber having only one regular member of the Tribunal amongst its members.

So far, two such chambers have been constituted, the *Chamber for Fisheries Disputes* and the *Chamber for Marine Environment Disputes*. The *Chamber for Fisheries Disputes* is available to deal with any dispute which the parties agree to submit to it on the conservation and management of marine living resources. The Chamber was first constituted in February 1997 when ITLOS decided by consensus on a list of 7 judges proposed by the President, following consultations. [94] It was reconstituted in October 1999. [95] The *Chamber for Marine Environment Disputes* was also constituted in February 1997 with 7 members proposed by the President. [96] It too was reconstituted in October 1999. [97]

So far, no cases have been referred to either of the two special chambers.

Chambers for particular disputes. Under Article 15(2) of the Tribunal's Statute, the Tribunal may form a chamber for dealing with a particular dispute if the parties so request. The composition of the chamber is determined by the Tribunal with the approval of the parties in accordance with Article 30 of the Rules of the Tribunal. Such a chamber was first established in December 2000 to deal with the *Case concerning the Conservation and Sustainable Exploitation of Swordfish Stocks in the South-eastern Pacific Ocean.* [98]

As the President of the Tribunal has noted, since the composition of such a chamber requires the approval of the parties, this type of special chamber may be of particular interest to States which generally prefer arbitration to other modes of dispute settlement. [99]

---

[92] Annex VI, Art. 15(1).
[93] Annex VI, Art. 17(2)-(6).
[94] ITLOS/Press 5, 3 March 1997.
[95] ITLOS/Press 31, 4 October 1999.
[96] ITLOS/Press 5, 3 March 1997.
[97] ITLOS/Press 31, 4 October 1999.
[98] *Annual Report of the International Tribunal for the Law of the Sea for 2000*, SPLOS/63, 6 April 2001, paras. 19-22 and 36-37. On this case, see further below, section 2.3.2, under *Conservation and Sustainable Exploitation of Swordfish Stocks in the South-Eastern Pacific Ocean, 2000 - Case No. 7.*
[99] Statement by P. Chandrasekhara Rao, President of the International Tribunal for the Law of the Sea, on the Report of the Tribunal at the Eleventh Meeting of the States Parties to the Law of the Sea Convention, 14 May 2001.

38   E.D. BROWN

Ad hoc Expert Panels under Fish Agreement. Finally, brief mention should be made of the *ad hoc* expert panels for which provision is made in Article 29 of the 1995 Agreement for the Implementation of the Provisions of UNCLOS on the Conservation and Management of Straddling Stocks and Highly Migratory Fish Stocks. Under Article 30, the provisions on settlement of disputes in Part XV of UNCLOS apply, *mutatis mutantis*, to disputes on the interpretation or application of the Convention. In addition, however, under Article 29, where a dispute concerns a matter of a technical nature, it may be referred to an *ad hoc* expert panel established by the States concerned. The role of the panel is to confer with the States concerned and endeavour to resolve the dispute expeditiously without recourse to binding procedures for the settlement of disputes. Following the entry into force of the Agreement on 11 December 2001, it will now be possible to employ this procedure.

2. Settlement of Disputes under UNCLOS

As has been seen, the new dispute settlement institutions called for by UNCLOS have now been established. This section broadens the picture and its purpose is to present a brief review of the operation so far of the UNCLOS system of dispute settlement. It begins with a description of the system and recalls questions which have been asked about the need for, and wisdom of establishing, such a complex system. Although it is still too early in the life of the new system to draw firm conclusions on its success, it may not be too early to attempt an interim, tentative evaluation. In drawing up a balance sheet, reference is made to the attitude of States Parties as reflected in declarations they have made (or not made) on choice of compulsory procedures; to the record of States in referring disputes to various fora since UNCLOS entered into force; and to the caseloads of the International Tribunal for the Law of the Sea and the International Court of Justice since the Convention entered into force.

2.1. THE DISPUTE SETTLEMENT SCHEME IN OUTLINE

The general scheme for the settlement of disputes is laid down in Part XV of UNCLOS and falls into three Sections dealing with: (1) the general obligation to settle disputes by peaceful means and preliminary steps to which all disputes are subject; [100] (2) compulsory procedures entailing binding decisions; [101] and (3) limitations and optional exceptions to such compulsory procedures. [102]

2.1.1. General Provisions
Article 279 of Section 1 is simply a restatement of the obligation upon members of the United Nations under Articles 2(3) and 33(1) of the Charter of the United Nations to settle disputes by peaceful means. It is followed by Article 280, which provides that

---

[100] Part XV, Section 1 (Arts. 279-285).
[101] Part XV, Section 2 (Arts. 286-296).
[102] Part XV, Section 3 (Arts. 297-299).

nothing in Part XV impairs the right of States Parties to agree at any time to settle a dispute between them relating to the interpretation or application of the Convention by any peaceful means of their choice. Such choice may be dictated by the terms of other agreements imposing particular binding procedures upon the parties to them, as recognised in Article 282, or emerge from the obligation to exchange views regarding settlement, as envisaged in Article 283.

2.1.2. Compulsory Procedures Entailing Binding Decisions
If no settlement is reached by recourse to Section 1, then, under Article 286, the dispute must be submitted, at the request of any party to it, to the court or tribunal which has jurisdiction under Section 2. Article 287(1) of this Section allows States Parties to make a written declaration choosing one or more of the four procedures specified: the International Tribunal for the Law of the Sea; the International Court of Justice (ICJ); an arbitral tribunal under Annex VII; or a special arbitral tribunal under Annex VIII (but only for categories of disputes specified therein). Where a party or parties to a dispute have made no such declaration or have not accepted the same procedure in their declarations, the dispute must be referred to arbitration under Annex VII.

2.1.3. Limitations and Exceptions to Applicability of Section 2
Section 3 includes three articles:
Article 297 embodies *general limitations* to the applicability of Section 2 procedures, those limitations being "general" in the sense that all States Parties are automatically entitled to invoke opting-out clauses in relation to categories of dispute referred to in Article 297.

Article 298 embodies further *optional exceptions* to the applicability of Section 2 procedures, these exceptions being "optional" in the sense that, if a State Party wishes to exclude any of the specified categories of dispute from the application of Section 2 procedures, it must make a written declaration to that effect; and, finally, *Article 299* recognises the right of States Parties, notwithstanding such limitations or exceptions, to settle disputes by any procedure they wish, so long as they have agreed upon it.

General Limitations on Applicability of Section 2 – Article 297. Article 297 refers to disputes concerning the interpretation or application of UNCLOS with regard to three separate areas: (1) the exercise by a coastal State of its sovereign rights or jurisdiction provided for in the Convention; (2) marine scientific research (MSR); and (3) fisheries. So far as MSR and fisheries are concerned, the relevant provisions – Article 297(2) and Article 297(3) – follow the same basic pattern: first, there is a statement of the general rule that disputes shall be settled in accordance with Section 2; and this is followed by a statement of the coastal State's right to exclude specified categories of dispute from submission to such settlement procedures.

Dealing first with MSR disputes, Article 297(2), after providing that MSR disputes will be settled in accordance with Section 2, allows the coastal State to opt out of any such procedure when the dispute arises out of: (1) the exercise by the coastal State of a right or discretion in accordance with Article 246 (governing MSR in the EEZ and on the continental shelf); or (2) a decision by the coastal State to order suspension or cessation of a research project in the EEZ or on the continental shelf under Article 253. If the researching State alleges that the coastal State has not exercised its rights under Articles 246 and 253 in a manner compatible with the Convention, the dispute may be submitted, at the request of either party, to conciliation under Part V, Section 2. However, the report of the commission is not binding [103] and, in any event, "the conciliation commission shall not call in question the exercise by the coastal State of its discretion to designate specific areas as referred to in article 246, paragraph 6, or of its discretion to withhold consent in accordance with article 246, paragraph 5". [104]

The pattern is similar in relation to fisheries disputes. Article 297(3), after specifying that fisheries disputes must be settled in accordance with Section 2, [105] allows the coastal State to exclude from this obligation "any dispute relating to [the coastal State's] ....sovereign rights with respect to the living resources in the exclusive economic zone or their exercise, including its discretionary powers for determining the allowable catch, its harvesting capacity, the allocation of surpluses to other States and the terms and conditions established in its conservation and management laws and regulations". Here too, where no settlement is reached by resort to Section 1, the dispute may be referred to conciliation, at the request of either party, in three specified cases. Once again, however, the recommendations of the conciliation commission are not binding [106] and in no case may it substitute its discretion for that of the coastal State. [107]

Optional Exceptions to Applicability of Section 2 – Article 298. Under Article 298(1), a State may declare that it does not accept any one or more of the Section 2 procedures for one or more of three categories of disputes:
(1) disputes over sea boundary delimitations or historic bays or titles;
(2) disputes over military activities and disputes over law enforcement activities concerning the exercise of sovereign rights or jurisdiction excluded from the jurisdiction of a court or tribunal under Article 297(2) or 297(3);
(3) disputes in respect of which the Security Council is exercising functions assigned to it by the United Nations Charter.

Such declarations in writing may be made when signing, ratifying or acceding to the convention or at any time thereafter and may be withdrawn at any time. Since, however, they are "without prejudice to the obligations arising under Section 1", such

---

[103] UNCLOS, Annex V, Art. 7(2), as applied by Art. 14.
[104] UNCLOS, Art. 297(2)(b).
[105] More detailed provision is made for the settlement of disputes concerning the interpretation or application of the Fish Agreement of 4 August 1995 (note 11 above) which implements the provisions of UNCLOS dealing with straddling fish stocks and highly migratory fish stocks. See further above, section 1.2.5, under *Ad hoc Expert Panels under Fish Agreement*, and E.D. Brown, "Dispute settlement and the law of the sea: the UN Convention regime", *Marine Policy*, Vol. 21, No. 1, 1997, pp. 17-43, at p. 41.
[106] Annex V, Art. 7(2), as applied by Art. 14.
[107] UNCLOS, Art. 297(3)(c).

declarations do not relieve States Parties of the general obligation to settle disputes by some peaceful means.

So far as "sea boundary delimitations" are concerned, Article 298(1)(a)(i) refers to disputes over the interpretation or application of Articles 15, 74 and 83, which deal respectively with delimitation of the territorial sea, the EEZ and the continental shelf between States with opposite or adjacent coasts. States making such opting-out declarations are obliged to accept nothing more than "compulsory conciliation" under Article 298(1)(a) of the Convention. This compels them to:

- "accept submission of the matter to conciliation under Annex V, section 2". [108] This would empower the Commission to proceed even if one of the parties declined to take part in the proceedings. [109]
- "negotiate an agreement on the basis of" the report of the Conciliation Commission. [110] That this is no more than a *pactum de contrahendo*, an agreement to negotiate in good faith with an intent to reach agreement, is evident from the fact that provision is made for the contingency that agreement is not reached in this way. Where this is so, the third element of compulsion is the obligation:
- "by mutual consent" to submit the question to a Section 2 procedure, unless the parties otherwise agree. [111] This again is no more than *pactum de contrahendo*. There is no guarantee that negotiation in good faith will enable the parties to agree upon a Section 2 procedure or any other form of binding settlement.

It may be noted that the obligation to submit to conciliation applies only "when such a dispute arises subsequent to the entry into force of this Convention and where no agreement within a reasonable period of time is reached in negotiations between the parties". [112]

Article 298(1)(b) deals with the optional exceptions in relation to disputes over military activities and disputes over law enforcement activities concerning the exercise of sovereign rights or jurisdiction excluded from the jurisdiction of a court or tribunal under Article 297(2) or 297(3). It allows States Parties to opt out of Section 2 procedures in two cases. First, they may declare any one or more Section 2 procedures unacceptable in relation to military activities by government vessels and aircraft engaged in non-commercial service. This would include, for example, disputes over military manoeuvres conducted by one State in another State's EEZ. The second potential opt-out follows from the provisions of Article 297(2)-(3), under which, as was seen above, States Parties may exclude from Section 2 procedures certain categories of disputes concerning marine scientific research or EEZ fisheries. Article 289(1)(b) ensures that disputes concerning the law enforcement activities of the coastal State in relation to such excluded matters may also be excluded from Section 2 procedures.

---

[108] Art. 298(1)(a)(i).
[109] Annex V, Art. 11(2).
[110] Art. 298(1)(a)(ii).
[111] *Ibid.*
[112] Art. 298(1)(a)(i).

Finally, States Parties may opt to exclude from Section 2 procedures disputes in respect of which the Security Council is exercising functions assigned to it by the UN Charter - "unless the Security Council decides to remove the matter from its agenda or calls upon the parties to settle it by the means provided for in this convention". [113]

2.1.4. Sea-bed Mining Disputes

So far as sea-bed mining is concerned, Section 5 of Part XI of UNCLOS makes provision for "settlement of disputes and advisory opinions". [114] The principal forum is the Sea-Bed Disputes Chamber of the International Tribunal for the Law of the Sea but certain types of dispute may, alternatively, be submitted to a special chamber of the Tribunal, [115] to an *ad hoc* chamber of the Sea-Bed Disputes Chamber, [116] or to binding commercial arbitration. [117]

The 1994 Implementation Agreement and Part XI of UNCLOS are to be interpreted and applied as a single instrument and, in the event of any inconsistency between them, the provisions of the Agreement are to prevail. [118] Given this integration of the two instruments, it may be assumed that the dispute settlement provisions of the Convention would apply to disputes arising from the terms of the Agreement, including disputes concerning any inconsistency between the two, and that the Sea-Bed Disputes Chamber would have jurisdiction over them [119] However, the Agreement also makes express provision for application of the dispute settlement provisions of the Convention in the following cases: disputes relating to the disapproval of a plan of work; [120] disputes concerning the provisions of GATT and its successor instruments in their application to the production policy of the Authority – but only in cases where one or more of the States concerned are not parties to these agreements; [121] and disputes concerning the interpretation or application of the rules and regulations for financial terms of contracts. [122]

---

[113] Art. 298(1)(b).
[114] Title of Section 5. Under Art. 191, the Sea-Bed Disputes Chamber "shall give advisory opinions at the request of the Assembly or the Council [of the Authority] on legal questions arising within the scope of their activities".
[115] Under Art. 188(1)(a), disputes between States Parties concerning the interpretation or application of Part XI may be submitted to such a chamber at the request of the parties.
[116] Disputes between States Parties concerning the interpretation or application of Part XI may be submitted to such an *ad hoc* chamber at the request of any party to the dispute under Art. 188(1)(b).
[117] See further E.D. Brown, *loc. cit.* in note 105, at pp. 26-27.
[118] Implementation Agreement, Art. 2(1).
[119] Art. 187 of UNCLOS, read with Art. 2(1) of the Implementation Agreement. Several States (Austria, Finland, Germany and Sweden) in making their choice of fora under Art. 287, expressly refer to disputes arising under both UNCLOS and the Implementation Agreement.
[120] Implementation Agreement, Annex, Section 3, para. 12.
[121] Annex, Section 6, para. 1(f)(ii). Where the States concerned are parties to these agreements, the dispute settlement procedures of these agreements apply (para.1(f)(i)).
[122] Section 8, para. 1(f).

2.2. AN OVERCOMPLEX SCHEME?

In considering whether the UNCLOS scheme for dispute settlement is too complex, it is useful to pose two fundamental questions:
(1) How far does UNCLOS succeed in providing a comprehensive dispute settlement system which places an obligation upon States Parties to resolve disputes by reference to compulsory procedures entailing binding decisions? and
(2) Is the system incorporated in the Convention well designed to achieve the two principal objectives of any such system: the definitive settlement of disputes and the development of a uniform and consistent jurisprudence?

2.2.1. *The Limited Scope of Compulsory Procedures Entailing Binding Decisions*
The answer to the first question is clear. The Convention permits States Parties to exclude disputes over several important areas from the scope of the obligation to submit disputes to compulsory procedures entailing binding decisions. As seen above, they include certain disputes relating to marine scientific research; fisheries; maritime boundaries; military activities and enforcement activities; and disputes where the Security Council is acting.

The degree to which States may exclude the application of the Convention's dispute settlement system varies from subject to subject. Thus, where, under Article 298, a State Party opts out in relation to military activities, enforcement activities, and disputes where the Security Council is acting, there is no consequential obligation to refer the dispute to conciliation under Annex V. On the other hand, where, under Article 297, a State Party opts out in relation to marine scientific research disputes or disputes over EEZ fisheries, there is a consequent obligation to refer the dispute to conciliation under Annex V. Similarly, where a State opts out in relation to boundary disputes, under Article 298, there is a consequent obligation to refer the dispute to conciliation. In this case, however, it is added in Article 298(1)(a)(ii) that, if negotiations based on the conciliation commission's report do not result in an agreement, the parties are subject to a *pactum de contrahendo* to submit the dispute to a Section 2 procedure, unless they otherwise agree (another *pactum de contrahendo*). The strongest of these obligations is that arising under Article 298(1)(a)(ii) but amounts to no more than an obligation to seek in good faith to agree upon a Section 2 procedure or to "otherwise agree". In all of these opting-out situations, there remains an obligation to settle the dispute by some peaceful means under Section 1 of Part XV but that is a very general obligation.

2.2.2. *Will the System Achieve the Definitive Settlement of Disputes and the Development of a Uniform and Consistent Jurisprudence?*
The answer to the first part of this question is that, no matter which of the various fora is employed, the system does offer States Parties the opportunity to settle disputes definitively. However, this is of course subject to the major qualification that this is so only if the dispute concerns an area of the Convention not subject to one of the opt-outs referred to in the previous section.

More difficult to answer is the question whether such a multifaceted system can achieve a uniform and consistent jurisprudence. [123] Much will depend on the pattern of choice of fora which will eventually emerge from declarations made by States Parties. As is noted below, 112 States Parties are deemed to have accepted arbitration under Annex VII simply because they have not so far submitted a declaration expressly choosing any other option. [124] If this practice is indicative of the final pattern, the prospects for the development of a uniform and consistent jurisprudence are poor. There will then be no limit to the number of different arbitral tribunals which may be set up and no way, therefore, in which uniformity and consistency may be guaranteed. Moreover, even if the pattern changes and ITLOS is able to establish itself as the principal judicial organ in law of the sea matters, it will still be difficult to ensure a uniform and consistent jurisprudence. As has been seen, if the parties to a dispute so request, the judgment of the tribunal may in fact take the form of one rendered by any one of the variety of special chambers or ad hoc chambers which ITLOS may form. [125] The need to provide a means to achieve harmonisation of the jurisprudence of the tribunal and its chambers was discussed in the Preparatory Commission and one delegation suggested the inclusion in the rules of the Tribunal of an article providing for a kind or harmonisation procedure when there were different views between either two or more chambers or between a chamber and the Tribunal. [126]

More recently, the former President of the International Court of Justice (ICJ), Judge Schwebel, has referred to this issue. He began by acknowledging that, "Concern that the proliferation of international tribunals might produce substantial conflict among them, and evisceration of the docket of the [ICJ]..., have not materialised, at any rate as yet", and noted that, "A greater range of international legal fora is likely to mean that more disputes are submitted to international judicial settlement". However, he was sufficiently concerned about the possibility of conflict to suggest a means of ameliorating it. He suggested that:

> In order to minimise ....significant conflicting interpretations of international law, there might be virtue in enabling other international tribunals to request advisory

---

[123] For differing views on the merits of introducing this new system of dispute settlement, see S. Oda (ICJ Judge), "Dispute Settlement Prospects in the Law of the Sea", *International and Comparative Law Quarterly*, Vol. 44, 1995, pp. 863-872, and, in response, J.I. Charney, "The Implications of Expanding International Dispute Settlement Systems: The 1982 Convention on the Law of the Sea", *American Journal of International Law*, Vol. 90, 1996, pp. 69-75. More generally, see L. Boisson de Chazournes *et al.*, "Implications of the Proliferation of International Adjudicatory Bodies for Dispute Resolution" (*ASIL Bulletin*, No. 9, 1995).
[124] See below, section 2.3.1, at note 144.
[125] Annex VI, Art. 15(4)-(5).
[126] The question was considered in Special Commission 4 during its examination of the draft Rules of the Tribunal in 1985. See LOS/PCN/SCN.4/L.3, 3 April 1985 (R. Platzöder, ed., *The Law of the Sea: Documents 1983-1989*, Vol. VII, 1990, pp. 29-42), at pp. 41-42, paras. 99-100. It was suggested that, "A conceivable way of meeting such a situation could be for the Chamber to relinquish jurisdiction to the Tribunal in order to achieve a continuous and reliable development in the interpretation of the Convention" (*ibid.*, para. 99). The *Provisional Report of Special Commission 4* (in Preparatory Commission, *Consolidated Provisional Final Report*, Vol. I, LOS/PCN/130, 17 November 1993, pp. 99-114) refers to this suggestion (at p. 105) but reports no further progress on the question. There is no such provision in the *Rules of the Tribunal* (ITLOS/8), adopted on 28 October 1997.

opinions of the International Court of Justice on issues of international law that arise in cases before those tribunals that are of importance to the unity of international law [127]

Judge Schwebel recognised that the case for such a mechanism was stronger – that there appeared to be no jurisdictional problem – in respect of international tribunals that are organs of the United Nations. [128] However, he added that:

> There is room for the argument that even international tribunals that are not United Nations organs, such as the International Tribunal for the Law of the Sea, ….might, if they so decide, request the General Assembly – perhaps through the medium of a special committee established for the purpose – to request advisory opinions of the Court. [129]

Judge Schwebel was clearly not yet persuaded of the wisdom of establishing new universal courts in competition with what he had earlier described as "the world's most senior international court" and what Article 1 of the Statute of the ICJ designates as "the principal judicial organ of the United Nations". He ended by advising that, "In any event, a certain caution in the creation of new universal courts may be merited in respect of inter-State disputes". [130]

The current President, Judge Guillaume, has recently expressed very similar views. Addressing the UN General Assembly on 30 October 2001, he contended that the proliferation of international tribunals, "….raises the risk of parties competing for courts – sometimes referred to as forum shopping – and overlapping jurisdiction". Again, he warned that, "The proliferation of international courts may jeopardise the unity of international law and, as a consequence, its role in inter-State relations". His advice was that:

> No new international court should be created without first questioning whether the duties which the international legislator intends to confer on it could not be better performed by an existing international court. International judges should be aware of the dangers involved in the fragmentation of the law and take efforts to avoid such dangers. However, those efforts may not be enough, and the International Court of Justice, the only judicial body vested with universal and general jurisdiction, has a role to play in this area. For the purpose of maintaining the unity of the law, the various existing courts or those yet to be created could, in my opinion, be empowered in certain cases – indeed encouraged – to request advisory

---

[127] Address to the Plenary Session of the General Assembly of the United Nations by Judge Stephen M. Schwebel, President of the International Court of Justice, 26 October 1999.
[128] *Ibid. E.g.* ,the international tribunals for the prosecution of war crimes in the former Yugoslavia and Rwanda.
[129] *Ibid.*
[130] *Ibid.*

opinions from the International Court of Justice through the intermediary of the Security Council or through the General Assembly. [131]

It seems unlikely, at least in the short term, that these judicial recommendations will have much impact upon States Parties to UNCLOS or upon ITLOS. Whether, in the longer term, they will commend themselves to the international community will very much depend upon the reputation which the Tribunal is able to establish.

2.3. A TENTATIVE EVALUATION

In attempting an interim evaluation of the success of the UNCLOS system of dispute settlement, it seems fair to gauge that success by referring to two indicators: (1) the declarations on choice of procedures made so far by States Parties; and (2) State practice in resolving law of the sea disputes since UNCLOS entered into force on 16 November 1994.

2.3.1. Declarations on Choice of Dispute Settlement Mechanisms

As at 6 February 2002, 40 States had included in declarations made on signing, ratifying or acceding to the UN Convention, or in separate declarations, statements relevant to dispute settlement matters. [132] In 6 cases (Bangladesh, [133] Brazil, [134] India, [135] Iran, [136] Pakistan, [137] and South Africa [138]), the States concerned simply reserved their right to indicate their choice in a later declaration. In 3 further cases (Algeria, [139] Cuba, [140] and Guinea-Bissau [141]), there is only a negative indication that the jurisdiction of the International Court of Justice is not accepted. Vietnam's declaration, [142] referring to peaceful negotiations as a vehicle for resolving a limited category of disputes, does not amount to a choice of fora under Article 287.

---

[131] Speech by H.E. Judge Gilbert Guillaume, President of the International Court of Justice, to the General Assembly of the United Nations, 30 October 2001.

[132] This analysis is based on declarations reproduced in *The Law of the Sea. Declarations and Statements with respect to the United Nations Convention on the Law of the Sea and to the Agreement relating to the Implementation of Part XI of the United Nations Convention on the Law of the Sea of 10 December 1982* (UN Publications, Sales No. E.97.V.3, 1997, referred to below as "E.97.V.3".) which contains declarations made down to 31 December 1996, updated by reference to UN website http://www.un.org/Depts/los/los_decl.htm, accessed 7 February 2002. On the position of the United States, see Transmittal Letter from the President to the US Senate, 7 October 1994 (34 ILM (1995), p.1396) and Submittal Letter from the Secretary of State to the President, 23 September 1994 (*ibid.*, p. 1397, at pp. 1399 and 1440-1443).

[133] LOS Bull. No. 46, p. 14.
[134] E.97.V.3, pp. 22-23.
[135] *Ibid.*, p.31.
[136] *Ibid.*, pp. 10-11.
[137] LOS Bull. No. 34, pp. 7-8.
[138] LOS Bull. No. 36, p.8.
[139] E.97.V.3, p.19.
[140] *Ibid.*, pp. 23-24.
[141] *Ibid.*, p. 30.
[142] *Ibid.*, pp. 45-46.

The analysis presented below is based on the remaining 30 declarations made by the following States: Argentina, Austria, Belarus, Belgium, Cape Verde, Chile, Croatia, Egypt, Finland, France, Germany, Greece, Hungary, Iceland, Italy, Oman, the Netherlands, Nicaragua, Norway, Philippines, Portugal, Russia, Slovenia, Spain, Sweden, Tanzania, Tunisia, Ukraine, United Kingdom and Uruguay. [143]

As has been seen, States Parties may include in declarations made on signature, ratification or accession, or later, *inter alia:*
(1) a choice of fora under Article 287;
(2) an indication that they wish to exclude issues referred to in Article 297(2) and (3) from the application of Section 2 binding procedures;
(3) an indication that they wish to exclude issues referred to in Article 298 from the application of Section 2 binding procedures; and
(4) a choice of forum to deal with disputes over the prompt release of detained vessels and crews under Article 292.

It has to be remembered too that, under Article 287(3), any State Party which has not made a choice under Article 287(1) and thus finds itself "a party to a dispute not covered by a declaration in force", will be deemed to have accepted arbitration in accordance with Annex VII. Pending any later submission of a declaration, this would apply to 112 of the 138 "States Parties" which had become parties to the Convention by 6 February 2002. [144]

The practice of the 30 States may be summarised as follows:
General Choice of Fora under Article 287. Of the 30 States surveyed, 5 expressly opted for arbitration under Annex VII (Belarus, Egypt, Russia, Slovenia and Ukraine). A further 4 States (France, Iceland, Philippines and Tunisia) made no express choice of fora under Article 287, but, under Article 287(3), will be deemed to have accepted arbitration under Annex VII for any "dispute not covered by a declaration in force". If, in addition, account is taken of the 112 States Parties which have not included in declarations any statement relevant to dispute settlement, [145] and are therefore also

---

[143] For the texts of these declarations made on signature (S), ratification (R) or accession (A), see the following sources: Argentina (R), E.97.V.3, pp.19-21; Austria (R), *ibid.*, p. 21; Belarus(S), *ibid.*, p.2; Belgium(R), LOS Bull. No. 39, p. 11; Cape Verde(R), *ibid.*, p.22; Chile(R), LOS Bull. No. 35, pp. 9-11; Croatia (Declaration of 4 November 1999 as successor State), LOS Bull. No. 42, p. 14; Egypt(R), E.97.V.3, pp.24-26; Finland(R), *ibid.*, p.26; France(R), *ibid.*, pp.26-27; Germany(A), *ibid.*, pp. 27-29; Greece(R). *ibid.*, p.30; Hungary (R), UN website, *loc. cit.* in note 132 above; Iceland(R), *ibid.*, p.31; Italy(R), LOS Bull. No. 34, p.7; Netherlands(R),E.97.V.3, pp.35-37; Nicaragua (R), LOS Bull. No. 43, p.13; Norway(R), *ibid.*, p.37; Oman(R), *ibid.*, pp.38-39; Philippines(R), *ibid.*, p.40; Portugal(R), LOS Bull. No. 36, pp.7-8; Russia(S and R), E.97.V.3, pp.14-15 and LOS Bull No. 34, p.9; Slovenia(Declarations of 11 October 2001), UN website, *loc. cit.* in note 132 above; Spain(R), UN website, *loc. cit.* in note 132 above; Sweden(R), E.97.V.3, pp.42-43; Tanzania(R), *ibid.*, p.44; Tunisia,(Declaration, 31 May 2001) LOS Bull. No. 46, p.14; Ukraine(R), LOS Bull. No. 41, p.14; United Kingdom(A), LOS Bull. No. 36, p.9; Uruguay (S and R), E.97.V.3, pp 44-45.

[144] That is, the 138 "States Parties" at 6 February 2002 minus the 26 States which had made a choice of fora under Art. 287 in their declarations. Under Art. 2 of the UN Convention, the term "States Parties" includes entities other than States "which become Parties to [the]... Convention in accordance with the conditions relevant to each ...".

[145] See text above, at note 144.

deemed to have accepted arbitration under Annex VII, the number committed to such arbitration rises to 116. Of the remaining States Parties, two States (Argentina and Chile) chose ITLOS and Annex VIII in order of preference. Two States (Austria and Germany) opted for ITLOS, Annex VIII and the ICJ in that order, while a third, Belgium, selected the same fora but in a different order: Annex VIII, ITLOS and the ICJ. Portugal chose ITLOS, the ICJ, Annex VII and Annex VIII without stating any order of preference. Three States (Greece, Tanzania and Uruguay) chose ITLOS alone (in the case of Uruguay, without prejudice to its recognition of the jurisdiction of the ICJ). Another six States (the Netherlands, Nicaragua, Norway, Spain, Sweden and the United Kingdom) chose the ICJ alone, though the United Kingdom was ready to consider submissions to ITLOS on a case-by-case basis. Finally, Finland, Italy and Oman opted for the ICJ and ITLOS without stating any order of preference, and Cape Verde, Croatia and Hungary chose ITLOS and the ICJ in order of preference.

It would seem, therefore, that the clear preference at the moment is for arbitration under Annex VII. It remains to be seen whether more States will opt for ITLOS once it has had more time to establish its reputation.

Choice of Fora for Specialised Disputes under Article 287. Special arbitration under Annex VIII has been chosen by ten States (Argentina, Austria, Belarus, Belgium, Chile, Germany, Hungary, Portugal, Russia and Ukraine) for disputes concerning fisheries, protection and preservation of the marine environment, marine scientific research and navigation, including pollution from vessels and by dumping.

Opting Out under Article 297. Egypt alone excludes from arbitration under Annex VII (its choice of forum under Article 287) disputes contemplated in Article 297. Mention should be made under this heading of the interesting declaration made by Spain to the effect that it interprets Article 297, without prejudice to its provisions on settlement of disputes, to mean that, under Articles 56, 61 and 62, the powers of the coastal State to determine the allowable catch, its harvesting capacity and the allocation of surpluses to other States may not be considered as discretionary. At first sight, it is not difficult to sympathise with the Spanish view of the powers of the coastal State over its EEZ fisheries. Although it has sovereign rights over these fisheries under Article 56(1), their exercise is limited by the obligation under Article 56(2) to have due regard to the rights of other States, and by the conservation and utilisation duties placed upon it by Articles 61 and 62. Given the detailed criteria to be observed by the coastal State in determining the allowable catch under Article 61 and the existence and allocation of surplus stock under Article 62, it might seem inappropriate at first sight to describe the coastal State's powers as "discretionary". However, in reality, the application of the criteria requires the coastal State to make a series of judgments and interpretations and it seems fair to describe the process of implementing these articles as the exercise in good faith of a discretion; and, as will be seen below, this is borne out by the terms of Article 297(3).

The more important question is whether the exercise of the discretion is open to binding third-party review and Article 297(3) gives a clear answer. The coastal State is not obliged to submit to a Section 2 compulsory-settlement mechanism any dispute relating to its sovereign rights over EEZ fisheries, *"including its discretionary powers for determining the allowable catch, its harvesting capacity [and] the allocation of surpluses to other States...."* What, then, is the effect of the Spanish declaration? Given the fact that it is expressed to be without prejudice to Article 297's dispute settlement provisions, it appears to be no more than an attempt to place a gloss upon Articles 56, 61 and 62 in the interest of maximising access for the large Spanish fishing fleet to the EEZ fisheries of other States Parties.

Opting Out under Article 298. Five of the 30 States surveyed exclude all disputes specified in Article 298(1)(a)(b) and (c) from the application of Section 2 compulsory procedures. Nicaragua accepts only recourse to the ICJ for such disputes and Norway declares that it does not accept arbitration under Annex VII for such disputes. In addition, 12 of the 30 States exclude specified issues from the application of Section 2 compulsory procedures. Maritime delimitation is excluded by 10 States; military activities are excluded by 9 States; 3 States exclude law enforcement activities in relation to marine scientific research in the EEZ or on the continental shelf, and EEZ fisheries; finally 8 States exclude disputes where the Security Council is acting. Reference is made to Article 298 in the Philippines declaration but its impact is unclear.

Choice for Disputes over Release of Vessels and Crew under Article 292. Three States (Belarus, Russia and Ukraine) have chosen ITLOS as their preferred forum for the resolution of disputes concerning the prompt release of vessels and crews detained by another State Party.

Lack of Discernible Trends in Choice of Fora. Given the considerable diversity in the choice of fora in declarations made so far, it is difficult to detect any marked pattern. One may well wonder if the fact that 81% of States Parties (112 out of 138 States Parties) have made no express choice reflects a conscious decision to accept arbitration under Annex VII by default, or is due rather to a lack of interest in the question. Whatever be its cause, it hardly reflects enthusiasm for the new dispute settlement scheme provided by Part XV.

2.3.2. State Practice in Resolving Disputes in UNCLOS Era

As has been seen, the dispute settlement scheme embodied in Part XV of UNCLOS offers a range of fora and mechanisms for the pacific settlement of law of the sea disputes and, just as it would be premature to predict the long-term pattern of choice of mechanisms on the basis of declarations made so far, so also it is too early to discern more than immediate trends in State practice on the reference of disputes to international courts and tribunals. This is particularly so in relation to the reference of disputes to arbitral tribunals established under Annex VII. For this reason, this brief survey concentrates on the two premier bodies – the International Court of Justice and the International Tribunal for the Law of the Sea.

The ICJ caseload. Since the Court was established as the principal judicial organ of the United Nations in 1945, 97 cases have been brought before it in contentious proceedings and 23 requests have been made for advisory opinions. [146] Of the contentious cases, 25 have been concerned with law of the sea or related issues [147] and, of the requests for advisory opinions, 3 have dealt in part with maritime questions. [148]

As regards the contentious cases, there was a very lean period between 1945 and 1973. During this period the Court dealt with only 3 law of the sea cases, though it has to be said that they were very important cases – the *Corfu Channel* case, which has had an important influence on the law relating to passage through straits and on environmental law; [149] the *Anglo-Norwegian Fisheries* case, which broke new ground on straight baselines; [150] and the *North Sea Continental Shelf* cases, which produced a seminal judgment on delimitation of the continental shelf between neighbouring States. [151] The pace quickened noticeably in 1974 when the Court gave judgment in the *Fisheries Jurisdiction* cases brought by the United Kingdom and Germany against Iceland, [152] and in the *Nuclear Tests* cases brought by Australia and New Zealand against France. [153] Since 1974, there has been a steady stream of law of the sea cases, the great majority of which have been concerned with maritime boundary questions. [154]

---

[146] Based on *List of all Decisions and Advisory Opinions brought before the Court since 1946* (http://www.icj-cij.org/icjwww/idecisions.htm, accessed 7 February 2002).

[147] This figure is rather inflated since it includes separately cases which dealt with virtually the same issues between different parties and cases which followed up earlier judgments. Thus, it includes the 2 *Fisheries Jurisdiction* cases brought by the UK and Germany respectively against Iceland (*I.C.J. Reports 1974*, p.3; *I.C.J. Reports 1974*, p.175); the 2 *Nuclear Tests* cases brought by Australia and New Zealand respectively against France (*I.C.J. Reports 1974*, p.253; *I.C.J. Reports 1974*, p.457); and 3 cases which were in a sense continuations of earlier proceedings (*Application for Revision and Interpretation of the Judgment of 24 February 1982 in the Case concerning the Continental Shelf (Tunisia v. Libya), Judgment, I.C.J. Reports 1985*, p. 192; *Request for an Examination of the Situation in Accordance with Paragraph 63 of the Court's Judgment of 20 December 1974 in the Nuclear Tests (New Zealand v. France) Case, I.C.J. Reports 1995*, p. 288; and *Land and Maritime Boundary between Cameroon and Nigeria, Request for Interpretation of Judgment of 11 June 1998, Judgment, I.C.J. Reports 1999*.

[148] *Constitution of the Maritime Safety Committee of the Inter-Governmental Maritime Consultative Organisation, Advisory Opinion, I.C.J. Reports 1960*, p.150; *Legality of the Use by a State of Nuclear Weapons in Armed Conflict, Advisory Opinion, I.C.J. Reports 1996*, p. 66 ; and *Legality of the Threat or Use of Nuclear Weapons, Advisory Opinion, I.C.J. Reports 1996*, p.226.

[149] *Corfu Channel, Merits, Judgment, I.C.J. Reports 1949*, p.4.

[150] *Fisheries, Judgment, I.C.J. Reports 1951*, p. 116.

[151] *North Sea Continental Shelf, Judgment, I.C.J. Reports 1969*, p.3.

[152] *Fisheries Jurisdiction (United Kingdom v. Iceland), Merits, Judgment, I.C.J. Reports 1974*, p.3; *Fisheries Jurisdiction (Federal Republic of Germany v. Iceland), Merits, Judgment, I.C.J. Reports 1974*, p.175.

[153] *Nuclear Tests (Australia v. France), Judgment, I.C.J. Reports 1974*, p.253; *Nuclear Tests (New Zealand v. France), Judgment, I.C.J. Reports 1974*, p.457.

[154] *Aegean Sea Continental Shelf, Judgment, I.C.J. Reports 1978*, p.3; *Continental Shelf (Tunisia/Libya), I.C.J. Reports 1982*, p. 18 and *I.C.J. Reports 1985*, p. 192; *Delimitation of the Maritime Boundary in the Gulf of Maine Area, I.C.J. Reports 1984*, p.246; *Continental Shelf (Libya/Malta), I.C.J. Reports 1985*, p.13; *Land, Island and Maritime Frontier Dispute (El Salvador/Honduras, Nicaragua intervening), I.C.J. Reports 1992*, p.351; *Maritime Delimitation in the Area between Greenland and Jan Mayen, I.C.J. Reports 1993*, p.38; *Passage through the Great Belt (Finland v. Denmark), I.C.J. Reports 1991*, p. 12 and *I.C.J. Reports 1992*, p.348; *Arbitral Award of 31 July 1989 (Maritime Boundary between Guinea-Bissau and Senegal), I.C.J. Reports 1991*, p. 53; *East Timor (Portugal v. Australia), I.C.J. Reports 1995*, p. 901; *Fisheries Jurisdiction (Spain v. Canada), I.C.J. Reports 1998* ; *Request for an Examination of the Situation in Accordance with*

THE LAW OF THE SEA AT THE MILLENNIUM 51

The current docket of pending cases (as at 6 February 2002) comprises 23 cases, of which 4 are concerned with maritime boundary questions. [155]

Capacity of the ICJ to cope with increasing caseload. Given the fact that the Court has made a considerable contribution to the development and interpretation of the law of the sea and continues to receive numerous applications in this area, the question arises whether there was any good reason for inventing the International Tribunal for the Law of the Sea. [156]

Some might argue that the establishment of ITLOS could only help the ICJ by relieving it of part of its burden. It is true that, in February 2000, the President of the ICJ, Judge Guillaume, admitted that there was a risk of delays in handling cases in the future [157] and, when he addressed the UN General Assembly in 2001, the Court still had 22 cases before it. [158] However, in recent years the Court has done much to improve its administration and procedures [159] and the increase in its budget for the biennium 2000-2001 will have enabled it to make further improvements. Nonetheless, its resources of about US$ 11 million per annum [160] are barely adequate and additional resources might have been more readily available if ITLOS had not been created. The Tribunal's budget for 2002 amounts to US$ 7,807,500, including US$ 1,808,100 for the remuneration, travel and pensions of the 21 judges (as compared with the 15 judges of the ICJ) and US$ 2,916,900 for salaries and related costs of 36 staff. [161]

The ITLOS Caseload. As at 6 February 2002, ITLOS had dealt with ten cases. The first two arose out of the arrest by Guinea of the *M/V "Saiga"*, flying the flag of Saint Vincent and the Grenadines, and were dealt with in three phases – proceedings for prompt release of the vessel, a request for provisional measures, and the merits phase.

Paragraph 63 of the Court's Judgment of 20 December 1974 in the Nuclear Tests (New Zealand v. France) Case, I.C.J. Reports 1995, p. 288; and *Kasikili/Sedudu Island (Botswana/Namibia), I.C.J. Reports 1999; Maritime Delimitation and Territorial Questions between Qatar and Bahrain, Judgment, 16 March 2001; Land and Maritime Boundary between Cameroon and Nigeria (Equatorial Guinea intervening), Provisional Measures, Order of 15 March 1996, I.C.J. Reports 1996, p.13; Preliminary Objections, Judgment, I.C.J. Reports 1998, p. 275.*

[155] *Maritime Delimitation and Territorial Questions between Qatar and Bahrain (1991-); Land and Maritime Boundary between Cameroon and Nigeria, Merits (1994 -); Maritime Delimitation between Nicaragua and Honduras in The Caribbean Sea (1991-);* and *Territorial and Maritime Dispute between Nicaragua and Colombia (2001).*

[156] On this question, see further above, section 2.2.2.

[157] *Press conference of Judge Gilbert Guillaume, President of the International Court of Justice,* ICJ Communiqué No. 2000/5, 16 February 2000.

[158] Speech by Judge Guillaume to UN General Assembly, 30 October 2001.

[159] Summarised in Judge Guillaume's speech (*loc. cit.* in note 158). See also *Report of the International Court of Justice, 1 August 2000 – 31 July 2001*, at paras. 19-22.

[160] According to Judge Guillaume's speech to the General Assembly of 30 October 2001, the Advisory Committee on Administrative and Budgetary Questions had recommended an increase in the Court's budget from US$ 20,606,700 for the biennium 200-2001 to US$ 22,873,500 for the biennium 2002-2003, an increase of 11%.

[161] Meeting of [UNCLOS] States Parties. *Report of the 11th Meeting of States Parties (14-18 May 2001)*, SPLOS/73, 14 June 2001, paras. 31-36.

The third and fourth cases were joined by the Tribunal and concerned proceedings instituted against Japan by Australia and New Zealand in relation to the exploitation of *Southern Bluefin Tuna*. The fifth and sixth cases – the *"Camouco"* case and the *"Monte Confurco"* case - were brought against France by Panama and the Seychelles respectively and were applications for the prompt release of vessels arrested by France. Case No. 7, between Chile and the European Community, concerned the *Conservation and Sustainable Exploitation of Swordfish Stocks in the South-Eastern Pacific Ocean.* The eighth case, the *"Grand Prince",* between Belize and France, and the ninth, the *"Chaisiri Reefer 2"* between Panama and Yemen, were two further prompt release applications. Finally, the *Mox Plant Case* was a case brought by Ireland against the United Kingdom, requesting prescription of provisional measures pending the establishment of an arbitral tribunal to consider the merits of a dispute over the alleged pollution threat likely to be posed by the operation of a nuclear fuel reprocessing plant. A brief review of each of these cases is given below.

The M/V "Saiga" Case, 1997 – Case No. 1 [162]
The first case to be decided by the Tribunal was concerned with the interpretation and application of Article 292 of the UN Convention, which embodies a novel procedure for effecting the prompt release of detained vessels and crews. [163] The case arose out of the arrest of the oil tanker M/V "Saiga", flying the flag of Saint Vincent and the Grenadines, by a Guinean Customs patrol vessel for allegedly supplying oil to fishing vessels and other vessels off the coast of Guinea contrary to Guinea's customs legislation. The Tribunal ordered the prompt release of the vessel and its crew upon the deposit of financial security. In a way, it is a pity that the Tribunal's first Judgment had to be delivered in a case such as this. Article 292 raises difficult questions of interpretation and it was no easy task for the Tribunal to produce a fully considered Judgment within the 3-week period prescribed by the Convention and the Rules of the Tribunal. [164] Nonetheless, the majority Judgment is open to criticism and was not perhaps the best possible advertisement for the services which the Tribunal has to offer. [165]

The M/V "Saiga"(No. 2) Case – Request for provisional measures, 1998 [166]
Following the Tribunal's Judgment of 4 December 1997, a Bank Guarantee was posted with the Agent of Guinea on behalf of Saint Vincent which subsequently declined a Guinean request, considered "unreasonable and either irrelevant or unacceptable" to

---

[162] *The M/V "Saiga" (Saint Vincent and the Grenadines v. Guinea* – International Tribunal on the Law of the Sea, Case No. 1), *Judgment,* 4 December 1997 (http://www.un.org/Depts/los/judg_1; *International Legal Materials*, Vol. 37, 1998, p.360).
[163] For an analysis of Art. 292 and a critique of the Judgment, see E.D. Brown, "The M/V 'Saiga' case on prompt release of detained vessels: the first Judgment of the International Tribunal for the Law of the Sea", *Marine Policy*, Vol. 22, Nos. 4-5, July-September 1998, pp. 307-326.
[164] UN Convention, Art. 292(3) and Rules of the Tribunal, Art. 112. This Article (and Art. 111) was amended on 15 March 2001 and now allows the Court a more extended period.
[165] See further Brown, *loc. cit.* in note 163, especially at pp. 319-325.
[166] *The M/V "Saiga" (No. 2) Case (Saint Vincent and the Grenadines v. Guinea* – International Tribunal for the Law of the Sea, Case No. 2: Request for Provisional Measures, Order of 11 March 1998 (http://www.un.org/Depts/los/ord1103.htm).

change its terms. [167] In the meantime, criminal proceedings were instituted against the Master of the *Saiga* in Guinea and a fine of about US$ 15 million (and, on appeal, a suspended sentence of six months) was imposed, the Court holding Saint Vincent civilly liable for the fine. [168] On 22 December 1997, the Government of Saint Vincent notified Guinea of its intention to refer the *Saiga* dispute to arbitration under Annex VII of the UN Convention. [169] As has been seen, States Parties which have not chosen another means of settlement under Article 287(1) are deemed to have accepted arbitration in accordance with Annex VII. This was the position of the two States. However, recognising that the constitution of an arbitral tribunal may take a considerable time, it is provided in Article 290(5) that, pending the constitution of a tribunal, ITLOS (or another court or tribunal agreed upon) may prescribe provisional measures if it considers that *prima facie* the tribunal to be constituted would have jurisdiction and the urgency of the situation so requires. Acting under this provision, Saint Vincent filed a Request for provisional measures on 13 January 1998. [170] The Tribunal was requested to prescribe that Guinea should forthwith comply with the Tribunal's Judgment of 4 December 1997 by releasing the *M/V Saiga* and its crew; suspend the application and effect of the judgments of the Guinean courts; desist from enforcing these judgments; desist from enforcing its customs law within the EEZ or at any place beyond that zone against vessels engaged in bunkering activities outside the Guinean 12-mile zone; desist from interfering with the rights of vessels registered in Saint Vincent, including those engaged in bunkering activities, to enjoy freedom of navigation and/or other lawful uses of the sea related to freedom of navigation; and to desist from undertaking hot pursuit of vessels registered in Saint Vincent except in accordance with Article 111 of the UN Convention. [171]

In a further twist to the proceedings, on 20 February 1998, the parties notified the Tribunal of an agreement requesting it to deal also with the merits of the case. [172] In accordance with that agreement, the Tribunal agreed that the Notification of 22 December 1997 instituting proceedings against Guinea under Annex VII should be deemed to have been submitted to the Tribunal on that date. [173] Similarly, it was to be deemed that the Request for Provisional Measures had been submitted to the Tribunal under Article 290(1). [174] The point of this latter stipulation was that, since there was no longer any question of constituting an arbitral tribunal under Annex VII, the authority to prescribe provisional measures under Article 290(5) no longer existed. Article 290(1) now provided the necessary authority. Under it, if ITLOS considered that, *prima facia*, it had jurisdiction under Part XV, it could "prescribe any provisional measures which it consider[ed] appropriate under the circumstances to preserve the respective rights of the

---

[167] ITLOS/Press 11, 13 January 1998.
[168] *Ibid.*, and ITLOS/Press 23/Add. 1, 1 July 1999.
[169] ITLOS/Press 11, 13 January 1998.
[170] Order of 11 March 1998 (*loc. cit.* in note 166), para. 21.
[171] *Ibid.*
[172] The text of the agreement is reproduced in *M/V "Saiga" No. 2, Order of 20 February 1998.*
[173] *Ibid.*
[174] *Ibid.*

parties to the dispute or to prevent serious harm to the marine environment, pending the final decision." [175]

On 28 February 1998, it was reported that the Judges had completed their deliberations and that the Order of the Tribunal would be read on 11 March 1998. [176] However, in the meantime, on 4 March 1998, the Tribunal was informed by Saint Vincent that Guinea had released the *Saiga* and its captain and crew. [177] Nonetheless, Saint Vincent maintained its Request for provisional measures in relation to the other aspects of the case. As a result of these developments, the unanimous Order for Provisional Measures of 11 March 1998 dealt only with the following four matters:

- *Guinea to refrain from enforcement measures.* The Tribunal ordered that:

> Guinea shall refrain from taking or enforcing any judicial or administrative measure against the M/V Saiga, its Master and the other members of the crew, its owners or operators, in connection with the incidents leading to the arrest and detention of the vessel on 28 October 1997 and to the subsequent prosecution and conviction of the Master. [178]

- *Parties to prevent aggravation or extension of dispute.* The Tribunal recommended that the parties endeavour to find an arrangement to be applied pending the final decision, and to this end they should ensure that no action was taken which might aggravate or extend the dispute. [179]
- *Parties to report on compliance with measures prescribed.* The Tribunal required both parties to report not later than 30 April 1998 on steps taken to ensure prompt compliance with the measures prescribed in the Tribunal's Order. [180]
- *Costs.* The Tribunal decided to rule on costs when deciding the merits of the case. [181]

The M/V Saiga(No.2) Case – Merits, 1999 [182]

The Tribunal's Judgment of 1 July 1999 dealt with a wide range of law of the sea issues, including registration and nationality of ships; [183] genuine link between flag State and ship; [184] exhaustion of local remedies; [185] nationality of claims; [186] the legal status of bunkering in the EEZ (that is the supply of fuel oil to vessels at sea); [187] hot

---

[175] UN Convention, Art. 290(1).
[176] ITLOS/Press 13, 28 February 1998.
[177] M/V "Saiga" No. 2, Order of 20 February 1998,para. 36.
[178] *Ibid.*, para. 52(1)
[179] *Ibid.*, para. 52(2).
[180] *Ibid.*, para. 52(3).
[181] *Ibid.*, para. 52(4).
[182] *The M/V "Saiga" (No. 2) Case (Saint Vincent and the Grenadines v. Guinea – International Tribunal for the Law of the Sea, Case No. 2: Judgment of 1 July 1999 (*http://www.un.org/Depts/los/Judg_E.htm).
[183] Judgment, paras. 55-74.
[184] *Ibid.*, paras. 75-88.
[185] *Ibid.*, paras. 89-102.
[186] *Ibid.*, ,paras. 103-109.
[187] *Ibid.*, paras. 137-138. See also Separate Opinion of Judge Anderson under "Arrest of the Saiga".

pursuit; [188] excessive force used in stopping and arresting a vessel; [189] reparation; [190] and costs. [191] However, the Judgment is of very limited interest on most of these topics simply because the case was relatively straightforward and, in most respects, required only the application of clearly established rules to the easily determined facts of the case.

Briefly, the Tribunal rejected Guinea's objections to the admissibility of Saint Vincent's claims; [192] found that Guinea had acted contrary to the UN Convention by applying its customs laws in a customs radius (rayon des douanes) which included parts of its EEZ; [193] held that Guinea's alleged lawful exercise of a right of hot pursuit failed to comply with several of the conditions laid down in Article 111 of the UN Convention; [194] and ruled that Guinea had used excessive force and endangered human life before and after boarding the *Saiga*. [195] Compensation of US$ 2,123,357 was awarded [196] but it was decided that each of the parties should bear its own costs. [197]

The Tribunal had very little opportunity in this case to break new ground and perhaps the only points deserving of comment in this brief review relate to the questions of registration and nationality of ships and to costs.

In challenging the admissibility of Saint Vincent's claim, the first objection raised by Guinea was that Saint Vincent did not have legal standing to bring the claims because the *Saiga* was "not validly registered under the flag of Saint Vincent and the Grenadines". [198] There was evidence that the ship was unregistered between 12 September and 28 November 1997. [199] Nonetheless, the Tribunal accepted Saint Vincent's argument that the *Saiga* was a ship entitled to fly its flag at the time of the incident giving rise to the case. The evidence adduced by St. Vincent included not only the provisions of its Merchant Shipping Act but also several indications of Vincentian nationality on the ship or carried on board. [200] The Tribunal took the view that this evidence was reinforced by St. Vincent's consistent conduct as a flag State during all stages of the proceedings before the Tribunal. [201] As additional grounds for rejecting Guinea's challenge, the Tribunal found that Guinea was estopped from raising its objection at the Merits stage of the case, having failed to do so until it submitted its

---

[188] *Ibid.*,. paras. 139-152.
[189] *Ibid.*, paras. 153-159.
[190] *Ibid.*, paras.167-177.
[191] *Ibid.*, paras. 181-182.
[192] *Ibid.*, paras. 55-109.
[193] *Ibid.*, para. 136.
[194] *Ibid.*, paras. 145-150.
[195] *Ibid.*, para. 159.
[196] *Ibid.*, para. 175.
[197] *Ibid.*, para. 182.
[198] *Ibid.*, para. 55.
[199] *Ibid.*, paras. 57-58.
[200] *Ibid.*, paras. 67-68 and 72-74.
[201] *Ibid.*, para. 68.

Counter-Memorial in October 1998, and because of its other conduct. [202] Interestingly, the Tribunal went on to observe that, in the particular circumstances of this case, it would not have been consistent with justice to decline to deal with the merits of the case. [203]

The only point on which there was a substantial difference of opinion in the Judgment relates to costs. On the other 12 findings the Tribunal was either unanimous (in 2 cases) or reached its verdict by majorities of 18 to 2 (in 9 cases) or 17 to 3 (in 1 case). [204] In deciding that each party should bear its own costs, however, the Tribunal was split 13 to 7. [205] Article 34 of the Tribunal's Statute provides that each party should bear its own costs unless the Tribunal decides otherwise and the majority saw no need to depart from the general rule in this case. [206] This was despite the fact that, in an Agreement of 1998 on which the Tribunal's jurisdiction was founded, the parties agreed that the Tribunal "shall be entitled to make an award on the legal and other costs incurred by the successful party...". [207] The contrary case is made in a Joint Declaration of the seven Judges who dissented on this question. [208] Their main argument for the view that costs should have been awarded to the "successful party" was that the parties were in agreement that the successful party should be awarded its costs and they found evidence of this alleged agreement in two places. First, they said that the two parties "requested the Tribunal to award costs to the successful party" in their Agreement of February 1998. [209] Secondly, they repeated the request when making their final submissions. [210] In fact, however, this is not wholly accurate. As noted above, the Agreement of February 1998 "entitled" the Tribunal to award costs; it did not "request" it to do so. It is true that both sides asked for these costs in their final submissions but this is quite normal and there was no repetition of the "entitled" formulation, let alone the "request" formulation referred to in the dissenting Joint Declaration. There was perhaps more force in the second argument deployed in the Joint Declaration – that it would have been consistent with the full achievement of the aim of wiping out the consequences of Guinea's illegal acts (the declared aim of the Tribunal's award of compensation) if the Tribunal had exercised its discretion in favour of awarding costs to Saint Vincent. [211]

Southern Bluefin Tuna Cases, 1999 – Cases Nos. 3 and 4 [212]

---

[202] *Ibid.*, paras.69 and 73(c).
[203] *Ibid.*, para. 73(d).
[204] *Ibid.*, para. 183.
[205] *Ibid.*, para. 183(13).
[206] *Ibid.*, para. 182.
[207] The text of the Agreement is reproduced in the Judgment at para. 4 and the reference to costs is in its para. 4.
[208] Joint Declaration by Judges Caminos, Yankov, Akl, Anderson, Vukas, Treves and Eiriksson on the Question of Costs (http://www.un.org/Depts/los/ITLOS/JD_Saiga.htm).
[209] *Ibid.*, second para.
[210] *Ibid.*
[211] *Ibid.* fifth para.
[212] *Southern Bluefin Tuna Cases (New Zealand v. Japan* – International Tribunal for the Law of the Sea, Case No. 3 and *Australia v. Japan* – Case No. 4), Order of 27 August 1999 (http://www.un.org/Depts/los/ITLOS/Order-tuna34.htm). The two cases were joined by Order of the Tribunal

This was the first case before the Tribunal to deal with a matter other than the prompt release of an arrested vessel under Article 292 of UNCLOS.[213] Like the earlier *Saiga* case, referred to above, it offers another example of how ITLOS may entertain requests for provisional measures even in cases where the merits are to be dealt with by another tribunal.

The facts of the case were as follows. Under the Convention for the Conservation of Southern Bluefin Tuna (SBT), 1993, the Commission for the Conservation of SBT established a Total Allowable Catch and distributed it among Member States. In 1998 Japan commenced an experimental fishing programme allegedly designed to obtain more reliable scientific evidence on the state of the stocks. Australia and New Zealand considered that such fishing was essentially for commercial purposes and increased the threat to the stock. Having been notified of the existence of a dispute, Japan proposed to refer it to mediation and, later, to arbitration pursuant to the 1993 Convention. Neither of these proposals was acceptable to Australia and New Zealand and they decided to commence compulsory dispute resolution proceedings under Section 2 of Part XV of UNCLOS. Although all three parties are States Parties to UNCLOS, none of them has chosen a means for the settlement of disputes under Article 287 and they are therefore deemed to have accepted arbitration in accordance with Annex VII of UNCLOS. Accordingly, Australia and New Zealand instituted proceedings under Annex VII on 15 July 1999. Since it would obviously take some time to establish an Annex VII arbitral tribunal and await its award, and since the applicants regarded the matter as urgent, on 30 July 1999, they submitted a request to ITLOS for the prescription of provisional measures. In so doing they were able to rely upon Article 290(5) of UNCLOS, whereby, as was seen above, pending constitution of an arbitral tribunal, ITLOS may prescribe provisional measures if it considers that, *prima facie*, the tribunal to be constituted would have jurisdiction and the urgency of the situation so requires.

In its Order of 16 August 1999, ITLOS prescribed a number of provisional measures. *Inter alia*, they requested the parties to refrain from taking any action which might aggravate or extend the dispute or which might prejudice the carrying out of any decision on the merits which the arbitral tribunal might render.[214] They were also required not to exceed early agreed annual catch limits and to refrain from conducting an experimental fishing programme, unless by agreement or within annual catch limits.[215] However, the measures prescribed also included positive measures. The parties were required to resume negotiations without delay with a view to reaching agreement on conservation and management measures;[216] and to make "further efforts" to reach agreement with other States and "fishing entities" engaged in fishing for SBT.[217]

---

of 16 August 1999.
[213] Though it is true that the Judgment in the *Merits* phase of the *Saiga* case went beyond the simple issue of prompt release under Art. 292. See further above, text, at note 182 *et seq.*
[214] Order of 27 August 1999, paras. 90(1)(a) and (b).
[215] *Ibid.*, paras. 90(1)(c) and (d).
[216] *Ibid.*, para. 90(1)(e).
[217] *Ibid.*, para 90(1)(f).

Moreover, each party was required to submit initial reports not later than 6 October 1999 and the President of ITLOS was authorised to request further reports and information as considered appropriate after that date. [218]

An Arbitral Tribunal established under Annex VII subsequently found that it lacked jurisdiction to decide the merits of the dispute and revoked the provisional measures prescribed by ITLOS. [219]

"Camouco" Case, 2000 – Case No. 5 [220]
This case was again concerned with an application under Article 292 of the UN Convention for prompt release of a vessel and crew. It arose out of the arrest of the Panamanian fishing boat "Camouco" by a French frigate, *Floréal*, for alleged illegal fishing in the EEZ of the Crozet Islands (French Southern and Antarctic Territories). The Tribunal ordered prompt release of the vessel and its master on deposit of a financial security of 8 million French Francs.

This was a relatively straightforward case, [221] though the Tribunal had to consider two novel points as regards the admissibility of the application. First, it found that the Convention does not require the flag State to file an application at any particular time after detention of the vessel and there was, therefore, no merit in France's contention that Panama's alleged failure to act promptly resulted in its losing its right under Article 292. [222] Secondly, referring to proceedings then pending before the court of appeal of Saint-Denis, the Tribunal observed that it was not logical to read the requirement of exhaustion of local remedies or any analogous rules into Article 292. It noted that that article provides for a quick independent remedy during which local remedies could normally not be exhausted. [223] The Judgment will also be referred to in future cases for its close consideration of the factors which have to be taken into account in determining the reasonableness of the bond or financial security required to be deposited under Article 292. [224]

"Monte Confurco" Case, 2000 – Case No. 6
This is yet another case on prompt release under Article 292 of UNCLOS. The *Monte Confurco*, registered in the Seychelles, was boarded by the French frigate *Floréal* in the EEZ of the Kerguelen Islands in the French Southern and Antarctic Territories and the Master was charged with failing to announce his presence in the EEZ and with illegal

---

[218] *Ibid.*, para. 90(2).
[219] *Southern Bluefin Tuna Case (Australia and New Zealand v. Japan). Award on Jursidiction and Admissibility, 4 August 2000.* Text on website of International Centre for Settlement of Investment Disputes, which administered the proceedings, at www.worldbank.org/icsid.
[220] *The "Camouco" Case (Panama v. France* – Case No. 5, *Judgment*, 7 February 2000 (http://www.un.org/Depts/los/ITLOS/JudgmentCamouco.htm).
[221] See however, the Separate Opinion of Vice-President Nelson and the Dissenting Opinions of Judges Anderson, Treves, Vukas and Wolfrum.
[222] *Judgment*, paras. 50-54.
[223] *Judgment*, paras. 55-58.
[224] See Judgment, paras. 64-76.

fishing in that zone. The Tribunal ordered the prompt release of the vessel and its Master upon the posting of a bond of 9 million French Francs, with a further 9 million FF security being provided by the monetary equivalent of the seized cargo of fish. [225]

As Lowe has noted, the Tribunal " ...laid down with great care the ground rules of its approach to prompt release cases, in terms likely to direct its approach to such cases for the foreseeable future". [226] In particular, developing its thinking in the *Camouco Case*, the Judgment provides useful guidance on the determination of the level of a "reasonable" bond in such cases, a level that reflects a balance between the law enforcement interests of the arresting State and the interests of the flag State, vessel and crew.

Conservation and Sustainable Exploitation of Swordfish Stocks in the South-Eastern Pacific Ocean, 2000 – Case No. 7
This case, between Chile and the European Community, is notable for at least two reasons. First, as noted above, [227] this was the first occasion on which ITLOS, at the request of the parties, established a Special Chamber under Article 15(2) of its Statute. [228] Secondly, this was the first time too that contentious proceedings between an international institution and a State had been instituted before a world court.

The Special Chamber was called upon to decide, *inter alia*, whether the EC had complied with its obligations under UNCLOS (especially Articles 116 – 119) to ensure conservation of swordfish in the fishing activities undertaken by vessels flying the flag of any of its member States in the high seas adjacent to Chile's EEZ, whether the Chilean Decree which purported to apply Chile's conservation measures relating to swordfish on the high seas was in breach of UNCLOS, and whether a "Galapagos Agreement" on fishing in the region, signed in 2000 without the participation of all interested States, was negotiated in keeping with the provisions of the Convention.

Whether the Special Chamber will have the opportunity to rule on these issues remains to be seen. In March 2001, the parties informed the President of the Special Chamber that they had reached a provisional arrangement concerning the dispute and requested that the proceedings before the Chamber be suspended. Accordingly, by an Order of 15 March 2001, the time-limit of 90 days for the making of preliminary objections will commence from 1 January 2004, though each party will have the right to request that the time-limit should begin to apply from an earlier date. [229]

Grand Prince Case, 2001 – Case No. 8

---

[225] The " Monte Confurco" Case (Seychelles v. France), Application for Prompt Release, Judgment, 18 December 2000.
[226] A. V. Lowe, "The International Tribunal for the Law of the Sea: Survey for 2000", *The International Journal of Marine and Coastal Law*, Vol. 16 (No. 4, 2001), pp. 549-570, at p. 564.
[227] See section 1.2.5, under *Chambers for particular disputes*.
[228] *Case Concerning the Conservation and Sustainable Exploitation of Swordfish Stocks in the South-Eastern Pacific Ocean (Chile/European Community), Constitution of Chamber. Order of 20 December 2000.*
[229] Order of 15 March 2001, para. 6.

This was the third application for prompt release under Article 292 to be concerned with an arrest of a vessel by the French authorities for alleged illegal fishing in the EEZ of one of its Southern and Antarctic Territories. The Tribunal found that it lacked jurisdiction to entertain the application because the evidence did not provide a sufficient basis for holding that Belize was the flag State for the purpose of making an application under Article 292. [230] It is a notable feature of this case that the majority judgment was supported by only 12 of the 21 Judges and a Joint Dissenting Opinion was given by the remaining Judges on the central question of the nationality of the vessel.

"Chaisiri Reefer 2" Case, 2001 – Case No. 9
The proceedings in this case were instituted by Panama on 3 July 2001 under Article 292 of UNCLOS in order to secure the prompt release by Yemen of the *Chaisiri Reefer 2* and its crew and cargo which had been detained for alleged violation of fishery laws on 3 May 2001. However, on 12 July 2001, the parties informed the Tribunal that, following the agreement of Yemen to release the vessel and cargo, they had reached a settlement of the dispute and agreed to discontinue the proceedings. The Tribunal accordingly agreed to remove the case from the Tribunal's list of cases. [231] The role of the Tribunal was thus limited to helping to concentrate the minds of the Yemeni authorities on the dispute and so bring about its speedy resolution.

MOX Plant Case, 2001 – Case No. 10
As in the earlier *Southern Bluefin Tuna* cases, these proceedings were concerned with a request for the prescription of provisional measures under Article 290(5) of UNCLOS, pending the establishment of an arbitral tribunal under Annex VII. The dispute stemmed from the British authorisation of the opening of a new MOX facility in Sellafield to reprocess spent nuclear fuel. Ireland's main concern was that the operation of the plant would contribute to pollution of the Irish Sea. Ireland notified the UK on 25 October 2001 of the submission of the dispute to arbitration under Annex VII and, as will be seen, it was material to the Tribunal's ruling on the request for provisional measures that there would be only a "short period before the constitution of the Annex VII arbitral tribunal". [232]

Before prescribing provisional measures under Article 290(5), the Tribunal had to satisfy itself that, *prima facie*, the Annex VII arbitral tribunal would have jurisdiction [233] and the UK argued that it would not have jurisdiction because, under Article 282, precedence should be given to procedures laid down in governing regional agreements – the OSPAR Convention, the EC Treaty and the EURATOM Treaty. [234] The Tribunal found, however, that Article 282 was not applicable to the dispute submitted to the

---

[230] *The "Grand Prince" Case (Belize v. France), Application for Prompt Release, Judgment, 20 April 2001.*
[231] *The "Chaisiri Reefer 2" Case (Panama v. Yemen), Order 2001/14, 13 July 2001.* See also ITLOS/Press 51, 5 July 2001 and ITLOS/Press 52, 16 July 2001.
[232] Judgment, para. 81. The time-limits for the constitution of an Annex VII tribunal are specified in Annex VII of UNCLOS, Art. 3.
[233] UNCLOS, Art. 290(5).
[234] See further Judgment, paras. 38-44. For the Irish response, see paras. 45-47.

Annex VII arbitral tribunal and that ITLOS was entitled to prescribe provisional measures. [235]

On the facts of the case, the Tribunal did not find that the urgency of the situation required the prescription of provisional measures in the short period before the constitution of the Annex VII arbitral tribunal. [236] Nevertheless, finding that, "the duty to co-operate is a fundamental principle in the prevention of pollution of the marine environment under Part XII of the [UNCLOS] Convention and general international law", [237] the Tribunal went on to prescribe measures different from those requested by Ireland. [238] They called for consultations, exchange of information, monitoring and measures to prevent marine pollution. Both parties were further required to submit reports to the Tribunal by 17 December 2001. [239]

2.3.3. Tentative Conclusions

It is clear from this brief survey of the recent practice of the ICJ and ITLOS that the ICJ has played an important part in the interpretation and development of the law of the sea and that it is continuing to do so in the UNCLOS era. It is less easy to evaluate the contribution made by ITLOS at this relatively early point in its development.

Reviewing the 10 cases summarised above, what has the Tribunal so far achieved? Of the 10 cases, 6 were concerned with prompt release under article 292 [240] and, of those, 1 was terminated for lack of jurisdiction [241] and another discontinued by the parties. [242] Useful though this specialised jurisdiction is, it must be said that it is not immediately apparent why a 21-judge tribunal is necessary to deal with the relatively simple issues which tend to arise in such cases.

In 3 of the Tribunal's cases, orders were made prescribing provisional measures [243] but, in one of these, the measures were later revoked by an arbitral tribunal; [244] in another, the measures prescribed, though helpful, were of relatively limited scope and not those requested by the applicant. [245] Finally, although ITLOS has been concerned with the merits in two cases, [246] in one of them, proceedings have been suspended following the conclusion of a provisional arrangement by the parties. [247]

---

[235] Judgment, paras. 48-53.
[236] Judgment, para. 81. The Tribunal's reasoning is set out in paras. 64-81.
[237] Judgment, para. 82.
[238] As permitted by Art. 89(5) of its Rules
[239] Judgment, para. 89.
[240] The M/V "Saiga" Case; the "Camouco" Case; the "Monte Confurco" Case; the "Grand Prince" Case; and the "Chaisiri Reefer 2" Case.
[241] The "Grand Prince" Case.
[242] The "Chaisiri Reefer 2" Case.
[243] The "M/V Saiga" Case; the Southern Bluefin Tuna Cases; and the MOX Plant Case.
[244] The Southern Bluefin Tuna Cases.
[245] The MOX Plant Case.
[246] The "M/V Saiga" Case and the Case concerning the Conservation and Sustainable Exploitation of Swordfish Stocks in the South-Eastern Pacific Ocean.
[247] The Case concerning the Conservation and Sustainable Exploitation of Swordfish Stocks in the South-

Given the fact that the Tribunal held its first session only about five and a half years ago, in October 1996, it is too early to reach any conclusion about the importance of its role and the contribution it is likely to make to the interpretation and development of the law of the sea in the longer term. It is true, of course, that the Tribunal has exclusive jurisdiction in relation to seabed mining cases arising under Part XI of UNCLOS. More generally, much will depend upon the attitude of States to the Tribunal and, as has been seen, a review of declarations so far made under Article 287 is not very encouraging. [248]

Looking beyond the ICJ and ITLOS to the wider picture, it is of course true that, in one sense, the scheme for settlement of disputes embodied in Part XV of the UN Convention has already been successful. The fact is that States, by ratifying or acceding to the Convention, are demonstrating that they are prepared to accept a conventional obligation to comply with a dispute settlement system which includes compulsory procedures entailing binding decisions. In that sense, Part XV has broken the mould and established a more positive attitude to compulsory third-party settlement than previously existed. However, this advance has been gained at a cost. As has been seen, the obligation to accept such compulsory procedures is subject to a variety of important exceptions. Moreover, the very feature which may have persuaded some States to accept Part XV – the proliferation of dispute-settlement mechanisms on offer – may lead to a fragmented system in which uniformity and consistence or jurisprudence will be seen to have been sacrificed to the primary objective of ensuring that disputes arising from the Convention may be finally and peacefully settled. However, if, over time, ITLOS establishes itself as the central tribunal to which law of the sea disputes are referred and the full Tribunal is employed when important points of interpretation are at stake, there is no reason why both certainty and uniformity cannot be achieved. In this context, it is perhaps pertinent to note that the Tribunal would certainly enhance its authority if its members could be persuaded to forego the indulgence of writing so many separate opinions and made greater efforts to reach more collegiate judgments.

Eastern Pacific Ocean.
[248] See above, section 2.3.1.

CHAPTER 3. MARITIME BOUNDARIES

Achievements to date and unfinished business

G.H. BLAKE

1. Introduction

Maritime boundary delimitation began in earnest soon after World War Two as the quest for hydrocarbon resources extended progressively offshore. In the half century since 1950 rather more than one third of the world's potential international maritime boundaries have been agreed. Estimates of the potential number vary somewhat, but in 2000 it was approximately 430. The number has risen in recent years with the creation of several new coastal states such as Estonia, Latvia, and Lithuania on the Baltic Sea coast. At the millennium there were some 160 formal agreements delimiting maritime boundaries between states, or 37 per cent of the estimated potential number. As the product of 50 years or more of delimitation endeavour this seems rather slow progress, averaging just over three agreements per annum. The number of annual agreements peaked during the 1970s while the Third UN Conference on the Law of the Sea was in progress. At the present rate it could take another 70 years before the bulk of the world's maritime boundaries are in place. By that time the international community might have devised new ways of managing ocean space, and there may also be a number of new states eligible for a share of the seabed.

Besides maritime boundaries between states, boundaries also have to be delineated between the outer limits of coastal state jurisdiction and the international seabed beyond (which is discussed below). As many as 60 states may be eligible to claim a share of continental shelf (CS) beyond their 200 nautical mile (nm) exclusive economic zones (EEZ). While this may not be politically as difficult as negotiating with a neighbouring state, such claims have to be formally presented to the Commission on the Limits of the Continental Shelf (CLCS) for approval., The process is technically complex and costly. A number of states have already invested heavily in preparing their continental shelf claims to go to the CLCS. The operation of the Commission and its judgements will attract a great deal of attention in the early years of the new century. Meanwhile there is no evidence that some optimistic forecasts of a "borderless world" emerging during the twenty-first century have diminished the eagerness with which coastal states are seeking to establish their offshore limits.

2. The 1982 UN Convention on the Law of the Sea (UNCLOS)

This Convention is arguably one of the most important treaties of the twentieth century.[1] From the maritime boundary perspective it is the most influential international agreement of all time, and it will set the agenda for offshore state sovereignty and maritime boundary delimitation for decades to come. UNCLOS has a high level of support from the international community. When it opened for signature in Jamaica in December 1982, 119 delegations signed on the first day. UNCLOS came into force in November 1994, 12 months after the 60th state had ratified it. Without UNCLOS we would be facing an anarchic scramble for state control of the oceans which would have been disastrous. It took from 1973 to 1982 to hammer out the 320 Articles and Annexes in UNCLOS. Nine negotiating sessions took place involving 120 states. Bearing in mind the difficulty of reaching agreement on several controversial matters, it is a miracle that UNCLOS was ever completed. Compromise was necessary, which perhaps explains why the text of the Convention is by no means perfect; it contains ambiguities and significant omissions, and there are no sanctions applicable to signatories who disregard some of its provisions. The burning questions which UNCLOS resolved, at least for the time being, are briefly outlined below.

2.1. BREADTH OF THE TERRITORIAL SEA

Article 3 of UNCLOS gives every coastal state the right to 12 nm (1 nm = 1.852 km = 1.156 statute miles) of territorial sea. While the right of innocent passage is preserved for other states, the territorial sea is subject to absolute sovereignty in other respects. Airspace sovereignty also extends to the outer limit of the territorial sea.

The idea of national sovereignty over a band of coastal waters did not gain much support until gunpowder created canons in Europe capable of controlling adjacent seas. The question then was how far out to sea should coastal state rights extend? Any form of coastal state control would clearly limit the activities of other states. A Dutch lawyer, Van Bynkershoek proposed a territorial sea of three miles in 1702, and this became generally accepted in Europe until the twentieth century. Many states however did not formally declare their claims to a territorial sea. After the Second World War many coastal states became increasingly preoccupied with extending their territorial seas because of growing interest in hydrocarbon resources, fisheries, and security concerns. Claims began to escalate to a degree where it was conceivable that the oceans could eventually be partitioned between coastal states (Table 1). An attempt had been made to resolve this and other sovereignty questions at the first UN Conference on the Law of the Sea, attended by some 60 states in 1958. Unfortunately, no agreement was reached about the breadth of the territorial sea, nor was it resolved at a second UN Conference on the Law of the Sea in 1960. Already the line-up was clear; in general, states with large commercial and naval fleets including Britain, the United States, and the Soviet Union argued for a narrow band of coastal sea, to ensure freedom of navigation. Developing states on the whole were more interested in controlling extensive adjacent waters and their resources.

Table 1: Territorial Sea Claims in 1980

Breadth (nautical miles)	Number of states
3	18
4	2
6	5
12	84
15	1
20	1
30	2
35	1
50	3
70	1
100	1
150	1
200	13

2.2. STRAITS USED FOR NAVIGATION

The territorial sea debate was complicated by concerns about freedom of navigation through international waterways. Under the old three mile regime some straits fell into the territorial waters of one or more coastal states, but extension of territorial waters to 12 nautical miles, as was being proposed, put something like 120 more straits used for navigation into territorial sea status. Many of these straits were in the hands of third world states with no great stake in freedom of navigation. To resolve this issue UNCLOS provides that navigation through narrow waterways customarily used by international shipping shall be protected by the right of "transit passage" (Article 38). This forbids the coastal state to interfere with shipping in international waterways even if it is passing through territorial seas.

2.3. THE EXCLUSIVE ECONOMIC ZONE (EEZ)

Coastal states are entitled to an Exclusive Economic Zone (EEZ) to 200 nm offshore measured from the same baseline as the territorial sea, a distance of over 230 statute miles offshore. The importance of the EEZ is very great indeed. Apart from anything else, it means that about one-third of the world's oceans fall into coastal state sovereignty, while two thirds have been saved from national sovereignty, and are to be managed for the common benefit of humankind. Rather more than 90% of fish landings and over 95% of oil and gas production comes from within 200 nm of the coast. Within their EEZs, coastal states have the exclusive right to exploit such resources, and to make any necessary arrangements to do this, for example by constructing platforms or other installations. They have the right to control marine research, and are obliged to undertake environmental conservation within the EEZ.

The EEZ concept emerged as a compromise between two former types of offshore sovereignty, both to do with resources. First, a number of states had unilaterally declared exclusive fishing zones. In some cases these were up to 300 nm offshore; several were 200 nm offshore. Since there was no consensus about a width for exclusive fishing zones they had given rise to considerable and growing international conflict. Secondly, states enjoyed continental shelf rights. After World War II a number of states claimed the exclusive right to the resources of the seabed in their own continental shelf areas. The Truman Proclamations of 1945 in which United States claims were set out, began the trend in continental shelf claims. Continental shelf rights covered seabed resources, notably oil and gas, but not fish. There was a certain logic in the idea that coastal states should have the right to the resources of the natural prolongation of their landmass under the sea. Nearly all the maritime boundary agreements in the post-war period from 1945 were continental shelf boundaries. Such agreements were naturally most common where there was some expectation of oil or gas resources, as in the Persian-Arabian Gulf, the Gulf of Mexico, and later in the North Sea.

2.4. DEFINITION OF THE CONTINENTAL SHELF (CS)

The problem with the Continental Shelf was that its outer limits were never satisfactorily defined. The 1958 UN Continental Shelf Convention thought it had solved the question by defining continental shelf rights as extending to 200 metres of water depth, or "beyond that limit, to where the depth of the superadjacent waters admits of the exploitation of the natural resources of the said areas" (Article 1). [2] In 1958 underwater technology was in its infancy and offshore exploitation of oil and gas beyond 200 metres of water depth seemed very unlikely. Before long however offshore drilling for hydrocarbons was going on in depths of over 1000 metres. This opened the door for coastal states to claim continental shelf rights far beyond 200 metres of water depth. With uncertainty over fishing zones and continental shelf rights, there was clearly the need to devise a new regime. In addition, and importantly for many Latin American and African states, C.S. rights did not necessarily give them much benefit since the physical extent of their natural C.S. is limited, especially compared with many developed states such as the United Kingdom, the former USSR, the United States, and Canada.

The introduction of a universal 200 nm EEZ was not the end of the matter, although many argue that it should have been. By gaining EEZ rights, coastal states were given much the same rights (plus a few more privileges) as would be available to them if they had both an EEZ and a CS claim in place. Some coastal states however argued for the continuation of their old CS rights in cases where the CS extended beyond the 200 mile EEZ. There was considerable controversy over this question, not least because it tended to be the rich states (not exclusively) who, by the accident of geography would gain most. In the end there was compromise. Under Article 76 (4), states in such a fortunate position are permitted to exploit resources in their CS up to an absolute maximum distance of 350 nm offshore, but they have to share any revenues accruing with the International Seabed Authority. If however their continental shelf terminates before 350 nm offshore, the 1982 Convention laid down two alternative methods whereby a line could be determined, taking into account the physical limits of the CS. This line is symbolically very important because it divides national sovereignty from the common heritage of humankind. In practice, exploitation of deep-sea resources has not yet become commercially attractive. The importance of the CS provision beyond 200 miles is that limits have been placed on coastal state expansion once and for all.

2.5. THE INTERNATIONAL SEABED

Arvid Pardo, the Maltese Ambassador to the United Nations sparked off the debate about ownership of resources of the deep seabed in a celebrated speech before the UN General Assembly in November 1967. Like many other observers he feared the progressive partitioning of the oceans between coastal states and the militarisation of the oceans, and argued in favour of the peaceful preservation of the deep seabed as the common heritage of humankind. At that time there was growing interest in the staggering mineral resources represented by manganese nodule deposits, especially in the Pacific and Indian Oceans. These potato-like formations contain nickel, copper, cobalt, chromium, and other strategic minerals in varying quantities, and the technology for their recovery is in place. There was particular interest in their potential when the price of land-produced minerals was high, or when their availability appeared to be threatened.

The 1982 Law of the Sea Convention (Article 136) established the principle of a common heritage beyond the outer limits of national jurisdiction, sometimes called "the global commons" or "the area". It is administered by the UN International Seabed Authority (ISA) which operates an agency called "The Enterprise" whose function is to grant concessions and exploit the minerals of the seabed for the benefit of humankind. Sadly, it was the arrangements for the exploitation of the deep seabed which proved to be among the most controversial provisions in UNCLOS. The objections to these arrangements were chiefly articulated by the United States, backed by Britain and Japan. As a result, three of the most important industrial countries with substantial interests in all the other provisions of UNCLOS failed to sign it. In the event, the world price of minerals has never justified the commercial exploitation of the deep seabed. This could change in the new century, and arrangements between the ISA and the multinationals may have to be reconsidered.

3. Boundary Delimitation

3.1. DELIMITATION PRINCIPLES

Maritime boundary delimitation has proceeded slowly for a number of reasons. First, in some parts of the world negotiations cannot begin because of political differences between the parties, often concerned with questions of territorial sovereignty or the ownership of islands. Secondly, the international law of boundary delimitation is both complex and difficult to apply. Until 1969 as a result of the 1958 UN Convention there was strong emphasis on looking for the equidistance line, to achieve equal division of seabed between states. In 1969 however the International Court of Justice (ICJ) ruled in the North Sea cases that states were not under any compulsion to seek lines of equidistance, but rather to determine that line which, considering all the circumstances, would lead to an equitable result. Thus geometry went out and almost everything else came in, making the quest for equitability an extremely complex process. In particular the ICJ emphasised the importance of the natural prolongation principle as one of the geographical circumstances to be taken into account. Thereafter, until the mid-1980s states began to look for the natural termination of their landmass under the sea using all the arguments of geology and geomorphology they could muster. Although very few maritime boundary agreements adopted the natural prolongation principle successfully, it added further complexity to maritime delimitation for 15 years or more. The collection and marshalling of evidence for boundary negotiations began to involve large teams of experts in many months or years of preparation.

UNCLOS (1982) incorporates the principle of equitability in Articles 74 [1] and 83 [1] which declare that EEZ and CS boundaries "shall be effected by agreement on the basis of international law, as referred to in Article 38 of the Statute of the International Court of Justice, in order to achieve an equitable solution." In the territorial sea however, failing agreement to the contrary, the applicable principle is the median line (Article 15). The sources of law cited in ICJ Statute 38 include international conventions, general principles of law recognised by civilised nations, jurisprudence, and above all state practice. The latter has become perhaps the dominant element in many of the cases presented by states in bilateral negotiations or in litigation before the courts. Most of the judgements of the ICJ refer to "state practice" as justification for at least some of the arguments used. The decisions of the ICJ are in turn very influential in shaping the principles of maritime boundary delimitation. The Libya-Tunisia (1982) and Libya-Malta (1985) cases were crucial in this respect. After the Libya-Malta case it was clear that natural prolongation had been abandoned in favour of *distance* criteria taking into account coastal length, shape, and direction. Geography, not geometry or geology was now the dominant criterion in maritime boundary delimitation. The objective remains an equitable result taking into account all the "relevant circumstances." *Proportionality* is regarded as a useful test of the equity of the proposed delimitation; the seabed to be allocated should be in rough proportion to the coastal lengths opposite the sea in question.

3.2. COMMON MARITIME ZONES

An alternative to seeking a maritime boundary delimitation is for the parties to agree on a *common zone* or *joint development zone* of some kind. This has the advantage of avoiding prolonged negotiation and quarrelling, and makes it possible to exploit the resources without undue delay. It is often assumed that Common Zones arise from the 1982 UN Convention Articles 74 (EEZ) and 83 (CS) which require states, pending delimitation agreements "in a spirit of understanding and co-operation ...... to enter into provisional arrangements of a practical nature and, during this transitional period, not to jeopardise or hamper the reaching of the final agreement." In reality, about half the existing Common Zones predate the 1982 UN Convention. The precise details of common zone arrangements vary considerably. Schemes can be devised to suit the parties involved. The agreement is generally regarded as temporary, pending a boundary delimitation, but a number have been in existence for many years, In 2000 there are at least 17 common zones in various parts of the world, most of which are reported to be highly satisfactory (Table 2). Common zones deserve far more attention than they receive. It would be no bad thing for them to be thoroughly reviewed early in the new century, perhaps with a view to more general application. The last major study of common zones was completed in 1989-90 by a group of British lawyers who were seeking to prepare a model agreement for joint development of offshore oil and gas. Although the model has never been adopted in detail, their two volumes are an excellent introduction to the subject.[3]

4. Unfinished Business

Although the 1982 Law of the Sea Convention is not perfect, it will continue to provide the basis for a maritime regime which commands a high degree of international support for the foreseeable future. More states continue to ratify it, and those few who have not already signed it may do so before long, including notably the United States. The Convention is universally regarded as the authority for state practice in relation to offshore jurisdiction and maritime boundary delimitation. While some of its provisions are open to various interpretations it has provided a framework for state behaviour which has undoubtedly made a significant contribution to peace. The scramble to partition the oceans which was gaining such frightening momentum in the 1970s and 1980s has been checked, and almost two thirds of the oceans have been secured as the common inheritance of humankind. Against these significant achievements of the last century, we need to consider some of the unfinished business in maritime boundary delimitation which will become dominant themes of the present century

Table 2. Joint Offshore Zones

As part of a full boundary agreement:

Bahrain-Saudi Arabia in the Persian Gulf (signed 1958)
Qatar-United Arab Emirates (Abu Dhabi) in the Persian Gulf (1969)
France-Spain in the Bay of Biscay (1974)
Norway-United Kingdom in the North Sea (1975)
Australia-Papua New Guinea in the Torres Strait (1978)
Iceland-Norway in the North Atlantic (Jan Mayen Island) (1981)
Australia-Indonesia in the Timor Sea (Timor Gap) (1989)
Denmark-United Kingdom (Faeroe Islands) (1999)

In lieu of/without a full boundary agreement:

Kuwait-Saudi Arabia in the Persian Gulf (1965)
Japan-South Korea in the Sea of Japan (1974)
Malaysia-Thailand in the Gulf of Thailand (1990)
Malaysia-Vietnam in the Gulf of Thailand (1993)
Argentina-United Kingdom (Falkland/Malvinas Islands) (1995)

Other joint zones:

Argentina-Uruguay in the Rio de la Plata (1973)
 Scientific research, fisheries, pollution control etc.
Saudi Arabia-Sudan in the Red Sea (1974) Metalliferous muds
Colombia-Dominican Republic in the Caribbean Sea (1978) Scientific research fisheries
Norway-USSR (Russia) in the Barents Sea. Fisheries (1978), Hydrocarbons (1996)

4.1. UNRESOLVED DISPUTES

Among the 270 or so maritime boundaries awaiting delimitation a number are already known to be in dispute. As more states embark on maritime boundary delimitations, more disputes are likely to emerge. Disputes are broadly of two kinds. First, where territorial sovereignty is contested by two (or more) states, maritime boundary delimitation cannot begin because it is land which gives title to the sea. Islands are by far the largest source of sovereignty disputes. Some islands may be prized for their own resources, or strategic location, but most are coveted because of their potential to bestow ownership of considerable areas of ocean. The Spratly Islands in the South

China Sea are a good example. Small and valueless in themselves, they are claimed altogether (or in part) by China, Taiwan, Vietnam, Philippines and Malaysia in the hope of winning sovereignty over thousands of square kilometres of seabed which are alleged to contain hydrocarbon resources. [4] Worldwide there are currently some 30 disputed islands or groups of islands, which between them are probably hampering up to 50 maritime boundary agreements. [5] Besides islands, there are a number of territorial disputes which inhibit maritime delimitations such as the status of Northern Cyprus, and the contested Egypt-Sudan land boundary on the Red Sea coast.

Second, there are disputes arising from overlapping claims by neighbouring states. These may arise from a variety of causes, including interpretations of the law, the weight to be given to islands, the historic behaviour of the states, contemporary diplomatic exchanges, the geography of the coasts including length of coast, and other circumstances. In some the seabed in contention may be quite small, but in other cases it may be very large. In most disputes however the parties are determined to gain maximum advantage, especially where oil or gas deposits can give colossal value to every square metre of seabed. Early in 2000 there were perhaps two dozen maritime boundary disputes worldwide, although there are problems in making a reliable inventory associated with questions of definition and scale. Taken together with island-related disputes, the total number of disputed maritime boundaries is approximately 70 to 80, or 26 to 30 per cent of the estimated potential. Although this may seem a high proportion, it is no cause for alarm. The majority of maritime boundaries are settled peacefully, albeit after long periods of contention, and sometimes only after recourse to expensive litigation. [6] As long as boundaries remain undelimited there is always the danger of serious incidents for example over access by fishing vessels, while oil companies are rarely willing to invest in exploring contested waters. There are therefore strong arguments for a speedy resolution of maritime boundary disputes.

Maritime boundary agreements are likely to proliferate in the next 10 or 20 years. There is now a considerable accumulation of state practice from over 150 successful agreements which is well documented and thoroughly analysed, notably in Charney and Alexander (1993, 1998). [7] Technical support for boundary delimitation is today better than ever, and tasks which once took great amounts of time can be completed relatively quickly. The availability of GPS is a colossal benefit, and a number of computer programmes can be acquired commercially to assist in the delimitation of straight baselines, archipelagic baselines, equidistance, and the outer limit of the continental shelf. In addition, extensive data is now obtainable on the internet which may be valuable in the delimitation process, such as bathymetry, chart availability, and fisheries information. Legal materials such as the proceedings of the ICJ can also be consulted via the internet. The preparation of materials prior to negotiation should thus be more thorough and more rapid than was conceivable 10 or 15 years ago. Nevertheless, a number of states, including some of the poorer island states, are deterred from proceeding with maritime boundary negotiation because of the cost, and the shortage of suitably trained technical experts and lawyers. In some cases the nautical charts currently in use are extremely old, and an accurate hydrographic survey will be necessary as a prelude to delimitation. Approximately half the maritime boundaries awaiting delimitation are associated with island states and island dependencies.

To date, the role of the ICJ and other Tribunals in the resolution of boundary disputes has been surprisingly small numerically, but their impact has been considerable., States which resort to litigation do so in the full knowledge that the process may take years, and will be extremely costly. Moreover, there can be no certainty as to the outcome. Nevertheless in the closing years of the last century the number of states using the courts to resolve their disputes was higher than ever, and the trend seems likely to continue (Table 3).

Table 3: Maritime boundary disputes taken to the ICJ and other Arbitral Tribunals 1951 - 2000

Bahrain – Qatar	1991	ICJ
Cameroon – Nigeria	1994	ICJ
Canada – France Arbitral tribunal (St Pierre and Miquelon)	1989-1992	
Canada – USA (Gulf of Maine)	1981-1984	ICJ
Denmark – Norway (Greenland-Jan Mayen)	1988-1993	ICJ
Denmark – Germany	1967-1969	ICJ
Dubai – Sharjah Arbitral tribunal	1981	
El Salvador – Honduras	1986-1991	ICJ
Eritrea – Yemen Arbitral tribunal	1996-1999	
France – United Kingdom (Minquiers and Ecrehos)	1951-1953	ICJ
France – United Kingdom Arbitral tribunal	1975 -1977	
Germany – Netherlands	1967-1969	ICJ
Greece – Turkey	1976 –1978	ICJ
Guinea – Guinea Bissau Arbitral tribunal	1983-1985	
Guinea Bissau – Senegal	1989-1991	ICJ
Honduras – Nicaragua	2000-	ICJ
Indonesia – Malaysia	1998-	ICJ
Libya – Malta	1982-1985	ICJ
Libya – Tunisia	1978-1982	ICJ

4.2. EXCESSIVE CLAIMS

Much has been written about the "excessive" claims of certain states, while the diplomatic protests of other states in response are well documented. [8] The problem however is far from being resolved, and excessive claims to maritime space continue to proliferate. Some arise no doubt from a generous interpretation of UNCLOS provisions made in good faith; others are made with cynical and blatant disregard for the spirit of the Convention. Abuse of the straight baseline provision in Article 7 is the most common cause for concern. Straight baselines may be drawn "where the coastline is deeply indented and cut into, or if there is a fringe of islands in the immediate vicinity...." (Article 7 [1]). Moreover "the drawing of straight baselines must not depart to any appreciable extent from the general direction of the coast...." (Article 7 [3]).

Unfortunately none of these geographical terms are defined in UNCLOS, and states interpret them in all kinds of geographical settings to their own advantage. Around Korea, Vietnam, Burma, and several other states large areas have been enclosed by straight baselines to become the internal waters of the coastal state. [9] Some courageous attempts have been made to set out specific rules but they have never been adopted. [10] If the process continues unchecked it will constitute a considerable reduction in freedom of navigation. By 1994 more than 60 states had declared straight baselines, and 10 had unpublished claims. Twenty seven are regarded as "excessive" by the United States. [11] While the United States arguably applies rather strict interpretations of Article 7, there can be no doubt that many claims are excessive.

Article 47 providing for archipelagic baselines has also been the subject of abuse, although in fewer cases than Article 7. An archipelagic state may draw straight baselines joining the outermost points of the outermost islands and drying reefs, provided that within such baselines the ratio of the area of water to the area of land is between 1 to 1 and 9 to 1 (Article 47 [1]). The length of such baselines must not exceed 100 nautical miles except in up to three percent of cases when the maximum length is 125 nautical miles (Article 47 [2]). Unlike Article 7, here the guidelines are clear and unequivocal, but some of the world's 17 archipelagic states have nevertheless declared baselines which do not conform with the Convention, thus expanding their archipelagic waters, and the seaward limits of their territorial seas. Certain continental states have also treated groups of offshore islands as archipelagic, contrary to the UN Convention, including Canada, Denmark and Ecuador. [12] The United States (and other states) make ritual protests, but no international action is taken. The signatories of the UN Convention, including the offending states, must grapple with the problem of excessive claims early in the new century, if only to revise or clarify Articles 7 and 47.

4.3. LIMITS OF THE CONTINENTAL SHELF BEYOND 200 NM

The international community is slowly waking up to the full implications of Article 76 of UNCLOS, which gives coastal states rights to their continental shelves beyond 200 nm. Those states fortunate enough to be in possession of such natural prolongations of their land mass must make a formal claim to the Commission on the Limits of the Continental Shelf. In 1998 Prescott (1995) identified 29 continental margins beyond 200 nm worldwide involving some 50 states. Since that time more potential margins have been identified and today some 60 states are preparing to submit claims to continental margins. The total area is larger than the Antarctic continent. [13] Several continental margins are being claimed in whole or in part by more than one state, and a number are claimed by up to seven states, for example off South West Africa. In submitting their claims to the CLCS states have to determine the outer limits of their continental shelves. Article 76.5 sets absolute limits to the continental shelf at *either* 350 nm from the baseline on the coast *or* 100 nm seaward from the 2,500 metre isobath. Short of these distances however, a boundary with the International Seabed Authority has to be determined. States can choose which of two methods set out in Article 76 [4] (a)(i) and (ii) will give them maximum benefit. Both require detailed scientific knowledge of the bathymetry and geology of the seabed, and the preparation of a submission to the CLCS as a result is costly and time-consuming.

It remains to be seen how the CLCS will operate, and how states with multiple claims will resolve their maritime boundaries in the continental margins. No state has yet submitted its claim but several are thought to be about to do so. There is some urgency because the deadline for submission of cases is 2004 for those states who had ratified the Convention in 1994. This may be extended to 2007. The Commission seems in danger of being overwhelmed by the sheer weight of submissions and the complexity of the evidence. Among the causes for concern is that some states will press for the inclusion of ridges and spurs associated with the continental shelf in their claims. There are undoubtedly oil and gas fields to be found in the continental margins, but whether (and when) they will be exploited remains uncertain. In the meantime we will become increasingly familiar with the emerging maritime political map depicting more boundaries, the 200 nm EEZ, and the continental margins beyond 200 nm. (Figure 1).

5. Conclusion

The second half of the twentieth century saw the beginnings of the political partitioning of the oceans between coastal states. The UN Convention on the Law of the Sea did not come into force until 1994, but it had profoundly influenced state behaviour for at least a decade before that. In the twenty-first century the provisions of the 1982 Convention will remain critically important in spite of misgivings about some of the detail. Maritime boundary delimitation seems likely to proceed until the process is largely complete some decades from now. In the meantime, as with land boundaries, the focus is likely to shift from delimitation towards management, from borderlines to borderlands, from preoccupation with limits to concerns for collaboration. Such a change of emphasis will be necessary to tackle urgent issues of conservation, resource exploitation, sea-use planning, and state security. States will be most inclined to embark on collaborative ventures if they feel confident that the regime for the oceans agreed in 1982 is really being made to work.

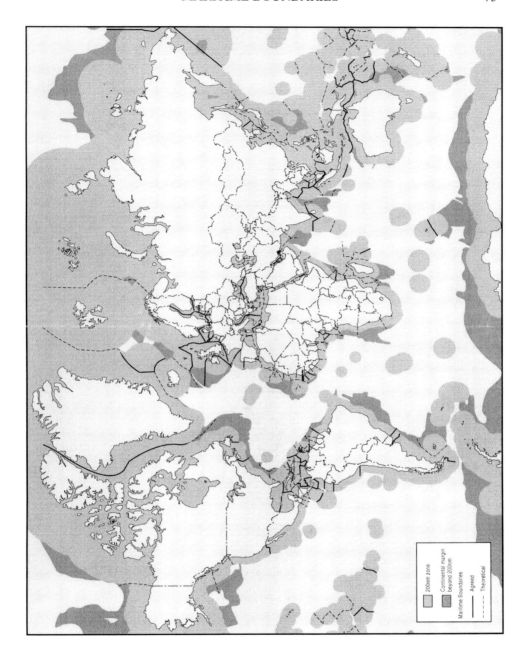

Figure 1. World maritime boundaries showing continental margins beyond 200 nm.

References

1. United Nations (1982) *United Nations Convention on the Law of the Sea*, United Nations Publications, New York. 192 pages.

2. Churchill, R.R. and A.V. Lowe (3rd edn. 1999) *The Law of the Sea*. Manchester University Press, Manchester. p. 147.

3. British Institute of International and Comparative Law (1989) *Joint Development of Offshore Oil and Gas : A Model Agreement for States,* B.I.I.C.L., London, 426 pages. and Fox, H. (ed) (1990) *Joint Development of Offshore Oil and Gas : Revised Model Agreement and Conference Papers*, B.I.I.C.L., London. 259 pages.

4. Dzurek, D.J. (1996). The Spratly Islands Dispute : Who's on First? *Maritime Briefing* Vol. 2(1). International Boundaries Research Unit, Durham. 67 pages.

5. Smith, R.W. and B.L. Thomas (1998). Island Disputes and The Law of the Sea : An Examination of Sovereignty and Delimitation Disputes. *Maritime Briefing* Vol. 2(4). International Boundaries Research Unit, Durham. 27 pages.

6. McDorman, T.L. and A. Chircop (2nd edn. 1991) The resolution of maritime boundary disputes in E. Gold (ed) *Maritime Affairs : A World Handbook*. Longman, Harlow. pp. 344-386.

7. Charney, J. I. and L.M. Alexander (eds) (1993) *International Maritime Boundaries* Vols. 1 and 2 (1993) Vol. 3 (1998). Martinus Nijhoff, Dordrecht for The American Society of International Law. 2,616 pages.

8. Roach, J.A. and R.W. Smith (2nd edn. 1994) *United States Responses to Excessive Maritime Claims*, Kluwer Law International The Hague..676 pages.

9. Scovazzi, T.G. Francalanci, D. Romano and S. Mongardini, (2nd edn. 1989). *Atlas of the Straight Baselines,* Guiffre Editore, Milan. 233 pages.

10. U.S. Department of State (1989) *Developing Standard Guidelines for Evaluating Straight Baselines.* Limits in the Seas No 106. Bureau of Oceans and International Environmental and Scientific Affairs, 37 pages.

11. *Ibid.*, p. 18.

12. *Ibid.*, p.23

CHAPTER 4. GEOGRAPHY AND GEO-STRATEGY OF THE OCEANS

A. VIGARIE

The relationship between geography and strategy has always existed; geography is at the beginning of the strategic objective which implies taking into consideration an environment, a setting, a place and the use of those characteristics integrated in operative decisions; but from the point of view here studied, there are also differences. Marine strategy deals with the nation's security and prosperity; it is important to mention what A.T. Mahan [1] wrote: "The aim of naval strategy is ... to create, to support, to increase the maritime power of the people during periods of war and peace." The geographer may have the same objectives, but does not focus his attention on the war like operations of the naval fleet: he has his own way of thinking; but the objectives are somewhat similar.

At the start, it is important to define, in geographical terms, what ocean geo-strategy is.[2,3] It is, in the first instance, an attempt at explaining behaviour of countries with reference to their relationship with the sea. It is also an analysis and an interpretation of the various interests relating to the independence of maritime orientated countries and their people. It centres on the study of *motives*, primarily economic motives and national attitudes, but it does not ignore the strong desire to expand ideological or political domination, with the sea as an enabling mechanism. These, however are mostly linked to economics. There is also a strategy of uniting the *means* for the implementation of defence policies with regard to legitimate interests.

Various consequences follow this brief definition. Ocean geo-strategy offers the scope to explain the complicated web of relationships which constitute the maritime sphere:

- This explanation can only be conceptualised on a global scale; the great mobility of men, goods, capital ideas and interests leads to the internationalism of all forms of exchange. Economy leads to geo-economics which again should be conceptualised on a global scale. Here, policy must also be interpreted as a geo-policy and strategy as a geo-strategy. The reflection must be a *Welpolitik* of countries.
- The relations of a state of medium power are also globalised, in other words it needs to have transport links, and in relation to this study, international maritime routes which themselves need to have strong institutional support, trustworthy middlemen, routes which operate with total safety and free access for traffic.

- In other words, we need to analyse conflicting aspects: diverse conflicts of interest with partners, differences or sometimes direct opposition, which can lead to an escalation of economic, political, diplomatic or military conflicts on various levels.
- As a consequence, ocean geo-strategy will be regarded here as the study of maritime inter-relationships surrounding a nation, a number of nations, or a navigation area and of the mechanisms linked to these relationships. Without doubt this is a vast problem because of the number of aspects that must be taken into consideration, and because of the range of attitudes likely to be encountered.
- Because the task is so complex, it is important to look for means of mastering and controlling the thought process. This is what is proposed below, starting from the premise that there is an inevitable need to utilise the sea in contemporary, national economies.

1. Theoretical Foundations of the Geo-Strategy of the Oceans

1.1. ALL MODERN COUNTRIES NEED EXTERNAL RELATIONS

In today's world, a situation of national self-reliance is not possible; it would lead to stagnation or a drop in production, the collapse of the standard of living and weaknesses in relations with the country's neighbours. This has been observed in the very few instances where this attitude has been at least partially adopted. Modern development implies the utilisation of raw materials and sources of energy, basic food supply, and manufactured or consumer goods. A country on its own, no matter how powerful it is, is totally incapable of looking after itself single-handedly. It is therefore stating the obvious that every state needs to import goods in order to develop its activities. The demands of the balance of payments imply financing imports by exports, or adding value to exports for the most advanced countries. All these facts are common knowledge.

Consequently, by analysing the balance sheet of a country, it is possible to establish its state of dependence regarding its needs; in other words, to estimate the degree of necessity of its trading links. This can be best done by looking at the number and diversity of trading partners. It is important to determine the degree of dependence, in other words the priority of those needs. As such, each year the United States imports 10,000 tons of cobalt, the content of a small bulk carrier, but this corresponds to 91% of their annual consumption; it also needs to purchase several tons of diamonds, graphite, mica, etc., but these amounts may well represent the whole of their annual consumption. The same applies, to varying degrees, to all countries, which need to resort to international trade. It is necessary to state that world trade is predominantly maritime as it involves 90% of the weight of goods and around 75% of their value. Foreign reliance of countries involved in international trade is therefore mostly a maritime reliance.

Some of the needs mentioned above which lead to trading with foreign countries, can be cut down, others cannot be or can be only partly reduced: there is a determinism to those needs, that is to say, the country in question is not free to give up the links without suffering consequences. France has virtually no oil; Italy has no coal. Their import is therefore absolutely necessary.

Consequently, it is necessary to establish a geo-economy of vital needs of nations: it is very important to present this in two forms:

- Static: This involves an assessment of the situation which surrounds these needs. This would include geology, geography, demography, production, type of society, etc., and the sea constitutes the only means of solving the problem by offshore production or by using maritime transport. This geo-economy, if established on a continental scale, reveals different aspects along the coasts and the surrounding areas. It is important to find out what is predetermined, what is inevitable; these factors form the main motives for orientation of nations towards the sea. For instance, it would be interesting to establish a geo-economy of China's needs in its current orientation phase towards the ocean.

- Dynamic: This determination of economic need must lead to coherent answers, in other words, to the drawing-up of commercial agreements and relations, or of opposition to other states with proposals to solve such conflicts.

1.2. THIS LEADS TO A GEO-POLICY TO SATISFY NEEDS

Part of a country's needs can be satisfied by exploiting the sea. Previous experience shows, however, that this is not sufficient, there is a compulsory system of exchange, of meeting up with other nations, which leads to certain behaviour, a policy of relationships.

- First of all, access to *essential material*: access to foreign resources is not always possible and presents difficulties; the most common problem is the strong concentration of reserves and producers, for various raw materials, energy or food; access to them can become difficult. When the USSR collapsed in 1993, the country owned jointly with South Africa 97% of platinum production, 94% of manganese, 78% of gold and 72% of vanadium. In 1995-1996, one single country held 63% of the world production of wheat and could therefore exercise pressure on its foreign customers. 68% of iron ore is exported by 2 countries, and 53% of coal by only 3 countries. Despite a certain degree of independence on the part of some industrial countries, OPEC continues to control 45% of oil sales. The possession of oil reserves in many cases provides economic 'weapons' which can have complex political repercussions.
- Henceforth, *a covert* war is being fought over access to resources, in particular with regard to "strategic" products. It is a war consisting not only of open or closed agreements between nations to satisfy their interests, but also of opposition bodies,

of struggles for or against sales, and of transport monopolies, embargoes and sanctions with political, ideological and economical motives closely entwined. South Africa faced a commercial blockade in the mid 1980s which reduced its fleet by 49% but the sea enabled the country to circumvent the sanctions.

This covert war can become open and lead to military conflicts. Problems of maritime economics took place in the Persian Gulf between 1980 and 1991 as witnessed in the Hormuz story.

This general scheme of maritime links resulting in access needs to basic resources involves the often-subtle role of governments but also of transnational corporations. They have their own policy towards finance or industrial groupings such as integrated fleets, in which they often give the sea the attention it deserves. They take part fully in covert wars sometimes in contradiction to their government's official position. Before 1988, Gulf Oil in Angola paid a substantial amount to the MPLA, who were very progressive and pro-Soviet , so that the company could continue to produce oil and export it by sea during the violent civil war. Meanwhile, the United States financed the other party, UNITA, which was pro-Western. It is also important to mention the attitudes of producers' associations (phosphate, coffee, bananas, etc.) who endeavour to dominate trade and transport and have often set up their own fleets.

Consequently, a complicated network of links has been set up across the oceans, links which vary according to the goods exchanged, each country becoming in turn buyer or seller. It is possible for one country to be in agreement with some and at the same time in conflict with others. Its network of maritime links are designed to protect its interests as far as possible.

- Then again, the export of goods, especially **manufactured goods** is not neutral; a commercial operation also frequently has a political meaning: Those goods bear the name of the company which manufactures them and which has its rules of external contracts and consumption. Those products, once sold, take with them a language, a culture, a mode of social integration; exchange can encourage deep interests, even the responsibility of partners. When considering factory delivery, manufactured high-tech products are integrated into a production line, weapons are bought abroad. If a breakdown occurs, it is necessary to contact a foreign technician. The need to get spare parts creates a state of dependence. Those exchanges lead to various degrees of intensity in the economic relationships, within which governments need to take a stand. And here again, the sea and ships are inevitably a means of facilitation.

- As a result, on an international scale, there is geopolitics of consumption involving agreements and opposition, and of movement of goods, most of which are transported by sea. Each nation has it own **system of routes** on which depends its provision of fresh supplies or exports. This system is carefully monitored and protected, and is used to cater for its commerce and military fleet. The sea routes for France are not the same as those for Russia, China or Japan. Each country would start a conflict to protect its own sea routes. The geography referred to above involves the control of sea routes on a national basis but also in a broader context of a group of countries which can have common needs.

Consequently the geo-economy of the needs and the geopolicy of corresponding solutions to those needs constitute the starting point of the ocean geo-strategy. Its interpretation is then done by country or by groups of countries.

These special forms of relations by sea increased in importance and diversity due to the growing maritimisation of the contemporary world. Henceforth, we can state as a first approximation that the growing link between the world economy and sea was through the exploitation of maritime wealth and the massive utilisation of ocean transport. It proved itself to a growing number of economic partners, due either to the independence of former colonial territories (the number of UN members went from 50 to 191) or to their recognition of the production functions and exchanges, or indeed of the new opportunities offered by the exploitation of the oceans and the new status of the sea. It needs to be centred on production and exchange systems largely because of new opportunities offered by the exploitation of the oceans and the new status of the sea.

The geo-strategy of the ocean reflects multiple interwoven interactions but it is not a vision of disorder. It is possible to observe a certain general organisation in the maritime space, so long as the underlying mechanisms, and the precise way in which nations and sea would interact, are analysed in detail.

2. General Methodological Approach of Research into the Geo-Strategy of the Oceans

It is clear that the method adopted is geographical but with a maritime approach, in order to measure the degree of intensity of links between the states and the sea. The aim of the following text is to provide settings in which a process of thought can provide research directions.

2.1. NATIONAL MARITIMISATION FACTORS AT THE BEGINNING OF THE 21ST CENTURY

It is not necessary to go back to problems already analysed at length. It is true that some populations are more sea-oriented, and are more aware of the sea, than others. Being situated by the sea can encourage insular economies; and the Convention of Montego Bay (1982) gives rights and duties to maritime countries, defining their source of interests. Questions over the delimitation of maritime boundaries are often a source of conflict. There are, however, landlocked countries such as Switzerland and Austria, without direct access to the sea, who adopt an active maritime policy.

It is important to consider carefully the impact of geographical and geological circumstances. A country or a group of countries exporting a product on an international scale do not necessarily adopt a strong maritime policy: The Middle East and the Near-East held in 1998, under national flags, only 1.8% of the world fleet and only 4.2% of ships under foreign flags. Demographic considerations are not an indicator of maritime power. The Indian sub-continent with 21.4% of world population is in charge of only 1.46% of global tonnage. These data have to be taken into account when geostrategic attitudes of certain countries are analysed.

There are other factors that influence attitudes and, for many countries, it is the legacy of half a century of events which have taken place that influence their maritime links. The transition between palaeo-technical economies (using primarily coal and iron and using corresponding maritime transport) and the neo-technical and post-industrial economies, which changed the pattern of consumption of energy and raw materials bringing about other forms of exchange, has now taken place. We can observe the impact of de-colonisation, especially towards the end of the 1960s, but also the attitudes of many states, including ex-mother countries, which led to a re-orientation of their links with the sea- and ex-colonies have not yet found the right balance in their external relations- with foreign countries within their maritime policy.

There are also ideological legacies. The doctrines of state controlled economics have not disappeared despite the fall of the USSR; socialist countries maintain purely continental attitudes beyond a certain degree of development. The USSR was a great sea power but as far as doctrine is concerned, maritime trade only served general evolution as opposed to being a source of new wealth (cf.: Lenin: "The Navy is a waste of people's money"). On the contrary, towards the end of the cold war, liberalism developed considerably, creating an immense expansion of trade and consequently new attitudes towards the ocean and also conflicts of interests as a result.

It is important to consider the place of each nation within these trends and it is not always perceptible through temporary and sometimes misleading attitudes expressed during economic recessions, or during local or regional crises.

It is possible therefore to envisage a *typology of nations*, which would result in the combination of the previous trends and which would direct research into geo-strategy. Each of the groups below has its own structure of maritime links.

- *Liberal industrialised countries* have advanced economies and a strong concentration of their population in the tertiary sector. This does not prevent them from extensively using international transport and they have the strongest interest in protecting the sea. They have set up the biggest volume of *manufactured goods* and use cellular fleets to carry containers. They own 28% of the world tonnage under their flags and even more than 28% of their ships are registered under foreign flags. They buy or sell bulky and heavy goods in large quantities, and are at the source of the gigantic size of oil tankers, ore-carriers and gas-tankers. The paradox is that national fleets have declined but these countries are the true owners of 85% of the world merchant navy.

- *The New Industrial Countries* refers to a small number of countries from South East Asia which have advanced rapidly since the 1960s and 1970s after Japan recovered from World War 2 the help of American aid. In the 1960s Western powers and Japan invested in technicians, capital and technology in Taiwan and Korea and this has stimulated rapid growth. In 1970s, those countries took part in the Western and Japanese investment in Hong Kong and Singapore. They did the same in the 1980s in Thailand and Malaysia and they contributed to some extent to the development of Indonesia and others. This process is know as 'Markov's Process Chain'. [4] The result is that these Asian countries with waterfronts and with real wealth constitute a true focus of maritime links that is growing despite periodic recessions. They have in common certain inequalities which have resulted in a spread of successive periods of growth. Japan belongs to the group of liberal industrial countries and remains a major player; the nearby newly industrialised countries account for 13.5% of the world's mercantile fleet and they host some of the most powerful shipowners in the world. Each 'maritime tiger' in South East Asia maintains its own distinctive style despite similarities of origin and evolution, tonnage of the fleets, demands for imported raw materials required for industrialisation, and export of manufactured goods produced by often poorly paid and poorly trained labour. In this context China has shown a specific evolutionary path and a maritime policy which generally brings the country into conflict with its neighbours. [5]

- The vast majority of *Developing Countries* (DCs) are previous colonies which gained independence in the 19th and 20th centuries. Their somewhat backward economies are outward looking and this integrates them into maritime activities generally with only a small fleet, but also with only a small number of adequately equipped ports. This situation often leads to an imbalance in the balance of payments, aggravated by periodic crises in the cost of raw materials, and also to serious debts and to an involvement of the IMF with privatisation measures and in deregulation of commercial ventures. When they are faced with these situations, such countries are not really able to cope and experience great difficulties in establishing a maritime policy. There has been an attempt, however, under the influence of the Group of 77 (this involves 77 developing countries who tried to organise themselves against major maritime nations) in UNCTAD. For 10 years, countries in West and Central Africa tried to maintain their indigenous shipping policies but were unable to. There are only local fleets and occasionally ships from other foreign countries exporting their raw materials and the countries maintain a few well-equipped ports, busy in most cases thanks to foreign capital.

- **Finally** there are the **LDCs** according to the UN terminology (Less Developed Countries). This leads one to think that there is a connection between a low standard of living and the absence of maritime links, thus implying that no sea strategy is adopted by their governments. It is important to point out that among

175 countries, 12 have the lowest level of development as defined by UNDP. Six of these are landlocked countries without access to the sea and the remaining, six states do not have a maritime policy. There are, however, exceptions as explained below:

It appears that belonging to a type of economy or having a particular form of behaviour reflects the way of thinking and understanding of their countries' strategic attitudes. There are differences within the same group, but this typology is a starting point.

The evolution of the world's 'sea order'

This expression deals with sea links evolving within the history of society and characterised by:

- The volume in weight and value of total world maritime trade;
- The nature of transported goods and the combination resulting from it: in bulk heavy goods, mixed general cargo, etc;
- The volume and composition of the fleets as a result (oil tankers, bulk carriers etc.);
- The type of links between commercial partners;
- The organisation of maritime traffic flows and sea routes and the problem of their protection.

During the second half of the last century many **changes regarding maritime supremacy** took place in backward sectors, but not in a synchronous way. The role of the oceans has evolved regarding their use for navigation. The order of navigation of the ***Atlantic Ocean*** lasted through World War 2, with seas regarded as borders. This ocean, in 1914, accounted for three-quarters of world trade with three-quarters of world ship tonnage. Afterwards a noticeable drop occurred; 57% in 1964, 48% in 1985 and today less than 40%. The Pacific Ocean represented 13.4% in 1964, 20.8% in the mid-1980s and today around 39% of freight. The main change took place in 1982 when, in value terms, United States trade across the Pacific overtook the American trade across the North Atlantic. This evolution, in other aspects, involves other oceans and other coasts: it is a movement of the centre of gravity of world navigation. The causes of the important changes are various: decolonisation, development of South East Asia, new routes for energy, international conflict and others. The impetus from the sea and from countries with a sea coast has evolved and the ocean geo-strategies of those countries have changed; the Pacific has a very strong centre of gravity: in the Spring of 1999 (but this has happened previously) several countries hurried to create new routes between North America and Asia involving 39 very productive ships serving 29 ports and several states, it is not a matter for shipowners: it is a sign of reinforcing the "order" in the Pacific offering new opportunities, even FESCO belonging to a penniless Russia followed this tendency.

With these general facts as a background, there is scope to examine the attitudes of governments towards the sea and, subsequently, to assess these attitudes.

2.2. THE MEANS: THE CAPACITY FOR INTERVENTION IN THE OCEANIC SPHERE

There are diverse forms of naval port facilities and equipment, various means of control including navigation routes, satellite tracking of ships and others. These need to be taken into account in order to summarise the situation, For simplification only fleets will be mentioned here.

Merchant and Fishing fleets and Offshore Infrastructure

The exploitation of sea wealth can be considered as a palliative to the lack of national terrestrial production especially when this wealth comes from territorial waters or from the Exclusive Economic Zone (EEZ).

As far as mineral production from the sea is concerned, it is important not to limit the study to hydrocarbons. The success of some developing countries has been particularly noticeable: for example India extracts 35 million tons of naphtha, and Gabon 18 million tons. Also significant is the extraction of dredged aggregates, which are important for European countries, of metallic ores such as tin from Malaysia, ilmenite from Sri Lanka, and other materials.

It is difficult to analyse the fishing power of a nation in precise terms for many reasons. Many small coastal or estuary boats play an important role but are not necessarily registered. Official vessel tonnages are often only partially relevant to assess the capacity for factory ships, which are also very important. The quantity of fish caught in official international statistics does not always convey a clear picture because of possible inaccuracies in spatial delimitation: How can sea fishing be differentiated from estuary fishing e.g. in China and Brazil and from river fishing? This classification approach is useful however, especially when it is completed with reference to the number of fishermen and consumption of fish per inhabitant.

While offshore activities and fishing remain useful indicators, the ***merchant navy*** is an interesting and partial indication of the interest a state can have in the sea.

It is true to say that there are countries which have become 'public transporters of the Ocean'. This is clear proof of their interest in the sea but is not necessarily confirmed by their foreign trade. Apart from countries of free registration, called "flag of convenience" (FOC) in the past, this 'public transporters' concept also applies to Greece and Norway. Other countries have, on the other hand, strong commercial links but with a ship tonnage that does not match. The second register flag and those mentioned above can hide the real role of a fleet in national ownership and the international character of armaments; it can also mask membership in a consortium.

As a result, looking at the number or capacity of merchant ships is interesting, and data need to be analysed taking into account fractions of merchant fleets, which show different compositions, variable productivity (for example, bulk, container ships) and different age profiles.

Taking these facts into account, it is suggested that joint indicators of dependence and equipment should be used. [6] These can also be inaccurate in part, but they have the advantage of establishing the between behaviour at sea of a nation and its demography. It is another way of assessing attitudes towards the sea and introducing more accurate statements about these indicators. China has become, without any doubt, a great sea power: its official fleet includes 16.5 mt dwt and the country is now ranked 6th in the world. It also has other vessels under different flags thus making China even more powerful. This is a sign of rapid maritimisation involving development from an early ocean strategy. This assessment, without being questioned, needs to be completed by a close look at technical indicators mentioned below: world equipment and dependence indicators are 0.077 Gt and 1,446 kg per inhabitant. Those of China are 0.012 Gt and 180 kg. This reflects the fact that the maritimisation which has taken place is not yet sufficient, that this country is still in an early phase, and that the expansion margin at its disposal is considerable.

Consequently, in order to assess the degree of orientation of a country towards the sea, and to study its participation in the geo-strategy of the oceans, the convergence and complementarity of the indicators need to be taken into account because their study in isolation would only enable incomplete conclusions to be drawn.

Naval Fleet

As the study is geographical, it is essential not to enter the realm of 'military strategy' which has its own experts. It is important to determine which form of global protection a naval power can offer in the general behaviour of nations towards the sea [7]

It is difficult to estimate the importance of this type of fleet. The greatest and most prestigious fleets are in possession of nuclear submarines (SNLE and SNA [8]) and airborne capability linked to aircraft carriers. At the forefront six countries only, including China, have this capability. Many other countries, including several developing countries, have traditional submarines but only the Indian and Brazilian navies has airborne capacity. These are criteria which help sustain an active maritime policy. But the possession of atomic weapons is on a different level: what is their use regarding conflicts in fishing zones, EEZ borders or sea coasts?

The protection of the most common maritime interests involves in most cases small units, such as frigates and corvettes. Besides these, the Gulf War has shown that the possession of deterrent-missiles is an important means of response not involving the use of heavy tonnage units and that it is easily possible to block a sea route. [9]

It is therefore obvious that naval forces of all kinds can take part in geo-strategy of defence, when a modem nation can have all sorts of interests at sea and must establish its independence whatever the area concerned. The position of the United States is interesting: they are the only nation with fighting forces scattered around the globe which can converge and exercise continuous military pressure. [10]

3. Research Perspectives and Directions for the Future

The geo-strategy of the oceans provides geographical research within a wider sphere.

3.1. THE DIVERSITY OF THEMATIC AREAS

This can encompass first of all the study of the behaviour of a nation towards the sea [11] or of a group of states to determine the volume, variety and nature of their maritime relations, solutions to their needs, and consequences of actions which may or may not involve conflict. There are so many aspects of international relations which are still unknown.

This can be followed by the grouping of problems on an *oceanic* scale [12] or on the scale of several oceans, and it involves the evolution of international relations; various aspects have been suggested before: world ocean order, international policy regarding weapon supply during conflicts and others.

It is interesting to note the possible existence of top level geostrategies, if there are such things as **models** of behaviour determined by common needs or difficulties. For example, this may be seen in certain developing countries, and it was observed in West Africa between 1975 and 1985. Is there a typology to discuss, to set up among those developing countries or among NICs?

It could then involve a maritime geo-strategy of **key products** which are considered as strategic. Jean Gottman suggested it in the case of oil and this has been outlined by Vigarie [13] and it can also be the case for natural gas, food grains, gravel or rare metals.

There are the problems of maritime routes and of "choke points", in other words, there are problems where there is a risk of restriction of trade, of food supply, of exports. The freedom of movement has been defined and guaranteed by the Montego Bay Convention but it remains fragile [9] and many countries have taken measures to over-ride it, especially in the event of a crisis. This is represented in the Chinese Law defining territorial waters with regard to the Spratly Islands, [5] the conflict of the Hanish Islands between Eritrea and Yemen; the pressing problems of Hormuz, still simmering, and the astonishing variety of potential conflicts surrounding the Japanese islands. [14]

This reminds us of the fact that there are many potential conflicts and that this has lead to numerous gentlemen's agreements to avoid serious risks. [15] They provide a wide field of research in a geographical perspective, and have proliferated under two influences. First of all, the Montego Bay Convention of 1982 with many beneficial consequences has prompted many countries to redefine their sea borders where this had not been done- in many cases a lot of conflicting interests were revealed. These were often solved in a peaceful manner without going to the International Court of Justice, but in other instances, a lot of fierce reaction followed, including the Icelandic Cod War, Greek-Turkish opposition in the Eastern Mediterranean, controversial allocations of oil concessions and others. A typology of conflicts, taking into account of regional peculiarities, still remains to be done. [16] The important development of naval capacities of Third World countries following the end of the cold war, and the sale of weapons by the major powers, have intensified these tensions.

3.2. A DYNAMIC VISION OF THE WORLD OCEAN

It has been shown above that ocean geo-strategy implies a progressive interpretation. The end of the cold war has lead to a rethinking of impending problems by modifying economic trials of strength and diplomatic and military relations in many parts of the world. The collapse of the USSR and its traditional customer base has brought major repercussions for maritime relations, fleets, ports in Africa, in Asia and in the West Indies-, the Black Sea and the far east of Russia. Those changes are the consequences of socialism and liberalism in the world, and have lead to major questions which have not yet been solved, such as the ocean position of India, of China and, more generally, of South East Asia in the 21st century.

From a different perspectives a dynamic approach also deals with changes in North-South maritime relations which can be identified: the recent opening of the Third World to maritime crises involving the cost of raw materials, the failure of UNCTAD policies, and the aggressive and overpowering attitude of the major shipping lines which impose themselves on developing countries. Having won the East-West routes, these lines now dominate the flows of the most valuable food products and the most sought-after cargoes in general.

Today's world is in a permanent state of change: ocean geo-strategy can only reflect this situation.

4. Conclusions

There is a wealth of options for ocean strategies: this brings about the purpose of the research outlined above. What is the intention of such studies?

First of all, they all lead to the acquiring of knowledge which could not otherwise be easily achieved. They also imply a geographical localisation, not only of the sea, but also of the knowledge which conveys the concept of multiple causality in a complex socio-political and commercial environment. They enable us to comprehend the world better in its state of constant change. It is important to reiterate that problems related to the sea do not always receive the attention they deserve. They help man, as a citizen and as an economic actor, to find his place in the current evolution of thoughts and facts.

The studies have also drawn the attention and interest of governments and politicians; these specific approaches have helped in the decision-making process. They precede the choice of response and provide information. They help to protect the maritime interests of nations, they enable the creation of a progressive ocean geo-strategy based first of all, but not exclusively, on the needs of humankind, secondly on political strategy and finally, if necessary, on naval policy.

References and Notes

1. Mahan A.T., (1989) *Influence de la Puissance maritime dans l'Histoire* (1660-1783). Traduction francaise, Paris 1989, 599pp.

2. Vigarié, A. (1990) *Géostrategie des Océans*. Ed. Paradigme, Caen. 399pp.

3. Vigarié, A (1995) *La Mer et la géostrategie des Nations*. Ed Economica. Paris 432pp.

4. The **Markov Chain** refers to a situation in which a cause C, creates consequence Cq 1; then the Aboelements add up to form a Cause C2 leading to consequences Cq2; they add up to form a new cause C3 leading to Cq3 and so on. This chain lends itself to the explanation of South East Asia's maritime growth.

5. Labrousse, H (Ed.) 1996) *Les iles conflictuelles des Mers de China*. Revue Maritime. 4th edn. P.74.

6. Dependence index: This is a theoretical measure of the weight of goods for each citizen of a given country which is involved in the maritime trade of that country. Equipment index: This is a theoretical measure of the official national fleet, in tonnes, which enables each citizen to benefit from foreign trade; it is measured as fleet tonnage per inhabitant.

7. Prezelin, B (1998) *Floyttes de Combat* (biannuel). Editions maritimes et d'Outre Mer, Paris

8. SNLE: Nuclear Submarine Launching Strategic Missiles (SSNB in British sources). SNA: Nuclear Submarine Attack (SSN in British sources)

9. It is important to reiterate that a few mines were sufficient to paralyse the traffic in the Red Sea in the Summer of 1984.

10. Vigarie, A (1999) *Elements Strategiques de l'Atlantique Nord et de la Mediterranee en 1995*, Atlas Permanent de la Mer et du Littoral, CNRS-Geolittomer, Nantes No. 4 Planches et textes 46 et 47

11. Vigarie, A (1991) *Approche d'une geostrategie des Pays Iberiques*. Colloque de l'Union Geographique Internationale, La Rabida-Huelva.

12. Op. Cit ref. 10

13. Vigarie, A (1993) *Echanges et transports internationaux depuis 1945*, ED Sirey, Paris, 3eme Edition, Chapitre 4

14. Pelletier, P. (1997) *La Japonesie*, Ed CNRS, Paris, 386pp.

15. Junnola, J.R. (1996) *Maritime Confidence Building in Regions of Tension.* The Henry L Stimson Center Report, Marsh, 1996.

16. Ridolfi, G (1988) *Il mare proibito*, Revista Geografia Italia, Giugno, p121.

CHAPTER 5. MARITIME TRANSPORT

J. MARCADON

World maritime traffic at the end of this millennium is more than 5 billion tons of goods transported by a merchant fleet of about 38,500 ships of more than 300gt. A real revolution in sea transport has seen over the past 40 years an increase in the size of ships, their specialised character including the unitisation of general goods by containerisation and roll-on - roll-off techniques. In the limited scope of this text, recent general trends and possible future developments regarding the world merchant fleet in various sectors, bulk transport, regular routes and passengers/freight, are discussed.

1. The Recent Evolution of the World Fleet and the Revolution in Sea Transport.

1.1. RECENT EVOLUTION ACCORDING TO FLAG.

The role of sea transport has changed step by step during recent history. European countries dominated world sea transport until the First World War. There is still an unequal distribution between nations and the rate of change is accelerating with the emergence of new extra-European sea powers playing an increasingly important role (Tables 1 and 2).

In 1961, 14 nations, mostly European, (10% of world states) owned a merchant fleet of more than 2 million deadweight tonnes. These 14 nations owned 92% of oil tankers, 83% of dry bulk carriers, and 86% of world tonnage overall. There were in 1981, 20 years later, 28 nations, twice as many sea powers. On 1st January 1998, 39 nations could be considered as commercial sea powers. The decline of some has been followed by the expansion of others: the United Kingdom represented 56% of world tonnage in 1900, 44% in 1914, 29% in 1938, 16% in 1964, 6.5% in 1980 and 1% in 1998 (18th place in the world). The French fleet held the 9th place in 1980 but the 28th in 1998.

These figures, however, do not take into consideration control or ownership of fleets. There is a constant movement of sale and transfer of ships to another flag without change of ownership. Maritime transport is a jungle of legal procedures which highlight the flexibility of the system of transport with the rationalisation of the operation of ships, even if men are too often forgotten in this state of affairs.

Traditional sea powers have taken measures to face the competition of other nations and have adopted, in certain cases, robust measures or have created a flag (for instance the Norwegian International Shipping Register or NIS, the Danish International Shipping or DIS) less costly for the ship owner regarding taxation and operating costs, involving crewing mainly from members of Third World countries. For example the cost of the crew on a ship (regular route) comprising 9 officers and 14 seamen under the French flag is about US$2 million per year and US$ 1.4 million under the French Arctic and Antarctic territories flag. It is US$ 700,000 for a ship equipped to the international standards of a flag of convenience.

New sea powers (Tables 1 and 2) can be grouped into various categories and there are different types of merchant navy in Third World countries. First are fleets of countries rich in energy and mineral resources. This includes the fleets of countries in the Middle East, which enjoy a lot of petrodollars and who invested in sea transport and ports in the 70s; and the fleets of major Third-World Countries such as India (13th rank in the world) or Brazil (26th rank, with their great wealth of minerals).

Then there are the fleets of newly industrialised countries in Asia such as Singapore (ranked 8th), Hong Kong (20th), South Korea (16th) and Taiwan (21st) which based their development on a maritime channel serving an upstream coastal steel industry, ship yards, and shipping within a state-controlled but basically liberal economy.

Next are the fleets of socialist countries (Republic of China) or ex-socialist (ex-USSR and Eastern Europe). The merchant navy of China in 1966 had a capacity of 700,000 gt, 6.6 million gt in 1980 and 14.9 million gt in 1998 (10th rank in the world). Leaders of the ex-USSR understood that it was essential to control the seas in order to increase their influence in the world, a great principle of geo-strategy. It is not sufficient, however, to own a military fleet - it needs to be backed up by merchant ships. This explains why the Soviet merchant navy was ranked 5th in the world in 1986. The tribulations of the economies of countries of the ex-Eastern bloc in 1998, led the Russian fleet to fall to 19th rank in the world, Ukraine to 40th, Romania to 33rd and Poland to 37th.

The emergence of flags of convenience is noticeable and now represents nearly half the world fleet. Therefore Panama, Liberia, the Bahamas, Malta and Cyprus hold respectively 1st, 2nd, 4th, 5th and 6th rank in the world as far as their fleet is concerned. The owners of these fleets (Table 2) are Greece, Japan, United States, Hong Kong and western European countries. Convenience flagging is a widely recognised phenomenon in tramping and bulk transport and now involves all sectors. For example, regular containerised routes involve 120 container ships operated by Evergreen and its subsidiary Uniglory and about one third include the Taiwan flag and the other two-thirds the Panama flag.

Table 1. Flags of the World's Top Merchant Fleets > 2,000,000 dwt (01/01/1998)

Country	Number of Vessels	'000 gt	'000 dwt
Panama	4,834	89,262	136,129
Liberia	1,599	58,714	95,698
Greece	1,199	25,093	43,162
Bahamas	1,070	24,852	38,333
Malta	1,312	22,581	37,476
Cyprus	1,533	22,996	36,099
Norway (NIS)	1,170	22,017	33,902
Singapore	968	18,472	29,031
Japan	3150	16,928	24,434
China	2045	14,947	22,399
Phillipines	935	8,675	13,314
USA	375	9,773	12,926
India	390	6,571	10,957
St Vincent	885	7,231	10,945
Marshall Is.	129	6,280	10,672
South Korea	742	6,951	10,587
Turkey	902	6,356	10,359
UK	460	7,314	9,943
Russia	1755	7,749	9,589
Hong Kong	268	5,754	9,543
Taiwan	232	5,812	8,913
Germany	631	6,547	7,883
Bermuda	96	4,555	7,349
Italy	623	5,810	7,304
Denmark (DIS)	502	5,549	7,083
Brazil	225	4,231	6,972
Malaysia	495	4,671	6,817
France	218	4,375	6,593
Iran	183	6,428	6,082
Netherlands	610	4,230	4,875
Indonesia	1034	2,781	3,929
Thailand	436	2,000	3,251
Romania	239	2,186	3,218
Kuwait	59	1,954	3,125
Australia	108	2,170	3,031
Antigua/Barbados	501	2,191	2,851
Poland	119	1,651	2,471
Belize	509	1,390	2,147
Sweden	251	2,652	2,048
Ukraine	365	1,774	1,953

Table 2. The Top ten Controlled Fleets (1998)

Liquid Bulk Fleet	Dry Bulk Fleet	Container Fleet
Norway	Greece	Germany
USA	Japan	Taiwan
Netherlands	China	Japan
Hong Kong	Hong Kong	USA
Greece	South Korea	Denmark
UK	Taiwan	China
Singapore	Norway	South Korea
Finland	Turkey	UK
Ex USSR	UK	Greece
Taiwan	India	Singapore

1.2. SHIP TECHNOLOGY, SIZE AND SPECIALISATION

1.2.1. The supremacy of the diesel engine and the arrival of fast ships.

The vast majority of the world fleet is diesel-powered except the oil burning steam mode. The other modes of propulsion are marginal: coal or diesel as a back-up for larger ships with sails like *'Wind Star'*, *'Wind Song'*, *Wind Spirit'*. *'Club Med 1'* and *'Club Med 2'*, launched between 1985 and 1993. Nuclear propulsion quickly showed its limitations in its commercial use. There have been only three examples and in each case the ship had to be denuclearised and provided with a new engine.

The HSSs (High Speed Ships) are the most striking. Merchant ships generally reach a cruising speed of 20 knots and for the past twenty years naval engineers have been designing ships able to reach 40 or even 50 knots.

A radical change took place in the 1960s with ships with a hull no longer in contact with the sea. These are jetfoils or hydrofoils, their hulls are supported by "skates", and hovercraft supported by air cushions. The hovercraft was launched on the Calais – Ramsgate route in 1958. By the end of 1994, there were already 600 fast ships throughout the world, 400 in Japan and South Asia, 70 in Italy and 60 in Scandinavia. This niche involves also passenger-freight ships which can turnaround quicker; this pleases customers and reduces the need for equipment and crew: a fast ship needs 14 crewmen whereas a normal ferry needs between 90 and 100. The use of HSSs for freight is gradually becoming a reality with the project of the American company Fastship (Transatlantic link Philadelphia-Cherbourg in 4 days) or the Japanese project of the "Techno Superliner". The project Fastship could operate from 2003 or 2004. The challenge is to increase speed by 20%. This also involves an increase of the propulsion power by 100% and has repercussions on the fuel consumption and on the fuel and commercial cost of the ship. There is also the problem of safety at 40 knots. Extreme forces act on the hull and the ship needs to be monitored more closely.

1.2.2. Size and Specialisation in the last 40 Years

The phenomenon of specialisation was highlighted by the construction in the 1960s of bigger and bigger oil tankers. After World War II, 25,000 dwt tonne vessels were normal, followed by Very Large Crude Carriers (VLCCs) up to 300,000 dwt (drawing 23m of water), then at end of the 1970s the Ultra Large Crude Carriers (ULCCs) to 565,000 dwt (drawing 28m of water). The oil crisis of the 1980s limited their value.

The ship is becoming bigger and more specialised with the development of optimum transport routes; unloading equipment becomes more adapted to the type of goods and type of ships. The stopover becomes shorter and the total cost of transport is lower. Ports are becoming a succession of specialised terminals dealing with specialised ships, for example cereal carriers, liquefied gas carriers, and container ships for general goods, all in different locations.

This phenomenon involves bulk carriers (trampships) as much as general cargo ships on regular routes. Thanks to them, it has been possible to increase port productivity considerably. On average a conventional quay deals with 100,000 tonnes of goods a year and a quay for containers one million tons. As ships get bigger and more sophisticated, they need bigger quays: 180m for the first generation container ships of 750 TEU (twenty foot equivalent unit) and 320m minimum for the fifth generation container-ships, post panamax, of 6,000 TEU capacity (8,000 TEU ships have been ordered). The very latest (August 1999) of the giant container carriers of Maersk shipping company, the '*Skagen Maersk*' is 347 m long, 47m beam and with a speed of 25 knots with a 12 cylinder diesel engine of 74,640 cc. The size of ships is still increasing and these ships will reach capacities of 13,000 TEU. Also notable is the transport of passengers in increasingly large ferries. The trans-channel ferry of the 1960s was 100m long and carried 950 people and 180 cars. At the beginning of the 1990s ships were 180m long and could carry 1500 passengers and 600 cars. At the end of the century, jumbo ferries were 200m long and could carry 2500 passengers and 650 cars.

So far as cruise ships are concerned, the "*Voyager of the Seas*" of Royal Caribbean Cruise Lines (RCCL) is the largest ship in the world. It started service in 1999, it is an over-panamax of 142,000 grt, 310.9m long, 48m beam with a capacity of 3,118 passengers (3,840 including upper bunk beds) and 1,181 crew members.

1.2.3. Company Mergers

The tendency regarding the organisation of sea transport is towards concentration. This is occurring both across a wide range of trades, as in the first case discussed below; and also more narrowly within container, car and passenger trades in the subsequent examples.

Japanese companies Mitsui-OSK Lines and Navix merged an April 1st 1999 and became the largest maritime enterprise in the world with a diverse fleet of 490 ships totalling 31.3 million tons capacity put under different flags and run under the name of Mitsui-OSK Lines. Mitsui-OSK itself was the result of a merger which took place in 1964 between Mitsui and the company Osaka Shosen. Navix was also the result of the merger in 1989 of Japan Line and Yamashita-Shinnihon. They are both major players in tramping (oil sector, dry bulk) and, regarding Mitsui, also the sector of scheduled services for containers and car transport.

The scheduled service sector of Safmarine was bought in February 1999 (comprising 50 private ships with a chartered capacity of 56,000 TEU) by Maersk for 240 million dollars. The reason for the sale was a major problem concerning the evolution of the trade. Safmarine joined the group A.P. Moller/Maersk and kept its trading name and this enabled the Maersk network, with a prime position in East-West links, to reinforce the North-South links especially in Africa. Better still, Maersk, leading transporter of containers in the world (Table 3) associated with SeaLand (6th place in the world) since 1996, in 1999 raised its capacity to 600,000 TEU, 10% of the world total, well ahead of other groups of ship owners (Table 3).

Table 3. Top Operators in the Containerised Sector (>100,000 TEU) (1998)

Shipowner	Country	Fleet Capacity (TEU)
Maersk	Denmark	378,000
Evergreen/Uniglory	Taiwan	297,000
P&O Nedlloyd	UK/Netherlands	263,000
Hanjin/DSR Senator	South Korea/Germany	233,000
COSCO	China	227,000
Sealand	USA	209,000
Mediterranean Shipping Co	Switzerland	199,000
NOL/APL	Singapore/USA	198,000
NYK	Japan	164,000
CP Ships & TMM	Canada/Mexico	139,000
Mitsui OSK Lines	Japan	129,000
CMA/CGM	France	116,000
Zim	Israel	114,000
Hyundai	South Korea	109,000
K Line	Japan	106,000
Ynag Ming Line	Taiwan	101,000
Hapag Lloyd	Germany	100,000

Two major European shipping companies, Wallenius from Sweden and Wilhelmsen from Norway merged in July 1999. The new entity, accounting for 23% of Roll-on Roll-off vehicle transport capacity, became the leading vehicle transporter in the world with fleet of 70 ships able to transport 1.5 million vehicles and 300,000 RO-RO units per year. 8 extra ships will join the fleet before 2000.

Regarding the transport of passengers, the merger of P&O European Ferries and Stena Line in March 1998 provides, in the Channel and North Sea area, a fleet of 10 ships concentrated on Calais-Dover (6 ferries, 30 departures every day in each direction) and Zeebrugge-Dover routes. The position of other (small) companies like Seafrance is becoming more precarious.

2. Fleet Developments and Organisation

2.1. BULK FLEETS AND INCREASED SAFETY AWARENESS

2.1.1. Ship Owners and Operators.

There are hundreds of ship owners of liquid bulk and dry bulk ships. Maritime hardware is constantly evolving so that figures quoted here and the composition of fleets are accurate only at a given moment.

Certain shipping companies involved in the hydrocarbon trades, for example, have a world wide impact with a fleet of a greater tonnage than the totality of the fleets of several maritime countries. Others operate locally in oil/oil products and have established transport in bulk liquid products except hydrocarbons and their flexibility is limited. It is impossible to be exhaustive here and only major companies playing an important international role are highlighted. Table 4 classifies the 10 top oil operators in the world taking into account only their oil fleet not their entire fleet and all types of ships. Anglo-Saxon and Scandinavian countries, with the exception of Greece, dominate the oil trade and trade with Asia and the Middle East. Independent oil tanker owners (that is to say independent from oil companies and state companies) are grouped in an association called INTERTANKO (International Independent Tanker Owners' Association) created in London in 1934 and sold off in 1970. It reappeared in 1972, grouping 75% of the world oil fleet owned by independent ship-owners.

The world dry bulk fleet (including ore carriers) included on 1st January 1998 more than 300gt, 6,139 ships for 275.9million dwt. The older elements in the fleet are still relatively important with 2,149 ships over 20 years old accounting for 72.6 Mdwt and the newer recent fleet of only 1,077 ships for 63 Mdwt. Four of the top ten dry bulk operators at the end of the 20th century were Chinese; this is a sign of their growing maritime strength (Table 5).

Table 4. Top Ten Tanker Operators (1998)

Shipowner	Country	Fleet capacity ('000 dwt)
Vela International	Saudi Arabia	7,314
Bergesen	Norway	6,466
World Wide Shipping Agency	Hong Kong	5,011
OSG Ship Management	USA	4,875
Chevron Shipping	USA	4,405
Acomarit UK	UK	4,257
Tanker Pacific Singapore	Singapore	4,229
Jahre-Wallem	Norway	3,996
Ceres Hellenic Shipping	Greece	3,820
Associated Maritime Hong Kong	Hong Kong	3,664

There is no such thing as a general mode fleet, but rather there are specialisations to suit the kind of product transported. The merchant fleet of each country and each region has a different functional and structural profile involving either oil tankers or dry bulk carriers, or, as we will see later, ships operating on regular routes.

Table 5. Top Ten dry Bulk Operators (1998)

Shipowner	Country	Fleet capacity ('000 dwt)
Cosco HK Shipping	Hong Kong	3,820
Cosco Qingdao	China	2,085
Shanghai Haixing Shipping	China	1,640
Pan Ocean shipping	South Korea	1,475
Iran Shipping Lines	Iran	1,461
Polstream Oceantramp	Poland	1,383
Guangzhou Maritime	China	1,150
Klaveness T	Norway	1,064
Marmaras Navigation	Greece	1,057
Navibulgar	Bulgaria	1,046

2.1.2. Fleet Diversity in Liquid Bulk

The transport of bulk liquid is by tankers. The holds of these ships consist essentially of waterproof tanks. Oil tankers are the most numerous. There are two types; crude carriers and product tankers. Crude carriers are noticeable by their sheer size with VLCCs and ULCCs. Product tankers carrying oil, gas-oil, fuel and kerosene can be small (100dwt), they are responsible for the food supply of islands, for instance, and can reach 60,000dwt. Chemical tankers belong to another category. They carry a whole range of chemical products derived from petro-chemicals or the chemical industry. They are polyvalent and carry vegetable oils, wines or alcohol. They are often small in size (1,000 dwt) but their tonnage can reach 50,000dwt. There are also monoproduct special tankers to suit the character of the product and the quality and hygiene requirements. This applies to sulphur tankers transporting sulphur at a fusion temperature of 135^0C,. The tanks need to be covered by insulation material. There are only 20 in the world and they do not exceed 20,000 dwt. On the opposite side of the spectrum, methane carriers transport liquid gas at $–161^0C$ (volume occupied by the gas is 600 times smaller). The largest reach capacities of $135,000m^3$. Food, like wine, can be carried by wine tankers or frozen concentrated orange juice tankers (FCOJ tankers), implying obvious hygiene measures.

The renewal of the fleet poses a problem for the oil fleet. After a decline caused by two oil crises in the 1970s, a low period in the mid 1980s, the world oil fleet, including gas tankers, progressed in tonnage (309.3 Mdwt on 1st January 1998 against 296.6 Mdwt two years previously) and in number (9,223 and 8,846 respectively). The fleet is getting old with 3,829 ships (totalling 121.5Mdwt) 20 years old on 1" January 1998. But the age of oil tankers still in service begs the questions as to how they will be replaced: there are only 1,049 oil tankers (totalling 46.4 Mdwt) which are less than 5 years old. Shipowners do not hurry to order new ships because the freight rates are not favourable. Renovating an oil tanker at the end of its reclassification period of 20 years is all the cheaper if the ship has been well built and well looked after. Changing structures, pipes and steel structure costs several million dollars, which is modest compared to the investment in a new ship.

2.1.3. Environment and Safety Measures.

Ships are changing following pressure from the environmentalists. New laws are modifying the ships' construction and ensure that operators are made responsible especially regarding the transport of hydrocarbons. The construction of bulk liquid ships has not changed since the *Gluckhauf* in 1886 (the first ship to transport oil in tanks integrated in the hull). It was only after the accident of the '*Exxon Valdez*' in 1989 that the design was modified. A double integral hull is now compulsory. It is not a new concept since certain specialised ships were already equipped with it.

Other maritime disasters have recently contributed to a modification of the law. The sinking of the '*Torrey Canyon*' (120,000dwt) at the tip of the Cornish coast contributed to the passing of the MARPOL convention in 1973, improved in 1978 following the '*Amoco Cadiz*' disaster on the coast of Brittany. It was amended in 1992 following American legislation.

All maritime transport sectors have been affected by an increased concern for safety, including the transport of passengers, following recent disasters *(Herald of Free*

Enterprise, Estonia and others). The International Safety Management (ISM) code has been in use since 1st July 1998, having been approved on 4th November 1993 by the IMO (International Maritime Organisation). Its aim is to improve- the safety of the operating of ships and to prevent pollution. This code is based on an independent body accredited by the state, generally a classification society, for company auditing and for monitoring of the condition of a ship at sea. The IMO also oversees the management and safety procedures between the shore and the ship. All maritime professionals (loading, chartering, insurance) will become increasingly more sensitive to the quality of the ship according to its conformity to the ISM code.

2.2. FLEETS ON REGULAR ROUTES.

About 30% of goods transported in the world are general cargoes, increasingly containerised (up to 11 %) and the majority follow networks of regular routes. They are expensive goods which generate more added value in ports and port communities than goods transported in bulk. In addition the transport of passengers constitutes a particular market fulfilling the increasing needs for the movement of people.

2.2.1. *The Conventional Freight Sector.*

This is the basis of maritime transport. A century ago everything loaded and unloaded in ports represented what is now called general cargo. This includes goods carried neither in bulk nor in containers or Ro-Ro. Four major categories of products constitute general cargo which has become increasingly containerised: manufactured metal products (tubes, barbed wire, sheet metals, etc.), food (flour, sugar, rice, fruit and vegetables, chilled products etc.), forest products (ranging from raw timber to reels of paper) which can be grouped as neo-bulks, and in the case of finished products includes heavy unit loads. Finally, conventional general cargo services still exist, and progress made in port and land logistics for goods of this type lend a particular perspective to this transport. Admittedly, although containers have killed conventional general cargo transport it continues, however, to endure in developing countries where lower prices in conventional freight transport are possible. Roll-on Roll-off shipping linked to truck haulage has also acquired a part of the conventional traditional trade. When tonnages are high, bulk transport replaces general cargo transport. There are various reasons for going back to the general cargo mode (lower costs, delays, safety, etc.). Contrary to preconceived ideas, the ordering of polyvalent ships capable of carrying conventional freight is as important as for container ships. In 1997 there were 132 ships in each category but the full capacity of container carriers is 3, 311,000 dwt but only 1,274,000 dwt for conventional ships. The 42 roll-on roll-off ships ordered that year represented a total capacity of 545,000 dwt.

An example of the perpetuation of conventional ships adapted to the modern types of transport and to the technological techniques is provided by ships with refrigeration capacity. The container has admittedly taken more than half of the trade (it is about 50

MT) and the trend is to continue this growth. The transport capacity of the world fleet of container carriers has reached 8.8 M m^3 against 8 M m^3 for the normal refrigerated fleet. The two modes of transport can coexist since they do not fulfil the same needs. The container is ideal for limited volumes moving throughout the year involving several ports of call for loading and unloading. The general cargo ship centred around a range of goods is suitable when a limited amount of specialised terminals is used. Five shipping lines have true world coverage with 50 ships each specialising in refrigerated carriers. Refrigerated ships constitute a pool of vessels belonging to different shipowners who manage them technically. It was Swedish originally, then the majority of the capital went in 1994 to the Norwegian group Leif Hoegh which in 1997 merged with the South African group Safmarine, leading to the creation of a new entity, Unicool, possessing a fleet of about 75 ships offering the greatest holding capacity. As far as the number of ships is concerned, the Dutch group Seatrade is the leading group in the world with a hundred units.

Other examples of co-ordination in refrigerated (reefer) ship ownership include the Lavinia Corporation Enterprise from the Greek Constantin Laskaridis which merged in 1984 with the German force Alpha Reefer Transport from Hamburg managing a pool of 35 ships and obtaining contracts with Soviet conglomerates. Lavinia took control in 1997 of the Latvian company Riga Transport Fleet and is a trading partner of Latvian Shipping Company and of the Ukrainian firm Yugreftransport. The pool runs in this sector about a hundred ships. The Danish group Lawritzen Reefers, the Anglo-German group Star Reefer (from the merger of Blue Star Line and a reefer subsidiary company of Hamburg-Sud) operate around 50 ships.

2.2.2. The Container Sector

This is the symbol of a great revolution in maritime transport in the last 30 years, mega-alliances between companies, emergence of new "hubs" and utilisation of over-panamax ships getting increasingly bigger (over 6,000 TEU). Containerised transport world-wide shows an over-capacity with, as a direct consequence, lower freight costs. This is holding back the smaller players, reinforcing the major maritime routes joining the world's major zones of production and consumption in North America, Western Europe and East Asia.

A concentration trend has been going on for three years in this sector of maritime transport; in 1996 the merger of P&O and Nedlloyd, in 1997 the take over by Hanjin of DSR Senator, the acquisition of APL by NOL and of Contship by CP Ships. In 1999 Maersk bought SeaLand and Safmarine. These are global transporters. Their world-wide organisation offers a multitude of routes generally in partnership with other players. Maritime transporters work most of the time within global or simply scheduled alliances. Great alliances like the New World Alliance or the Grand Alliance operate especially the East-West routes but North-South ship owners pursue policies of co-operation.

Table 5 shows the top container operators in 1998 in millions of TEU. It illustrates the division of the fleet in groups of countries with Asia representing 52% of the world capacity, ahead of Europe (32%) and North America (6%). However, this type of classification is relative; it is difficult to have figures for the fleet of each ship owner because of the variable short time chartering which cannot be taken into consideration. In fact, because of the alliances, each line or group has access to a fleet of ships much greater than its actual fleet, but the classification given is relevant and comprises owned ships and freighters on long or medium term leases.

Among the economic conditions which affect the life of shipping lines or groups, some very contrasting situations have been highlighted along certain routes according to economic and political factors. The Asian financial crisis in the summer of 1997 affected the ability to fill containers on some routes; load factors are also affected by differences in legislation, for example between the EU and the American Federal Authorities. Increasingly, the large port operators such as Hutchison Port Holdings, P.S.A. Corporation, Stevedoring Services of America, Eurogate (Bremen Lagerhaus, Gesellschaft and Eurokai) fit in with the strategy of the big shipping lines.

The future seems favourable to the development of containerised maritime flows despite the over-capacity on certain routes and the political and economic problems. Experts have planned for at least a doubling of containerised exchanges over the next 15 years (187.8m TEU in 1998) with traffic expected to exceed 400m TEU.

2.3. MARITIME PASSENGER TRANSPORT

Maritime transport has been an essential component of the populating of new countries since the nineteenth century. The recent history of the populating of the Americas and Australia is well known. Millions of poor immigrants from Europe have taken to the sea for a better future: 72 million Europeans crossed the Atlantic between 1820 and 1920.

Those maritime routes have practically disappeared since the expansion of the Airlines starting in the1950s and the transport of passengers has become dual with the cruise phenomenon and with links across narrow sea passages such as the English Channel or the Baltic.

The phenomenon of short sea passenger traffic is a crucial component of regional economies. Each year Dover, like Calais, has over 20 million passengers; the European shoreline includes several ports handling millions of passengers per year across the Channel and Mediterranean. This phenomenon exists on many coasts especially in Asia and America. As in other sectors, a number of important companies dominate the market, for example including P&O European Ferries, Stena in the Pas de Calais and Buquebus in the Rio de la Plata.

A similar evolution is taking place in the cruise market where there is a concentration of players and again a move towards bigger ships, although this does not exclude the existence of small operators in the niche segment of the market. The concept of cruises, developed in the 1960's, made the ship an industrial product. The over-panamax vessels have been popular for a number of years. Apart form the 'Norway' the first of this type was launched in 1996 by the most powerful cruise group in the world: Carnival Cruise Lines controlling since 1997 five trade names totalling 33 ships offering 50,250 berths. This includes 'Carnival Destiny', 101,000 dwt, 262 metres long, 38m beam, accommodating 3,360 passengers in 1,320 cabins with a crew of 1,058. This ship is huge as is the latest one which came into operation in November 1999: the 'Voyager of the Seas' of RCCL. In 1999 there were 260 cruise ships in operation in the world offering 228,.000 berths. By 2003, there will be a further 52 new ships with a capacity of 92,000 berths and beyond that 23 options are confirmed, with ships costing between 300 and 400 million US dollars each. The cruise industry is very buoyant at the beginning of the 21st century.

The phenomenon of cruise shipping evident at different levels apart from the handful of large groups (Carnival, RCCL, P&O and the new carrier, Malaysian Star Cruises based in Singapore), which are able to finance increasingly bigger ships to follow the market expansion ". There is a whole range of small cruise operators using the niches created by this traffic and aimed at a particular group of customers. There are companies like Special Expedition with three or four ships (71m long) for 120 passengers offering cruises in inhospitable areas on the equator or at high latitudes.

Also of note are sailing liners such as the Company des Isles du Ponant (based at Nantes, France with a "Three-Master" based in Pointe a Pitre and able to accommodate 67 passengers. Large groups are also interested in those routes since Carnival Cruises launched their three-masted "*Wind Star*" "*Wind Song*" and "*Wind Spirit*" (built in the workshops and ship yards of Le Havre) with 485 berths in total. The line is run by the Arison Family who founded Carnival Cruises in 1972. They took over the two Club Med ships in 1998, both five-masters also built in Le Havre.

3. Conclusion

The impact of maritime transport on the world coastal zones and on the geo-strategy of the oceans is profound. Maritime transport is indispensable to the activities of nations; the effects of maritime traffic on coastal zones and continental interiors are considerable. Ships have become increasingly large and increasingly more specialised, needing large scale and specialised quay equipment. The necessity to provide space to accommodate greater tonnages of goods has modified port morphology and links between the urban areas and ports. Decisions taken by shipping companies in particular locations involving in some cases world-wide strategies taken far away from their headquarters have repercussions in ports situated on the other side of the world. This involves not only economic aspects with social consequences but also geostrategic aspects. Fundamentally there is a geography of the needs of nations in terms of raw materials, energy, food and consumer goods. These demand patterns lead to a geoeconomy with oceanic flows and also to a global dimension which encompasses the measures taken to satisfy those needs. Maritime transport, in the context of globalisation and its increasingly stronger influence remains, and will remain, at the heart of development of the planet during the twenty-first century.

CHAPTER 6. OFFSHORE OIL AND GAS AT THE MILLENNIUM

H. J. PICKERING

1. Introduction

This chapter explores the offshore oil and gas industry at the turn of the millennium, identifying the major geographical patterns and trends. It discusses the major reasons behind these trends and what lies in store for the first decade of the next century. A temporal perspective is adopted, with the chapter not only representing a 'snap-shot' in time, but also drawing on an analysis of past trends and predictions of the future, as appropriate. The first part of the chapter sets the scene in terms of the market and price trends that have significantly influenced the pattern of offshore oil and gas development over the 1990s and the first few months of the new millennium. The chapter then turns to the evolution of global production and the contribution of offshore supplies from different regions of the globe. To this are added current patterns of offshore exploration and development activity, which will serve to modify the geography of the industry offshore over the next few decades. For a longer-term perspective, estimates of future offshore reserves are also drawn upon.

Given that the geography of offshore oil and gas activities is a manifestation of a number of factors not only the size and characteristics of the resource base, in both the aforementioned sections of the chapter and those that follow other aspects of the industry's 'operating environment' are drawn into the discussion. The influence of global economics, market demand and budgetary challenges on the geography of offshore hydrocarbon exploitation are drawn out. The opportunities presented by developments in technology and institutional restructuring are explored and the ongoing challenges posed by intra- and inter-national politics highlighted. Finally, the role of growing environmental awareness in guiding offshore activities is addressed.

2. Market and Price Trends

Figure 1 demonstrates the pattern of global demand for oil over the 1990s. It demonstrates a pattern of overall growth over much of the decade, but also a dramatic decline in demand in the countries of the former Soviet Union.[1] The decade continued the trends of previous decades, with rising demand spurring a growth in upstream exploration, development and production. Global crude oil production over the 1990s increased by 11% (between 1990 and 1998), with marked growth in production from

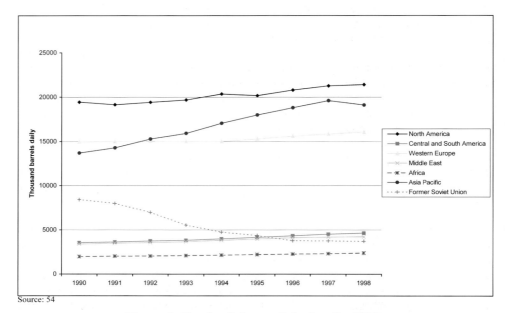

Figure 1: Crude oil demand during the 1990s

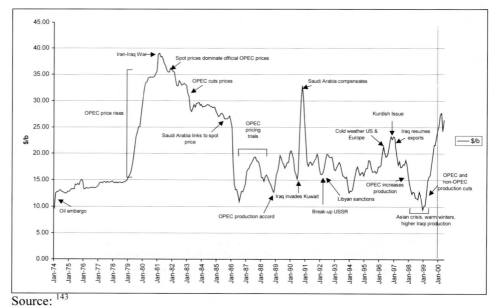

Figure 2: Refiner acquisition cost of imported crude oil, USA

the American continent and Western Europe. Global natural gas production over the same period increased by 14%, with all regions of the globe experiencing an increase, apart from the Former Soviet Union. However, the East Asian plunge into economic crisis in 1997 undermined this upward trend, sending the industry into somewhat of a roller-coaster ride extending into the new millennium. The crisis had significant consequences for the offshore hydrocarbon industry through its effect on oil prices and product markets. Prior to 1997, East Asia fuelled the growth in world oil demand. However, in 1998 and early 1999, the combination of economic slowdown, energy deregulation [2] and currency depreciation in relation to the US Dollar (increasing oil import prices) resulted in a fall in oil consumption and demand in most countries in the region. China was the notable exception, seeing oil demand rise from 3.79 million bbl/d in 1997 to 3.98 million bbl/d in 1998. The crisis had serious manifestations for oil exporters in the region, but also had much wider connotations geographically and economically.

As figure 2 illustrates, while the global oil price has been somewhat erratic over the last decade, there was a marked collapse in 1997 and 1998. The decline in the fortunes of the Asian economy and the associated rapid weakening of Asian crude demand, in combination with global over-production, resulted in a fall in the oil price from $20 plus bbl in 1997 to a low of less than $10/bbl by late 1998. Global oil demand grew by less than 1% during 1998 (including Asia), while global oil production increased by 1.2%, to 67.17 bpd. The Asian economy was expected to recover with time. However, with the growing non-OPEC production and the influence of financial markets on the price of crude over recent decades, the oil price has become highly volatile, responding rapidly to political, market and supply issues, exacerbating reactions to economic fortunes. [1] As a consequence, both oil producing countries and companies were to see their budgets seriously affected and the economic viability of their plans and operations shift unfavourably.

The effects were also felt in terms of natural gas, where short-term growth in regional demand was curtailed (Figure 3) and many exploration and development plans were scaled back for fear that projects would be left without markets. With Japan and South Korea representing 71% of world liquefied natural gas (LNG) imports in 1997, the effect on world-wide LNG trade, in particular, was substantial. [3] The LNG trade had been growing over the decade, with Asia, Africa and the Middle East the main producers, Indonesia alone accounting for 40% of LNG exports and Algeria 21%. [4] The effects were, therefore, dramatic for several countries and regions, but the effects were not isolated. Markets elsewhere in the globe were also affected by the economic slow down in Asia and several mild winters. Natural gas prices fell alongside those of oil in the major natural gas consuming countries of the world, both importers and those with domestic supplies (Figure 4). Traditionally the distribution and marketing of natural gas has been operated by highly regulated monopolies and in many countries still is. In these countries, the price of natural gas is set either: relative to that of competing fuels (notably oil); using a cost-plus formula (i.e. the acquisition cost of the gas plus a mark-up for non-gas costs and a return on capital, the mark-up often being influenced by the price of competing fuels); or a combination of the two methods. Similarly, inter-fuel competition plays a significant role in short-run price determination for natural gas in countries that have moved to more competitive pricing mechanisms. [5,6] As such, the collapse in oil prices has had knock-on effects for the price of natural gas globally.

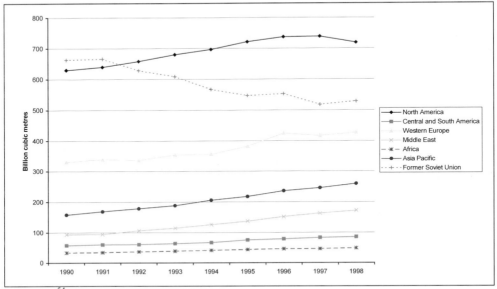

Source: [54]

Figure 3: Natural gas demand during the 1990s

In March 1999, in an attempt to counter the glut of oil on the market, the OPEC oil ministers crafted an agreement to cut 1.7 million bopd of production, which unlike the two previous output reduction agreements, achieved 90% compliance. [7] In addition, unlike previous agreements, four non-OPEC nations: Mexico, Norway, Russia and Oman (some of the largest producers) also imposed production constraints. Oil prices subsequently improved. It was a degree of international co-operation that heralded a break in strategy for a number of countries, both OPEC and non-OPEC nations: a change from maximising oil production and market share towards greater international production management. The change in strategy by these countries reflected the scale of the financial pressures caused by the low oil price and significant international pressure.

The ability to sustain the oil production cuts set in March 1999 along with a growth in oil consumption in 1999 of over 1 million barrels per day, saw prices re-bound during 1999 and early 2000, outstretching many expectations. Demand in south-east Asia had started to recover, which with strong growth in consumption in industrialised countries (equating to 60% of the growth in demand in 1999) and economic growth in the Russian economy, the balance of global supply and demand had shifted. [1] The shut down of exports by Iraq in November 1999 in protest of the United Nations' terms for renewed oil exports tightened markets further. [1] By March 2000 the price had soared, the Brent Blend breaking the $31/bbl mark, levels not seen since the invasion of Kuwait. [8] After which, despite falls in April and August to $20.8/bbl and $25.4/bbl, respectively, the price continued to climb, with the Brent Blend reaching a peak of $37/bbl on 7th September. [9]

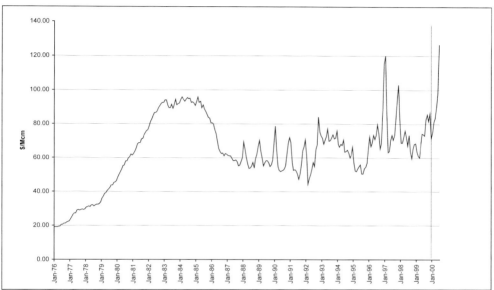

Source: [143]

Figure 4: Natural gas wellhead price, USA

During the first half of 2000 oil demand continued to grow, with Asia continuing to account for the majority of the growth with China playing a key role. [10] Even though global production increased over a 12 month period by 3.8 million b/d (June 2000 over June 1999) [8] crude oil prices remained strong, pushed up by short-term demand, relatively low crude and product inventories and speculation on the financial markets. Natural gas prices were likewise strong (Figure 4). The NYMEX price for natural gas in June 2000 averaged $4.30/mmbtu, versus $2.34/mmbtu in June 1999, and prices were expected to continue to climb through the remainder of 2000. [10]

The very high levels to which oil and gas prices soared, while favourable for short-term production economics, were viewed by many as unsustainable. The oil price, in particular triggered international calls for action to reduce the oil price to more reasonable levels. In response, OPEC increased its production targets in June 2000. However, the prices remained high (Brent exceeding $31 per barrel). [8]

The last few years of the old millenium were, therefore, somewhat of a roller-coaster ride for the industry, which had to face falling, collapsing and then booming markets all in the space of a few years. It is a situation that has significantly influenced the pattern of offshore activity and will continue to do so over the next few years (as will be come evident throughout this chapter).

Over the longer term, offshore activity will continue to be driven by demand, although field development will ultimately depend on how supply is offset against this and the resulting oil and gas prices. Assuming the continuation of pre-Kyoto conference energy policies and "business as usual", the International Energy Agency [11] predicts that world energy demand will continue to grow (by 1.9% per annum over the next 20 years), with oil maintaining its market share and natural gas increasing its share significantly. Oil is expected to account for 38% of energy consumption in 2020 (down 1% of 1997 levels), and natural gas 29%, up from the 22% in 1997. [1] Although, the complexities and uncertainties of economic growth, energy prices, technology and consumer behaviour will determine actual levels of growth [12], and energy intensity [13] is expected to fall with developments in technology, economic and industrial restructuring and fuel substitution, offshore activity looks set to continue. [1]

3. Contribution of Offshore Production

Two decades ago the vast majority of the world's oil and gas supplies came from onshore reserves – 84%. (Figure 5 and Figure 6) Since then, with new discoveries, technological developments and favourable economics, offshore reserves have become an increasingly important source of supply. While onshore reserves still dominate natural gas supplies (accounting for between 78% and 85% of global natural gas production over the period 1975 to 1995), the contribution of offshore reserves to crude oil production has almost doubled over the same time period. By 1995 offshore oil production had increased its contribution to global oil production, to 30% from 16% in 1975.

As offshore production has expanded so has its geographical distribution (Figure 7). In 1975 the Middle East, Venezuela and the USA dominated offshore crude oil production. By 1995, although the Middle East retained a significant portion of global offshore production (18%), this had fallen from 34%, and the Asia-Pacific region (20%) and Western Europe (21%) were now dominant. The reduced contribution to global production from the mature provinces off the USA (8%) and Venezuela (4%) also reflected growing production off Africa and non-Venezuelan Latin America, reaching 16% and 12% (the latter from 3% in 1975), respectively, by 1995.

Global offshore natural gas production has also demonstrated a regional expansion over time. (Figure 8) In 1975, the USA accounted for 55% of global offshore natural gas production, with Western Europe and the Middle East both contributing around 18.5%. By 1995, the contribution of the USA had fallen as a percentage to 25% and that of the

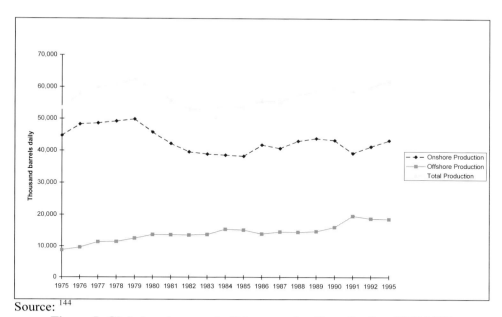

Source: [144]

Figure 5: Global onshore and offshore crude oil production 1975-1995

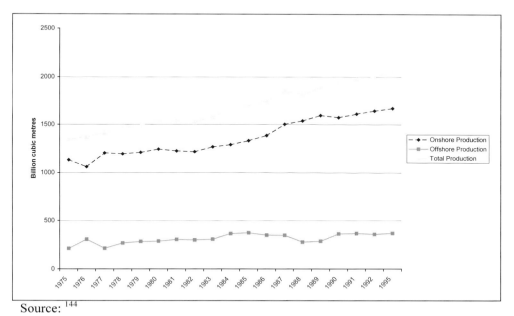

Source: [144]

Figure 6: Global onshore and offshore natural gas production 1975-1995

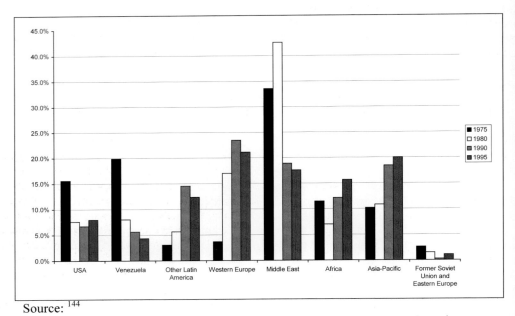

Source: [144]

Figure 7: Percentage of global offshore oil production from each region

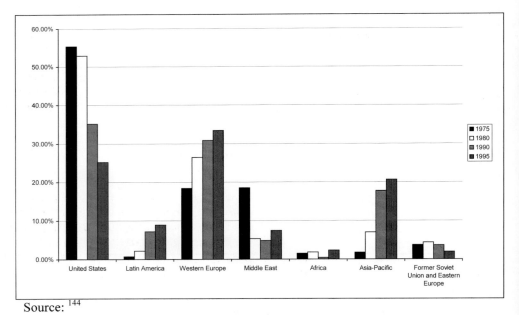

Source: [144]

Figure 8: Percentage of global offshore natural gas production from each region

Middle East to 8%. In contrast, Western European production increased over the period, rising to 34% of global production. The remaining production came from the growing number of offshore fields in the Asia-Pacific region and Latin America, which in 1995 contributed 21% and 9% of global offshore natural gas production, respectively.

As the traditional provinces have been maturing, the hydrocarbon companies have been searching for new reserves to add to their portfolios in 'frontier' areas, hoping to find larger fields and more favourable development and production economics. With countries eager to develop secure domestic supplies, developments in technology permitting greater depth and environmental capabilities and an increasingly mobile offshore support sector, global expansion has been ensured.

In terms of the balance between fuels there has also been a shift in emphasis, with natural gas and natural gas liquids taking a more prominent role, both as an exploration target and in terms of production.

Over the 1990s in particular, natural gas gained credibility as a valid exploration and production target. This reflected a number of factors. The relative stability exhibited by natural gas prices, in contrast to the high volatility exhibited by crude oil prices, the growing perception of natural gas as a 'clean' fuel, and the international trend towards the deregulation and privatisation of natural gas and electricity generation markets has made it a more attractive fuel. [6] Its share of the electricity generation market alone increased from 12.1% to 15.5% globally between 1973 and 1997. [14] Over the last decade many of these forces accelerated, giving rise to current levels of activity. Operators have been focusing on natural gas developments to compensate for lower oil prices, with certain smaller independents aiming to carve a market niche for themselves with natural gas. In support of offshore activities, new ways have been explored to develop gas resources and pipeline infrastructure developments are underway in many countries to facilitate greater market penetration (e.g. in West Africa a major pipeline is destined to export natural gas from Nigeria to Ghana and the pipeline infrastructure between Canada and the USA is being improved to accommodate U.S. import demand). [1,15]

4. Offshore Exploration and Development

As current patterns of production are the product of past exploration and development, future patterns of production, into the new millennium, will be the product of recent and current exploration and development activities. As the international active rig count would appear to indicate, offshore reserves are likely to contribute significantly to this production over the next few decades. During the 1990s offshore rig activity represented a significant proportion of rig activity globally (Figure 9), with the waters off all continents represented (Figure 10). As a result of the dramatic swings in the oil price of recent years however, the tale end of the 1990s saw the level of activity somewhat tempered. With the collapse in the oil price and reduced revenues, many oil companies (both national and international) tightened their budgets and reassessed their exploration and development plans. [16] The result was reduced investment in exploration and development globally and a geographical concentration of activity. The companies focused in on particular regions and exploration targets, and put other targets and regions on hold. [17] Offshore rig activity fell generally relative to 1998 levels, but relative to rig activity as a whole largely maintained its contribution, with the exception of North America and Central and South America (Figure 9).

In terms of exploration activity, Eastern Europe and the Former Soviet Union (not shown in the figure) were particularly hammered, with an estimated 17% decline in drilling activity in 1999, the effects of which were expected to potentially linger for some time, consolidated by political and economic instability in the region. [17,18] African activity also lost some of its gains of previous years, with drilling projects in the promising prospects offshore West Africa cut back extensively, albeit not as dramatically as in other areas. [18,19] Activity in the South Pacific, a popular target between 1995 and 1997, also fell by 31% in 1999. [18,20]

The Middle Eastern countries (notably, Saudi Arabia, Oman, Qatar) saw exploration and development drilling drop by 14% in 1999 (although it still registered as a relatively good year), with Qatar one of those countries particularly affected. [18,21] Due to previous estimates of oil revenues failing to materialise and reduced LNG demand from Asia, Qatar faced significant financial problems in servicing the debt incurred by previous hydrocarbon associated infrastructure commitments. [16,22] Even the United Arab Emirates, which had the most diverse of the economies in the region felt considerable pressure from the 1998 oil price crash. With oil and derivative products accounting for approximately 78% of exports, the country experienced a slow down in its economy and a serious reduction in government revenues. [16,23] With government revenues a major source of industry investment in these countries, exploration and development budgets were inevitably hit.

In Europe, whilst oil production continued at record levels, drilling was slack and mostly associated with routine appraisals and efforts to extend field life spans. [23] In June 1999 only four drilling rigs were left active in the UK sector and a number of projects had been delayed. [23,24] The low oil price, however, was not only to blame. The Norwegian government imposed delays on a number of field developments and in the United Kingdom changes were proposed to the fiscal regime and the power generation market, making the sector nervous. [25,26] The industry was similarly uneasy about proposed legislative changes in the Netherlands.

In North America, offshore Gulf of Mexico rig demand had declined steadily since early 1998. Demand for shallow water rigs was the hardest hit, although some deepwater work was also delayed in the Gulf. [19] There were even cases of production cut backs and workers being laid off. [16] However, with its existing infrastructure and the ability to make economies of scale, exploration in the Gulf did well relative to other offshore areas. [15,27,28,]

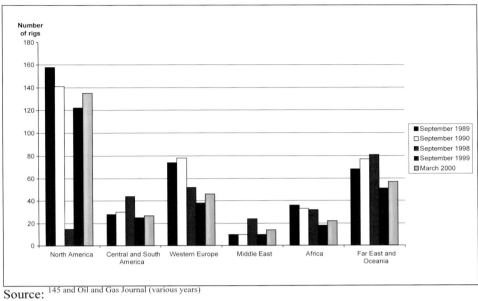

Source: [145] and Oil and Gas Journal (various years)

Figure 9: Active offshore drilling rigs as a percentage of all active drilling rigs in region

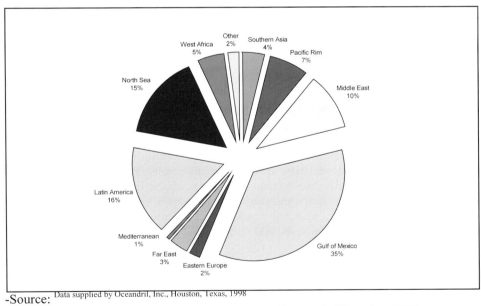

-Source: Data supplied by Oceandril, Inc., Houston, Texas, 1998

Figure 10: Regional distribution of offshore drilling rigs 1998

Despite this reduction in activity globally, offshore areas did not suffer as badly as onshore areas, partly due to the extended planning horizon associated with offshore fields, especially deepwater fields. [29] Deepwater field and gas reserves remained key exploration targets throughout the period, even though expensive deepwater projects represented sizeable cost-cutting opportunities. The attractiveness of the potential returns from investment apparently outweighed the benefits of short term cost savings. World-wide offshore rig utilisation rates generally fell by less than 20% of capacity. [1]

With the strengthening oil prices, activity started to take a more positive turn upon entering the new millennium. [30] However, while offshore rig activity increased slightly in all areas, activity was slow to take-off. [19,30,31,32] December 1999 showed little positive movement over earlier in the year, with the exception of the Gulf of Mexico: activity in the Gulf being buoyed by strong domestic natural gas prices. [19,31,32] With high levels of oil price volatility expected to continue, the industry elsewhere remained cautious, preferring share buy-back programmes and debt reduction to increased exploration and development spending. [29] Overall exploration and development spending during early 2000 remained down on 1997 levels. It should be noted, however, that this reflects in part the number of large corporate mergers that took place in the industry, and the integration of their exploration and development programmes. [33]

Nevertheless, despite the slow start, there were signs of improvement. Industry surveys indicated that the industry was anticipating and planning higher levels of activity. As the strong oil price persists the economics of drilling were being viewed more favourably. [34,35] For example, companies in Canada were looking to supplement their capital budgets, the deep waters of the Gulf of Mexico were attracting attention and high levels of drilling were being maintained off Mexico. Western Europe was also seeing renewed activity, aided by developments in government policy and regulation. [15] The United Kingdom licensing rules had been streamlined and proposed changes to the fiscal system put on hold and in Norway, the government was investing in project related technology as an incentive for ongoing development. [23,36,37]

Further east, the waters of the Former Soviet Union were attracting more cautious attention, notably off Sakhalin Island and the Caspian Sea. The latter has been one of the most closely watched areas of the world, potentially holding similar reserves to the Middle East. [15,38] However, the sizeable investment and major exploration contracts that were entered into during the 1990s had still to yield results matching expectations and two international consortia had already pulled out of the region after disappointing drilling programmes. [15,39] Other consortia remained. One of the key problems faced was the lack of seismic data, hopefully now to be overcome after a major feat of engineering, which successfully shipped one of the latest generation seismic vessels to the landlocked sea.[40]

In South America, there was a more positive note, with notably Brazil and Trinidad attracting active interest. Several major oil companies made Brazil their top investment priority, despite nervousness amongst investors over economic and political developments in the region.[41] The main areas of interest were, however, Africa and the Asia/Pacific region. With West Africa accounting for more discoveries (all offshore) than any other region in 1998, half of which were in deep water, and with significant unexplored potential in the region, West Africa represented a key exploration target. Almost every country had a major offshore play, including Cote d'Ivoire, Ghana, Nigeria, Gabon, Equatorial Guinea, Congo and Angola and those that hadn't were trying to get in on the action with licensing rounds, notably Senegal, Namibia, The Gambia and Guinea.[15,42] Similarly, the Asia-Pacific region was proving a popular target, just as over the previous decade:[43] India and Thailand attracted more development drilling and New Zealand is demonstrating good exploratory interest. Other countries in the region also opened-up to industry interest, building on licensing rounds held during 1999.[18]

As the level of exploratory activity continued to recover it was anticipated that development work would do likewise. Development projects had suffered alongside exploration activity under the low oil price. In November 1999 there were 515 offshore field development projects underway world-wide, down 142 on 1998, with certain smaller marginal projects cancelled and larger projects suspended or postponed.[44] With a resurrection of confidence and an upturn in the oil price, many of these developments were incorporated within upward revisions in corporate budgets. The majority of the large and deep water finds, however, continued to be pursued. The most active regions, reflecting the pace of exploration were the Asia/Pacific region and North America.[44] The popularity of the Asia/Pacific region was manifest in it accounting for 29% of offshore installation projects planned for the millennium (Figure 11). North America and Europe followed with 24% and 18% respectively. The Middle East and Former Soviet Union between them accounted for 11%, and offshore Africa and offshore South America accounted for 10% and 8%, respectively.

Over the next 20 years, to supply the estimated growth in demand, this pattern of world-wide development is expected to continue despite price fluctuations. In contrast to the pattern over the last 20 years, however, an increasing dependence on supplies from the Middle East is anticipated (unless unconventional sources of liquid fuels are developed to compensate). Non-OPEC provinces are expected to mature and discoveries in these areas are likely to become more marginal, given the high depletion rates and close to capacity production of non-OPEC reserves. Non-OPEC supply will nevertheless remain important, with the major increments in oil supply expected to come from offshore resources, especially the Caspian Basin and deepwater West Africa.[1,45] Oil production increases, albeit less significant, are also expected from Pacific Rim producers through the use of enhanced exploration and extraction technologies – deepwater fields again featuring - the major area of growth being within the Timor Sea.[33] Central and South American production also has the capacity to expand, through foreign investment under an improved economic and political climate.[1] The new large deepwater fields off Brazil will make a significant contribution.[33]

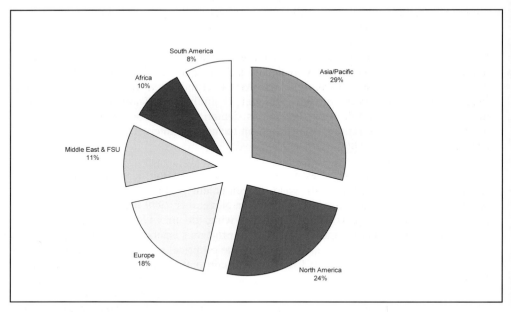

Source: [44]

Figure 11: Installation projects planned for the Millennium

In contrast, production from the North Sea is expected to peak around 2004 and 2005, with the UK continental shelf starting to decline earlier. Similarly, production from the USA is expected to decline, although estimates of the rate of decline have been revised recently with the discovery of reserves in the deeper waters of the Gulf of Mexico. [29] As of the end of 1998, 72 fields had already been discovered in the Gulf of Mexico in water depths exceeding 1,000 ft and more are anticipated, [29,46] making the Gulf one of the world's production hotspots of the future and a key source of future American production. [33] Canada and Mexico are also expected to help offset any decline from the North American region as a whole, with production increases in the former principally coming from the Atlantic offshore [1,33] and for the latter, from the Bay of Campeche. [33]

As noted above, deepwater reserves are expected to play a particularly significant role in future global production additions. [47,48] As of December 1998 there were already 52 deepwater fields in the Gulf of Mexico, 20 off Brazil and 17 off West Africa, reaping almost 23 billion boe, but this is only a fraction of the potential of deepwater areas, especially off West Africa. [47] Other deepwater provinces include the north-west Atlantic frontier and deepwater Norway. [49] One estimate, in 1999, placed deepwater reserves as representing approximately 14% of the World's total offshore reserves awaiting development between 1999 and 2007. [45] World-wide, operators currently have almost 100 field development projects under study in waters in excess of 1,500ft and 36 deepwater rigs able to drill in 3,000ft of water under construction. To meet requirements, the 1999 fleet of six ultra-deepwater drilling rigs, able to operate in 7,500ft of water, is expected to be tripled over the next few years. [50]

In terms of gas, the fortunes of production are integrally tied into onshore infrastructure and local markets. As such, an expansion of infrastructure will need to be instigated to match the growing market demand and desire to expand production. This pipeline and infrastructure expansion has already been initiated in many parts of the world, running in parallel with offshore development activity. In synergy, this development work, plus current foci of exploration, will lead to offshore gas reserves playing an increasingly significant role in the future energy mix in many regions. For example, natural gas production is expected to show substantial growth from the federal waters of the Gulf of Mexico, from both known and estimated, yet to be discovered reserves. The estimates for the Gulf equate to the 1997 estimates of recoverable gas resources for the USA as a whole. [51] The role of Egyptian waters is also expected to dramatically increase, with the Nile Delta being on the verge of becoming a world-class gas province. The Delta experienced a boom in exploration and development over the latter half of the 1990s. [52]

5. Reserves

The outcome of this activity is ultimately more accurate estimates of global hydrocarbon reserves. Reserves take on several definitions, with the estimates quoted and probabilities attached reflecting the interaction of a large number of factors. Some of these factors reflect down-hole conditions and knowledge of the reserves themselves, while others reflect external forces, such as evolving technology, the economic environment, the availability of investment funds and international and national politics.[53] The level of knowledge of down-hole conditions and hydrocarbons in place increases with field production, more often than not leading to an upward revision of estimates of proven reserves over time.

At the end of 1998 global proved reserves of crude oil (onshore and offshore sources combined) lay around 1053 thousand million barrels and for natural gas 146 trillion cubic metres. By the year 2000, gas reserves had more that doubled over levels 20 years ago, while current oil reserves represent an increase of 62%. [1] At current rates of production, the average reserve to production ratio globally for natural gas lies at 63.4 years, compared with 41 years for oil, [54] the ratio being highest outside the mature provinces of North America and Europe. [1] Table 1 gives the regional breakdown.

In 2000, the USGS completed a study "USGS World Petroleum Assessment 2000" [55] projecting forward over a 30 year time frame (1995 to 2025) the quantities of conventional oil, natural gas and NGLs outside of the USA that could be added to these existing reserves. In terms of combined onshore and offshore production, they estimated that 75% of the world's grown conventional oil endowment and 66% of the world's grown conventional gas endowment had already been discovered (for areas outside the USA regarded as priority or boutique areas) by 1995. In addition to which, they estimated that by the end of 1995, 20% of the former and 7% of the latter had already been produced.

The areas the USGS expect to contain the greatest volumes of undiscovered conventional oil are the Middle East, the north-east Greenland Shelf, the western Siberia and Caspian Sea areas of the Former Soviet Union and the Niger and Congo delta areas of Africa. Significant new, currently undiscovered oil resources are also expected to exist off Suriname, which along with the north-east Greenland shelf, has no significant production history to-date. Offshore and coastal provinces are prominent in the projections.

In terms of natural gas, the greatest volumes of undiscovered conventional gas are estimated to lie in the western Siberia basin and the Barents and Kara shelves of the Former Soviet Union, [56] the Middle East and in the offshore Norwegian Sea. Several areas are expected to reveal significant additional undiscovered gas reserves even though they have already been targets of exploration and development, notably the north-west shelf of Australia and Eastern Siberia.

Figures 12 to 14 indicate the range of estimates provided by the USGS for offshore reserve additions over the period 1995 to 2025 relative to the mean estimate of reserve additions from combined onshore and offshore sources. Note that the figures for North America exclude the USA and include Greenland. In terms of the mean estimates for crude oil (Table 2), the offshore areas of Central and South America, North America and Africa are expected to contribute the greatest volume relative to other offshore sources, and in each of these regions offshore sources are expected to account for the majority of reserve additions. Europe is also anticipated to be highly dependent on offshore sources for its oil reserve additions over the time period.

In terms of the mean estimates given for natural gas reserve additions from currently undiscovered offshore sources (Table 3), the Former Soviet Union dominates. Estimates for the other regions are much lower, but more evenly spread than those for oil. This is a pattern somewhat mirrored by the estimates given for NGL reserve additions. Offshore sources of natural gas and NGLs are expected to be fundamental for North America (excluding the USA), Europe and Central and South America and also expected to play a significant role in future reserve additions in the Former Soviet Union and the Asia-Pacific region.

Globally (excluding USA), the estimated mean offshore reserve additions equate to 22% of known global oil volumes (i.e. cumulative production plus known remaining reserves), 54% of known natural gas volumes and 86% of known NGL volumes. The regional breakdown is given in Table 5, which when compared to the current spatial distribution of reserves (Table 4) indicates the regional and global significance of the contribution of offshore sources.

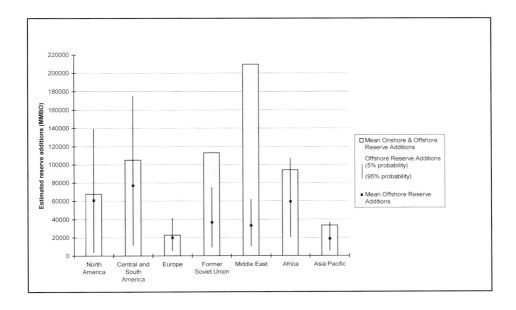

Source: [55]

Figure 12: Estimated oil reserve additions (1995-2025) from undiscovered offshore resources

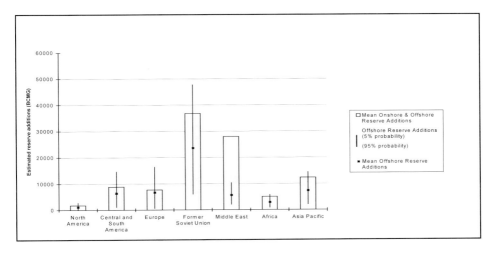

Source: [55]

Figure 13: Estimated natural gas reserve additions (1995-2025) from undiscovered offshore resources

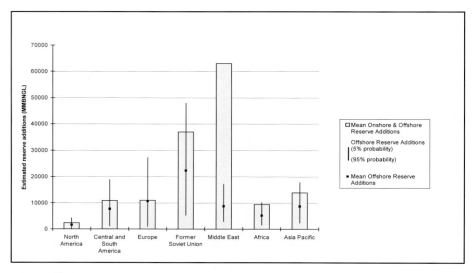

Source: [55]

Figure 14: Estimated NGL reserve additions (1995-2025) from undiscovered offshore resources

These estimates are based purely on geological factors. They do not take into account the many socio-economic and technological parameters that will ultimately determine what portion of these reserves is economically retrievable. Changes in the economic environment, manifest through capital and operating costs, product prices and access to investment funds, affects the profitability of exploration, development and production, leading to revisions in the estimates of reserves that can be extracted under commercial operating conditions. The development of new exploration and production technology enhances the discovery, definition and extraction of in-field hydrocarbons and, thereby, estimates of reserves recoverable. Similarly, the fiscal policies of the state and their policies towards energy production, resource depletion and the environment serve to influence the costs of production, the lifetime profitability of a field and exploitation rights, facilitating or constraining both the discovery of reserves and the ability to recover them. Several organisations have produced statistical projections extrapolating from past performance to estimate the effects of such factors. However, the evolution of the industry has shown, that the consequences of such factors are not easily predicted.

Table 1. Proved reserves from combined onshore and offshore sources at end of 1998 and percentage held by countries with significant reserves.

	North America	Central and South America	Europe	Former Soviet Union	Middle East	Africa	Asia-Pacific
Proved reserves of crude oil (thousand million barrels)							
	Total = 85.1	Total = 89.5	Total = 20.7	Total = 65.4	Total = 673.7	Total = 75.4	Total = 43.1
	Mexico 56% USA 36%	Venezuela 81%	Norway 53% UK 25%	Russia 74%	Saudi Arabia 39% Iraq 17% UAE 15% Kuwait 14% Iran 13%	Libya 39% Nigeria 30% Algeria 12%	China 56% Indonesia 12%
Proved reserves of natural gas (trillion cubic metres)							
	Total = 8.4	Total = 6.2	Total = 5.2	Total = 56.7	Total = 49.5	Total = 10.2	Total = 10.2
	USA 57% Canada 22% Mexico 22%	Venezuela 65%	Netherlands 34% Norway 22% UK 15%	Russia 85%	Iran 46% Qatar 17% Saudi Arabia 12% UAE 12%	Algeria 36% Nigeria 34%	Malaysia 23% Indonesia 20% China 13%

Table 2. Estimated (mean) crude oil reserve additions (1995 to 1025) from currently undiscovered offshore sources as a percentage of total additions (both onshore and offshore) in region

North America	Central and South America	Europe	Former Soviet Union	Middle East	Africa	Asia-Pacific
Greenland 68	Brazil 44	Norway 56	Kazakhastan 12	Iran 6	Nigeria 24	Indonesia 17
Mexico 19	Suriname 12	UK 28	Russia 10	Saudi Arabia 5	Angola 15	Australia 14
Canada 3	Venezuela 4	Rest 4	Turkmenistan 6	Rest 4	Gabon 8	India 5
	Falkland Islands* 6		Azerbaijan 5		Congo (Brazzarille) 6	China 5
	Rest 7		Rest 1		Rest 10	Brunei 5
						Malaysia 9
						Rest 1

* Considered to be one of the last pure frontier plays, but so far there have been no commercial successes. [55]

Table 3. Estimated (mean) natural gas reserve additions (1995 to 2025) from currently undiscovered offshore sources as a percentage of total additions (both onshore and offshore) in region

North America	Central and South America	Europe	Former Soviet Union	Middle East	Africa	Asia-Pacific
Greenland 52	Brazil 39	Norway 58	Russia 56	Iran 12	Nigeria 15	Australia 21
Mexico 11	Venezuela 10	France 7	Azerbaijan 4	Saudi Arabia 4	Egypt 7	Indonesia 12
Canada 2	Trinidad Tobago 9	UK 6	Rest 4	Rest 4	Angola 6	Malaysia 9
	Rest 14	Spain 6			Gabon 5	Bangladesh 4
		Italy 5			Eritrea 4	Myanmar 4
		Rest 5			Rest 21	Rest 8

Table 4. Regional distribution of known volumes (i.e. cumulative production plus known remaining reserves) from combined onshore and offshore sources

Region	Known Oil (MMBO)	Known Gas (BCFG)	Known NGL (MMBNGL)
North America (excl. USA)	5%	3%	5%
Central and South America	8%	5%	3%
Europe	4%	7%	9%
Former Soviet Union	19%	38%	22%
Middle East	48%	31%	39%
Africa	8%	7%	14%
Asia Pacific	7%	9%	8%

Table 5. Estimated (mean) reserve additions (1995 to 1025) from currently undiscovered offshore sources as a percentage of known volumes from combined onshore and offshore sources

Region	Offshore oil reserve additions	Offshore natural gas reserve additions	Offshore NGL reserve additions
North America (excl. USA)	94%	65%	48%
Central and South America	67%	72%	310%
Europe	34%	87%	152%
Former Soviet Union	14%	64%	135%
Middle East	5%	20%	30%
Africa	51%	58%	51%
Asia Pacific	19%	59%	137%

6. Opportunities for the Millennium

Given that the geography of offshore oil and gas activities is the manifestation of the interaction of a number of factors, not only the size and characteristics of the resource base, it is pertinent here to discuss some of the other factors that have and will shape the development of offshore oil and gas exploitation over time. The influence of global economics, market demand and budgetary challenges on the geography of offshore hydrocarbon exploitation has already been discussed to some extent. The chapter will now concentrate on the opportunities presented by developments in technology and institutional restructuring, the ongoing challenges posed by intra- and inter-national politics and the issues generated by the global environmental agenda. These factors are both unique to the offshore environment and universal. They are characterised by dramatic developments having already been made, facilitating the nature and scale of offshore activity to-date, and the necessity for further developments to be made if offshore hydrocarbon resources are to utilised as effectively and 'sustainably' as possible.

6.1. TECHNOLOGICAL DEVELOPMENTS

With the low price of oil and the associated budgetary constraints on exploration and development, the growing upkeep and work-over costs as producing facilities and reservoirs age in the mature provinces and the mounting liability for field shutdown and environmental risks, the 1990s saw producers face significant pressure to shave costs wherever possible. They responded on a number of fronts, notably, through adopting technological improvements and standardisation practices, out-sourcing and contracting, and reducing staffing levels and multi-skilling their personnel. [57,58,59]

The main technological research and development (R & D) strategies adopted were to: improve the accuracy of exploration and appraisal well targeting; locate, quantify and access bypassed oil; raise drilling efficiency; and reduce offshore platform and onshore plant costs. They were tasks that had to be undertaken with reduced R & D budgets and pressures for extremely short lead in times. [60] Over the 1990s, a number of significant technological developments were made. For example, the introduction of 3D seismic surveying techniques [61] increased the potential accuracy and speed of the exploration process, while such as swath sonar systems reduced the vessel time necessary to collect bathymetric data over large areas. [62,63] With decreasing profit margins, the oil companies were investing heavily in refining data collection and processing with the intention of minimising the subsequent exploration and development costs. Using 3D seismic surveying, for example, the risk of dry wells is significantly reduced, the number of appraisal wells required is reduced and the accuracy of positioning production wells improved. [57,64,65,66] In an exploration and production company poll carried out in 1995, 75% of oil company respondents cited 3D seismic techniques as having the 'most impact on their business' of all technological developments. [67] A view which is maintained today. [2]

In terms of platform design, new technologies have been developed to cut construction costs further and reduce retrofitting costs, so to optimise the commercial viability of marginal fields. The world's designers and builders have been looking to meet ever more demanding criteria: flexible designs, reusable structures, minimum cost, short lead-in times, and high safety, reliability, and environmental sensitivity. [68] The era of custom designed, or over-designed platforms has given way to standardisation and streamlining to cut costs [69] with floating production systems and sub-sea satellite developments receiving significant interest. [70] For the former, the attraction lies with the ability to re-deploy a substantial element of the development costs and the potential to increase flexibility and reduce downside risk. This flexibility is matched by the latest generation of purpose-built floating, production, storage and offloading (FPSO) platforms offering gas processing, gas export or re-injection, water injection and chemical treatment as well as dynamic risers and dynamic position keeping systems for use in deeper waters.[47,70,71]

There have also been developments in drilling technology, principal among which has been the evolution of horizontal drilling technology and multilateral completions. [63,66,72,73] A single horizontal well, tracking precisely through a thin reservoir can replace several vertical wells and the associated costs, prolonging the project lifetime rate of return. [33,57,65,74] Operating from a central producing facility, it can also reduce the number of surface installations required. [65] The technology also offers the opportunity to add reserves through side-tracking within existing fields. As such it was heavily utilised during the 1990s in the Gulf of Mexico and the North Sea, accounting for a sizeable proportion of drilling activity in these mature areas. [29] Between 1990 and 1995 alone, horizontal and extended reach drilling and multilateral completions generated savings exceeding £1 billion in north-west Europe and additional production of £54 billion (at August 1996 prices). [75]

As a result of the many technological developments that have been made over the last few decades, it is now possible to retrieve potentially 70% or more of a field's reserves (subject to the resource and reservoir characteristics), compared to 40% in the 1970s. [76] In addition, the percentage of exploratory wells successful and completed rose during the 1990s from 25% to 45%, due to cost savings and reduced risk. [77] Further technological developments are in the pipeline, [78] which should result in further cost savings as they are disseminated throughout the industry. [33] It should be noted, however, that technological advances alone do not determine costs. Between 1996 and 1998 technology-driven cost savings were more than offset by market-based upward cost pressures. The market for drilling rigs and other oilfield services tightened, increasing costs dramatically. In contrast, between late 1997 and early 1999 the market for oilfield services loosened, as demand fell. Consequently the rates charged fell. [33] Development ultimately depends on whether the cost of capital plus an acceptable rate of return (often targeted at 15%) is covered by the current and anticipated price for the product. [33]

The technological developments made over the last decade have not solely been concerned with cost cutting. Major reformulations in design theory have been needed to cope with the environmental forces and water depths of many of the deepwater and Arctic developments underway around the world. [47,56,79,80] Some of the more interesting developments in platform technology have been towards subsea completions and unmanned structures, [81] supported by developments in, *inter alia*, unmanned sub-sea intervention and equipment reliability. [82] As of 1998, the industry had the capacity to drill in water depths up to 7,700ft, and to produce from water depths of up to 5,300ft. [47] Developments, however, are still needed to make the most of deepwater reserves. It can still take ten years from spudding the first exploration well to being ready with a production solution in such frontier areas. [47]

6.2. STATE OWNERSHIP TO PRIVATE INVESTMENT

The financial realities of the decline in the oil price since 1996 have seen technological developments complemented by institutional restructuring. Significant changes have occurred in the pattern of ownership and resource exploitation in many countries, in terms of both oil and natural gas. Notably, there has been a global trend towards greater private sector involvement. Certain non-OPEC nations (e.g. Mexico and China) have in recent years been moving away from their traditional state-ownership and state operated industrial models of hydrocarbon exploitation towards greater private sector involvement. [83] The change reflects shifting political ideologies; serious reductions in oil revenues; global credit shortages; and the need for capital and technological know-how to facilitate reserve expansion programmes, bring non-producing fields on-line and enhance production from currently producing fields. [84,85,86,87,88,89] Various models have been adopted, often alongside each other: foreign ownership; joint ventures, where investment costs and production are divided according to each party's share in the venture; or production sharing agreements, where the foreign partners act as subcontractors, providing the investment funds needed for exploration and development and recovering their investments once production begins.

The shift to private sector involvement, however, has not been confined to the non-OPEC nations. Since the late 1980s, several OPEC nations (e.g. Algeria and Nigeria) have been looking to private industry to secure vital investment, expertise and access to market opportunities. [74,88,90,91,92,93] With the decline in the oil price this process has been accelerated, with several Middle Eastern OPEC nations also opening their doors in an attempt to increase investment in domestic reserves and infrastructure. Iran, for example, has been looking to the private sector to, amongst other things, kick start exploration: there having been almost no new exploration activity in the country for thirty years.

Some of the most notable shifts to private sector involvement have been by communist nations, notably China and the Former Soviet Union. The former has been following a process of reform since 1978, gradually opening its economy and hydrocarbon industry to foreign investment. With the Asian economic crisis, this reform was accelerated. By 1999 China had signed contracts with over 70 foreign petroleum companies. [94] Following the break-up of the Former Soviet Union, the Russian Federation likewise initiated (1993) a process of reorganisation and privatisation within its hydrocarbon industry. The country needed both critical investment and expertise to modernise and replace equipment and infrastructure, and to restart exploration in the country. [95] While foreign ownership was initially limited to 15%, [95] this restriction has now been abandoned. Eleven private vertically integrated companies and a small number of regional independent producers have now been established through this process. [96] Foreign investment, however, remains limited due to long term economic and political problems in the country and constraints on export opportunities. [15] The governments of other former republics in central Asia have also been actively restructuring their oil and gas sectors, creating "joint stock companies" and opening up to and actively seeking international investment. [97]

On the American continent, moves to increase private sector involvement in the industry have been seen in Canada, Ecuador, Peru, Argentina, Brazil, Colombia and Venezuela. Liberalisation of Argentina's oil sector has been underway since the early 1990s, along with Peru's. In Brazil the state oil company Petrobras lost its constitutional monopoly in 1995 and the government has moved to reduce its stake in the company. Petroecuador's monopoly was ended by constitutional reforms in 1998 and in

Venezuela, while the sell off of the state oil company has not been favoured, private investment in the sector has been encouraged. [98] It is a pattern of reform that has global manifestations. Governments in several countries in the Asia Pacific region, for example, are similarly moving to attract in international oil and gas producers and investment. [99]

Opportunities are, thereby, opening up in many parts of the world previously closed to the oil companies, with countries competing to attract in foreign investment and corporate involvement. [100,101,102] With much of the oil majors output coming from mature fields reaching their peak or already in decline and new fields in the traditional provinces being marginal and relatively high cost, the need to increase reserves has acted as a major incentive to explore such opportunities. [103] The expansion of existing reserves through extending the life of fields and exploiting fields previously inaccessibly can only be relied on so far, leading to a high cost portfolio and involving the use of expensive technology. The discovery of larger, cheaper reserves elsewhere offers the potential to adjust this position. Production costs in Iran, Iraq, Saudi Arabia and Kuwait, for example, are less than \$2/b while in the North Sea they are over \$10/b, [103] making the latter one of the most expensive hydrocarbon basins in the world and the former one of the cheapest. [104] In the hunt for new reserves, there has been increasing competition for new acreage in such as Venezuela and Columbia, growing interest in reserves in the Former Soviet Union and a positive response among the majors to invitations from the OPEC nations. [103] The international oil and gas companies now have a wider choice of investment and asset acquisition opportunities globally and consequently evaluate the relative financial return to be gained and the contribution of each to their asset portfolio. As a result the geography of hydrocarbons is changing.

However, Western governments were not necessarily so keen for the oil companies to take-up the invitations of some nations. With respect to Cuba, American oil companies have been bound by the scope of the U.S. economic boycott, and others potentially affected by U.S. legislation designed to penalise third country companies who profit from properties confiscated under Fidel Castro. [105] Iran's attempt to modernise its industry, expand offshore production and add to its reserves by inviting in foreign investment has been complicated by U.S. economic sanctions and diplomatic pressure (manifest through the United States Iran and Libya Actions Act 1996), [106] which has prevented both US companies and other companies with business interests in the USA taking up the opportunities presented. US-based Conoco, for example, was obliged to abrogate a \$550-million contract to develop several of Iran's offshore oil and gas fields. [103,107,108,109] A sanctions waiver, however, was given to firms of the European Union investing in Iran, which has led to European countries consolidating on a first mover advantage in the country. [110] Iraq and Libya have also been affected by U.S. energy sanctions. Iraq remains under sanctions imposed after its invasion of Kuwait in 1990 and Libya fell to the same sanctions as Iran. UN sanctions against Libya due to its failure to release for trial the two men suspected of the bombing of Pan Am flight 103 over Lockerbie, Scotland, were however lifted in April 1999, when the men were extradited. The opportunities on offer do, however, remain attractive despite sanctions and several non-American companies have enthusiastically taken them up.

6.3. CORPORATE RESTRUCTURING

The economic challenges of the 1990s were not only reflected in technological developments and increased global opportunities for the oil companies, they were also been marked by significant changes in the structure and management of the industry. Mergers, adjustments in asset portfolios, the thinning of professional staffs, corporate downsizing and the increased use of contracted services, under both fixed rate and risk/reward sharing, characterised the decade. [58,64,111,112,113] It was a pattern manifest in both international and national companies in the desire to maintain profitable operations.

While many of these developments were initiated during the 1980s and spanned both decades, the 1990s may be best remembered for the significant shifts in the pattern of reserve ownership that took place during the decade. Corporate ownership saw many changes throughout the decade. At the field level, asset portfolios were revised as the companies repositioned themselves on the global scene to make the most of newly expanding global opportunities and to reduce their costs. [114] 1996, for example, saw Chevron sell several of its more costly North sea interests to Oryx and Ranger [115] and BP sell off four of its least profitable fields (Beatrice, Buchan, Clyde and Thistle). [116] The sale of existing assets served to release cash and expertise for use in new ventures, optimise fiscal liabilities and spread risk, whilst allowing new companies to buy into the sector or existing companies purchase assets relatively cheaply (i.e. relative to the investment necessary to find and develop new reserves). [48] Asset acquisition also proved to be easier to finance, relative to exploration and development. With the lower risk attached to asset transfer, the finance companies made finance for asset transfer activities more readily available and under preferential conditions relative to that for exploration and development. The finance sector was eager to avoid a repeat of the considerable losses [117] it took when many loans to small operators turned sour under a collapsing oil prices during the 1980s. [118] Towards the end of the decade, revisions in asset portfolios took on a more marked cost-cutting focus. For example, Saga Petroleum, Norway's largest privately-owned offshore oil and gas group, not only cut its exploration budget in 1998 and delayed drilling on its deepwater Atlantic margin investments, but also sold off certain domestic and foreign assets. In the United Kingdom, Shell U.K. Ltd invited bids for certain of its Central North Sea interests and BP and Amoco sold several North Sea oil and gas fields to enable them to concentrate on bigger younger fields in newer provinces (such as to the west of Shetlands). [119]

The oil price collapse, on top of desires for reserves, cost cutting and higher refining and market shares, also prompted corporate level developments, with the latter years of the decade seeing a flood of large corporate mergers, which affected the majors as well as the independents. [120] 1998 and 1999 saw Arco take-over Union Texas Petroleum, BP and Amoco merge, Exxon and Mobil combine in a take-over/merger, Total move to acquire Petrofina and the culminating Totalfina enter into an agreement to merge with Elf Aquitaine. Arco's Chairman also approached BP Amoco's Chief Executive for a possible merger, which has been completed. Several defensive mergers also went ahead amongst smaller independent oil companies. [108] Together these developments constitute a major shake-up in the corporate structure of the industry.

7. Challenges for the Millennium

7.1. CIVIL UNREST AND POLITICAL UNCERTAINTY

While the asset base, exploration programmes, research and development and corporate restructuring has equipped the industry to cope with the roller-coaster ride of market forces and the evolving geography of hydrocarbons that has marked the turn of the Millennium, there are still challenges to be met. Many new reserves are located in countries and regions where western firms have difficulty doing business. Nationalist feelings, civil unrest and political uncertainty affects or threatens to affect offshore operations in many parts of the world, as in Angola, Colombia, Nigeria, Indonesia, Guyana, Surinam and French Guiana. While deepwater and other offshore prospects off Africa are a significant attraction to the hydrocarbon industry, and the operators are competing for the associated exploration licences, the volatile and diverse politics of the region still make the industry cautious in its involvement in the region. [121] 1998 saw Angola plunge back into a civil war that had already spanned three decades. With Angola's economy highly dependent on crude oil production of which almost 70% comes from the Cabinda offshore basin, the consequences of the violence extending to offshore facilities or the onshore support infrastructure could be substantial had albeit to-date focused on hostage taking and demonstrations. [109,122] In Nigeria civil unrest and political uncertainty has undermined the safety and cost of operations. Ethnic unrest and violence in the Niger Delta region, one of the key areas for production in the country, has resulted in disruption and damage to installations and personnel. In addition, the frequent changes of government has generated uncertainty and disruption to both domestic and foreign operations, as with the cancelling of crude oil sales contracts and exploration agreements set up by previous governments (including 11 awards in Nigeria's deepwater area) and debates over the propriety of previous awards procedures. [109] While Nigeria hopes to double its oil reserves to 40 billion barrels by 2010, many field operators have postponed exploration and field development, looking to neighbouring Angola instead. [121,123,124] Confidence has not been helped by the failure of the Nigerian National Petroleum Corporation (NNPC) to pay its share of capital spending under joint venture agreements on time. [121,124]

In Colombia, offshore (gas) reserves have in contrast become a focus of exploratory activity, in a country where guerrilla and paramilitary action (including the bombing of oil pipelines), difficult geology and the low price of oil has reduced onshore exploration significantly and resulted in companies scaling back their operations or pulling out of the country altogether. While Colombia's oil production in 2000 was at an all-time high, Colombia faces the prospect of becoming a net importer by 2005-6 unless additional reserves are found and developed. [125] On the same continent, political unrest in the region of Guyana, Surinam and French Guiana has hampered the spread of offshore exploration south east from Trinidad and Tobago and the Caribbean. [126]

In Indonesia, the consequences of civil unrest and political uncertainty have extended even to the international legal basis of offshore hydrocarbon exploitation. The economic crisis in Asia and its knock on effects for the national economy spurred civil unrest, nationalistic feelings, violence and riots in certain parts of the country: manifest, *inter alia*, in the moves towards independence by the East Timorese from Indonesia. In addition to the blood shed caused by the split, the legal status of the Timor Gap Treaty also became a victim. The treaty covers oil and gas exploration in the waters between Australia and Indonesia, including the Bonaparte and Browse Basins. On September 7, 1999 Indonesia's Energy Minister announced that Indonesia would revoke the Timor Gap Treaty leaving the Government of East Timor to pick-up the pieces and resolve the legal implications of its revocation. The Timor Gap is an exploration target for a number of international oil companies, who anticipate sizeable discoveries. [109] The basins are believed to contain natural gas reserves of the order of 40 trillion cubic feet, rivalling the natural gas reserves of Western Australia's north-west shelf. [127]

The precise effect of civil unrest on the industry depends on the level of risk and cost incurred relative to existing assets in the region, the level of expenditure already committed to the region and expectations as to how the dispute will pan out. The outcome is, therefore, somewhat country and company dependent, but is always unsettling.

7.2. BORDER DISPUTES

International disputes have also influenced the pattern of offshore activity. While many offshore boundary disputes have been resolved globally, offshore exploitation is still constrained by such issues in certain parts of the world, either long-standing disputes or ones of more recent origin, as with the Timor Gap issue. The prospect of hydrocarbon in an area can often precipitate and aggregate such disputes, if not spur their resolution.

One of the key areas in relation to offshore hydrocarbon exploitation is the Caspian Sea. As a consequence of the break-up of the Former Soviet Union, the status of the Caspian Sea has changed from internal waters to international waters and as such the rights of the newly independent states to exploit the reserves beneath the Caspian Sea requires clarification. The question faced is on what basis the dividing lines between the states should be drawn up. Until this issue is resolved, offshore exploitation is effectively constrained or the subject of dispute, as with the seismic surveys and exploration contracts initiated by Iran in areas claimed by Azerbaijan. [128] Several solutions have been proposed, with the position of the governments' of the respective countries changing over time in terms of the solution they prefer. In July 1998 Russia and Kazakhstan signed an accord to divide up oil and gas development in the north Caspian region using the median line principle, giving each country full jurisdiction of its offshore resources. Russia has also favoured a treaty on joint jurisdiction so that each country would have a voice over trans-Caspian pipelines and navigational issues (export routes being the principal constraint on exploitation in the area). [96] In contrast, Iran has supported operating under the agreement signed with the U.S.S.R. in 1921 and 1940. [128] However as noted, three stances have changed over time.

Elsewhere in the globe, as exploration moves out from the traditional provinces and into deeper water, the number and complexity of the claims increase, with previously low profile issues gaining in importance. Both Cameroon and Nigeria, for example, are claiming the Bakassi peninsula in the Gulf of Guinea, which is believed to contain significant reserves of oil both onshore and offshore. Several oil discoveries have been made but operations in the area have been suspended due to the dispute. The dispute was sent to the International Court of Justice for resolution in 1994, and formal hearings were started in March 1998. The boundary between Nigeria and Equatorial Guinea is also in dispute, with the ownership of the Zafiro field in question. Nigeria contends that the oil structure straddles the national boundaries, while Equatorial Guinea claims sole ownership. [124]

In Asia, 1999 saw the government of Cambodia active in trying to resolve a twenty-five year old dispute with Thailand over acreage in the Gulf of Thailand, estimated to encapsulate 10-11 tcf of gas. Exploration under licences issued by Cambodia in 1998 has been hampered by its non-resolution. [99,129] Exploration in parts of the East China and South China Seas has likewise been hampered by the ongoing sovereignty issue over a small uninhabited group of islands (including the Spratlys) between Vietnam and the People's Republic of China, the subject of naval conflict in 1988. Other parties are also contending ownership of these islands: Brunei, Malaysia and the Philippines. [130] Further north, hydrocarbon deposits have been located underneath the Diaoyu/ Senkaku islets, the subject to another sovereignty dispute, between Japan, China and Taiwan. [130]

In Europe, the indication of large undiscovered oil reserves in the Faroe-Shetland Basin and growing corporate interest in the area, brought the longstanding and stalled [131] negotiations between the Faroe Islands and the United Kingdom over their North Atlantic boundary to a head. As reputedly stated by Faroese officials, the low price of oil and the need to find additional reserves, were a major financial incentive for the partners to resolve the dispute, [119] which was finally achieved in early 2000. In the Persian Gulf, the long term boundary dispute between Qatar and Bahrain over the divisions of rights in the Persian Gulf and ownership of the Hawar Islands also showed signs of easing, with the parties referring to the International Court of Justice for arbitration and resolution. [22] It is hoped that the resolution of the boundary will reap similar benefits to the recent resolution of the Libyan/Tunisian offshore boundary in the Gulf of Gabes (Persian Gulf).

7.3. ENVIRONMENTAL IMPACT MITIGATION

Another challenge that still faces the offshore oil and gas industry is that of environmental impact mitigation. While significant developments have been made over recent decades, governments continue to extend the range of environmental policies that affect the energy sector: both acting directly on upstream activities and indirectly, through down-stream activities. One of the major sources of uncertainty at the turn of the Millennium is the ultimate form that Kyoto Declaration driven policies will take.

With rising concentrations of greenhouse gases in the Earth's atmosphere, the 1990s saw several international conferences targeted at addressing the issue of greenhouse gas emissions. These culminated in the representatives of over 160 countries meeting in Kyoto, Japan in 1997 to agree limits on greenhouse gas emissions. The outcome was the

'Kyoto Protocol', which established emissions targets for the participating developed countries. For the industrialised countries, which are expected to account for approximately 30% of the incremental world-wide energy use between 1997 and 2010, the Protocol's targets could mean a reduction in consumption of between 15 and 30 million barrels of oil equivalent per day, depending on the strategies and fuels used to achieve the targets. [1] Although the energy mix that will be used to achieve this is still the subject of discussion, along with some of the more fundamental aspects of the protocol, the growing role of natural gas in the energy mix world-wide is likely to be accelerated, along with the trend towards improved energy efficiency. The improvements being made to energy efficiency in the industrialised nations are expected to spread to the developing nations, the former Soviet Union and Eastern Europe over the next twenty years, resulting in energy efficiency improvements here too, of at least 1% per annum. [1]

As noted by the International Energy Agency, with energy demand estimated to increase by 65% between 1995 and 2020 and fossil fuels expected to supply 92% of energy needs even at the end of this period, the "business-as-usual" scenario will inevitably result in significant further increases to CO_2 emissions: an estimated 70% increase in energy related CO_2 emissions from OECD countries alone. [132] This is recognised as a problem by both industry and governments. [132] However, many countries have yet to clearly define how they intend to achieve the objectives set, if they intend to at all, which leaves the industry in an environment of uncertainty. [133]

The current approaches used to reduce CO_2 emissions with implications for the industry include the Norwegian carbon dioxide tax on field production, which proved unpopular in the face of the financial challenges already facing the European industry during 1998 and 1999. [23] Another approach has been to increase the efficiency of resource usage. In line with the objectives of the Kyoto declaration, but also reflecting the desire to increase the commercial use of natural gas, a number of countries (e.g. Angola and Iran) who traditionally flared gas have been trying to reduce gas flaring. Angola has estimated reserves of 1.6 Tcf of natural gas but approximately 85% of that in 1999 was being flared. [122] Additional strategies will no doubt be needed, in addition to these, to achieve the desired objectives, but it is not so clear what these should or will be.

In terms of more regional and national issues, the traditional provinces have seen the growth of onshore and offshore infrastructure matched by a growing emphasis on environmental issues in the regulatory regimes that govern them. Rising concern for the environment in society in general, concern generated by the incidence of major oil spills and greater use of coasts for leisure and tourism has led to increasingly stringent requirements for environmental assessments and the imposition of more demanding performance and equipment standards. Standards have been developed both nationally and internationally for the main operational biproducts and impacts of offshore hydrocarbon exploitation. Approvals procedures have also been tightened in terms of environmental criteria, with technological, process and emission standards stipulated as mandatory conditions of licences for exploration, development, production and decommissioning. Moves closer inshore, with their sensitive habitats and coastal populations, have seen the most demanding developments, with supplemental environmental conditions attached to contracts, if development is permitted at all.

In the developing regions, many governments are just starting to face these issues and develop regulatory frameworks to address them. The experience and management frameworks of the more mature provinces are proving important as models of reference and are being drawn on, with a country often taking and mixing approaches from several regimes. For the Caspian Sea, the issue of most prominence for which a solution has recently been sought is that of pollution management and liability for past environmental damage. The Caspian Sea is a fragile ecosystem. It has limited self-cleaning potential due to its closed nature and yet hosts a number of species particularly sensitive to water pollution, including three quarters of the world's sturgeon. While terrestrial non-hydrocarbon associated industry probably accounts for most pollution to date, upstream hydrocarbon activities have contributed, with Azerbaijan a major culprit, being more developed than its neighbours. Recognising the need for environmental protection, there are a number of initiatives underway among the Caspian Sea nations, albeit in their early stages. Azerbaijan and Kazakstan have successfully reached a bilateral agreement on environmental standards for the Caspian Sea and Azerbaijan has been working on definitions for environmental limits for future consortia, comparable to those in the North Sea and Gulf of Mexico. In addition, Azerbaijan has included strict environmental provisions into production sharing agreements with foreign companies. To assist these efforts, the United Nations Environment Programme, World Bank, and European Union have been helping with monitoring equipment and personnel to ensure enforcement. There is, however, no multilateral agreement for the Caspian Sea in place at present, initiatives being on a country-by-country basis. [134]

In addition to governmental environmental regulations, the global industry has over the last decade faced a new challenge, that of growing and increasingly effective public lobbying. In the USA there have been local campaigns and lawsuits against offshore hydrocarbon activity for the last 20 years on environmental, aesthetic and economic lines. [135] However the 1990s saw this achieve the creation of effectively industry no-go zones for significant areas of the continental shelf off California, Alaska, the Atlantic and the Florida coast of the Gulf of Mexico. [22,108,135,136] In Europe, the industry has faced a particularly active Greenpeace. In July 1998 it said that it would seek an injunction to prevent drilling for gas in the Waddenzee area of the Southern North Sea. It submitted applications for licence blocks during the 17th UK licensing round in an attempt to turn 22,000 square miles of blocks on offer into a marine wildlife sanctuary and in 1995 it protested over Shell and Esso's plans to sink the Brent Spar decommissioned oil storage unit. The latter action resulted in militant and consumer action against the companies concerned across Europe, with the effect that the disposal plans were changed and abandonment was thrown into the public spotlight. [137]

The question of abandonment has become a particularly emotive issue for many countries. As offshore provinces mature the question of abandonment has to be addressed. For approximately 97% of the world's offshore platforms this is not a significant problem, falling within the size and weight limits for which complete removal is stipulated under international law and generally achievable. Many platforms within this category have already been decommissioned globally [138] and the technology for decommissioning and disposal is relatively well developed. The remaining 3% (approximately 450 platforms) are those over which the decommissioning issue centres. [138,139,140] These larger platforms, given their size and weight, are eligible for partial removal under international law, in recognition of the technological challenges, safety and cost issues faced with their disposal. The vast majority lie in the northern parts of the Australian continental shelf and the northern North Sea (UK and Norwegian sectors), where such platforms were the state-of-the-art solutions to the deep waters and hostile environments found in these provinces at the time they were developed. The mobile and reusable platform concepts developed over the last decade for deepwater operations are not accompanied by similar concerns. For the UK, while the peak of decommissioning will fall between 2006 and 2011, this will predominantly be comprised of small satellite developments and sub-sea tie-backs in the central North Sea. The larger structures will require decommissioning and disposal solutions to be in-place by 2011-2016. Environmental impact assessment techniques, removal and disposal techniques, toppling technology, underwater steel cutting and stability assessments have been a key focus of research and development programmes over the last decade to prepare for this challenge and research is ongoing. [138,141] With an agreement of the European members of the OSPAR Convention for the North East Atlantic, banning the dumping of all North Sea oil and gas platforms over 10,000 metric tons, [119,142] the stakes in one part of the globe have recently been raised.

8. Conclusions

The turn of the millennium, therefore, sees the industry faced with a number of challenges, both current day extrapolations of challenges that have accompanied the industry since its first moves offshore, and challenges of more recent origin. Disputes over international legal boundaries, civil unrest and political uncertainty, which have beset the industry since its early days, still constrain offshore operations in several parts of the world, although the regions and countries affected have changed over time. Many of these challenges have come to the fore with the discovery of new reserves and the opening up of regions previously closed to the industry. However, the discovery of new provinces and the opening-up of resource opportunities, with the reserves they contain, means that these challenges are likely to be met head-on, just as with the challenges of the past: The balance of costs versus benefits continuing to favour investment. The prospect of sizeable reserves remaining undiscovered globally and the level of investment already sunk or underway in frontier regions means that offshore hydrocarbon exploitation will span well into the new millennium.

With petroleum products expected to continue to dominate the energy mix for the foreseeable future, additional supplies will need to be brought on line over the coming decades to meet demand. These will come from an increasingly international portfolio of reserves, some currently identified and others not, with offshore reserves playing a significant role. While the known reserves in the mature provinces of the North Sea and the shallow waters of the Mexican Gulf may be facing decline, the waters of the former Soviet Union, West Africa, Asia Pacific and Central and South America are set to become the major offshore provinces of the new millennium, with deep water reserves featuring heavily. Even in provinces already developed, significant exploration and development expenditure continues to be committed with ongoing expectations of sizeable reserve additions. Downstream investment is also being made, particularly in terms of natural gas processing and transportation infrastructure, drawing on the growth of natural gas as a exploration target of choice.

The 1990s represented an era of increasingly constrained industrial economics, with the globalisation of both industry and commodity markets and increasing diversity of ownership of production, which undermined the ability of the hydrocarbon industry to manage both supplies and prices. However, the industry is in a healthy state going into the new millennium. The culmination of the 1990s would appear to have offered up to the industry a particularly bright new year. With the low product prices experienced during the 1990s, the arena of operation of the private oil companies has expanded beyond expectations, as shortfalls in oil revenues spurred countries to court foreign investment to develop or rejuvenate their industries and economies. Along with shifts in political ideologies, these forces have given rise to a global pattern of privatisation that has opened up some of the most exciting offshore reserve concentrations of the globe to private exploitation.

Within the industry itself, the price challenges have underwritten significant technological developments over the decade. These have provided the industry with the capacity to target reserves more accurately, reach deeper and extract more than ever before, while at the same time they have facilitated improvements in operational safety and reductions in costs, both financial and environmental. Institutional restructuring, also low oil price inspired, has produced a streamlined, global industry, whereby a few fully integrated national companies have given way to a complex international network of oil companies, contractors and sub-contractors. It is a flexible and relatively low cost formula, suited to the exploitation of the opportunities opening up. This is not to say, however, that further technological and institutional developments are not needed. Continuing technological and process improvements will be required in all areas of operations to sustain the industry throughout the coming decades and meet the ever more stringent standards required of it. However, it is a healthy industry going into the new millennium, with its offshore activities likely to remain an important feature of the human presence at sea.

References

1. Energy Information Administration (EIA) (2000) *International Energy Outlook 2000*, U.S. Energy Information Administration, Washington. http://www.eia.doe.gov/oiaf/ieo/index.html, 15 July 2000.

2. notably the removal of consumer subsidies in several countries (including Indonesia and Thailand).

3. Energy Information Administration (EIA) (1999d) *East Asia: The Energy Situation*, U.S. Energy Information Administration, Washington D.C. http://www.eia.doe.gov/emeu/cabs/eastasia.html, 27 October 1999.

4. Energy Information Administration (EIA) (1996) *International Energy Outlook 1996*, US Energy Information Administration, Washington D.C.

5. Priddle, R. (1998) *Natural Gas Pricing in Competitive Markets*, International Energy Agency, Paris.

6. Radetzki, M. (1999) European natural gas: market forces will bring about competition in any case, *Energy Policy* **27**, 17-24.

7. in part attributed to improved relations between Saudi Arabia and Iran

8. International Energy Agency (2000c) *Monthly Oil Market Report July 2000*, International Energy Agency, Paris.

9. Energy Information Administration (EIA) (2000b) *Selected Crude Oil Spot Prices*, U.S. Energy Information Administration, Washington. http://www.eia.doe.gov/emeu/international/crude1.html, 28 September 2000.

10. Radler, M. (2000) World oil demand rises despite higher prices as production struggles, *Oil and Gas Journal* **98**(30), http://ogj.pennwellnet.com, 24 July 2000.

11. International Energy Agency (2000b) *World Energy Outlook,* International Energy Agency, Paris.

12. Energy projections are subject to uncertainty caused by a number of factors: economic output and structure; population growth; technical change and capital stock turnover; human behaviour; fossil fuel supplies and extraction costs; energy market developments; energy subsidies; and, *inter alia,* changing environmental objectives and policies.

13. Energy use per unit of economic activity.

14. International Energy Agency (2000d) *Key World Energy Statistics 1999*, International Energy Agency, Paris.

15. DeLuca. M. (1999) Recovering industry focusing on West Africa, deepwater Gulf of Mexico, natural gas, *Offshore* **59**(5), http://os.pennwellnet.com, 28 October 1999.

16. Feld, L. and MacIntyre, D. (1998) OPEC nations grappling with plunge in oil export revenues, *Oil and Gas Journal* **96**(38), http://os.pennwellnet.com, 28 October 1999.

17. Obut, T. and Sarkar, A. (1999) Comparing Russian,Western major oil firms underscores problems unique to Russian oil, *Oil and Gas Journal* **97**(5), http://os.pennwellnet.com, 28 October 1999.

18. Anon (1999b) World trends: some positive signs are emerging, *World Oil* **220**(8). http://www.worldoil.com/archive/archive_99-08/99-08_world.staff.html 27 October1999.

19. Offshore Data Services (1999) *Offshore rig demand hits lowest point since 1992*, press release 9 March 1999, http://www.offshore-data.com, 22 December 1999.

20. Bradshaw, M. T, Foster, C. B., Fellows, M. E., and Rowland, D. C. (1999) Patterns of discovery in Australia: Part 2, *Oil and Gas Journal* **97**(24), http://os.pennwellnet.com, 28 October 1999.

21. Some international credit rating firms have downgraded Qatar's debt.

22. Energy Information Administration (EIA) (1999g) *Qatar*, U.S. Energy Information Administration, Washington D.C. http://www.eia.doe.gov/emeu/cabs/qauar.html, 27 October 1999.

23. Beckman, J. (1999) Record UK-Norway production, initiatives over-shadow slow drilling, *Offshore* **59**(8), http://os.pennwellnet.com, 28 October 1999.

24. Cresswell, J. (1998) delays push back North Sea start-ups, *Petroleum Review* **52**(618), 14-15.

25. Beckman, J., Terdre, N. and Potter, N. (1998) Politics, low oil prices destabilise European operations, *Offshore* **58**(8), http://os.pennwellnet.com, 28 October 1999.

26. Knott, D. J. (1998) Northwest Europe's offshore activity still brisk despite mandated slowdown in Norway, *Oil and Gas Journal* **96**(33), http://os.pennwellnet.com, 28 October 1999.

27. Gurney, J. (1998) Going for gold in the Gulf, *Petroleum Review* **52**(619), 40-42.

28. Miller, N. W. (1998) Exploration pace picks up in Canada's Maritimes, *Oil and Gas Journal* **96**(39), http://os.pennwellnet.com, 28 October 1999.

29. Cochnener, J. (1999) Prospects for Gulf of Mexico gas remain bright despite dismal oil market, *Offshore*, **59**(6), http://os.pennwellnet.com, 28 October 1999.

30. Marsh, T. (editor) (1999) Rig market outlook: better times ahead, *The Offshore Rig Newsletter* **26**(8), 16pp. http://www.offshore-data.com/news/samples/r9908.htm, 2 November 1999.

31. Offshore Data Services (1999d) Rig market outlook: better times ahead, *The Offshore Rig Newsletter* **26**(8), http://www.offshore-data.com/news/samples/r9908.html, 2 November 1999.

32. Offshore Data Services (1999e) *Outlook mixed for world's rig markets*, press release 17 December 1999, http://www.offshore-data.com, 22 December 1999.

33. Matsuda, T. (2000) World oil supply outlook to 2010, paper presented at *The 5th Annual Asia Oil and Gas Conference, 28-30 May 2000, Shangri-La Hotel, Kuala Lumpur, Malaysia*, International Energy Agency, Paris.

34. Anon (1999e) Industry E&P spending plans for 2000 reflect renewed optimism, *Oil and Gas Journal* **97**(51), http://ogj.pennwellnet.com, 4 February 2000.

35. Anon (1999f) Lehman: World E&P outlays to rise 10% in 2000, *Oil and Gas Journal* **97**(51), http://ogj.pennwellnet.com, 4 February 2000.

36. Anon (1999c) UK announces plan to boost offshore E&D, *Oil and Gas Journal* **97**(15), http://os.pennwellnet.com, 28 October 1999.

37. Anon (1999d) Norway acts to boost blighted E&P sector, *Oil and Gas Journal* **97**(20), http://os.pennwellnet.com, 28 October 1999.

38. Grace, J. D. (1998) Caspian production, export, investment outlooks sized up, *Oil and Gas Journal* **96**(34), http://os.pennwellnet.com, 28 October 1999.

39. Wilson, D. (1999) Russian Far East holds huge resource potential, *Oil and Gas Journal* **97**(29), http://os.pennwellnet.com, 28 October 1999.

40. Anon (2000) Moving greater seismic capability to the Caspian Sea, *Offshore* **60**(5), http://os.pennwellnet.com, 28 July 2000.

41. DeLuca, M. (1999b) Brazil providing foothold as producers expand South America presence, *Offshore* **59**(10), http://ogj.pennwell.com, 4 February 2000.

42. Offshore Data Services (1999c) *West Africa offshore oil and gas activity poised for recovery*, press release 27 August 1999, http://www.offshore-data.com, 22 December 1999.

43. Delmar, R. (1999) Asia/Pacific, *Offshore* **59**(4), http://os.pennwellnet.com, 28 October 1999.

44. DeLuca, M. (1999c) 515 field development projects underway worldwide. *Offshore* (Pennwell) **59**(11), http://os.pennwellnet.com, 21 December 1999.

45. Knight, R. and Westwood, J. (1999) Long-term prospects very bright for deep waters off West Africa, *Oil and Gas Journal* **97**(3), http://os.pennwellnet.com, 28 October 1999.

46. Ray, P. K. (1998) US Gulf oil production could double as early as 2002: Thick submarine fans harbor stacked plays, *Offshore* **58**(6), http://os.pennwellnet.com, 28 October 1999.

47. Furlow, W. (1998) Ultra-deepwater plays still face stiff challenges, *Offshore* **58**(12), http://os.pennwellnet.com, 28 October 1999.

48. Quinlan, M. (1998) Deepwater discoveries promise sharp rise in oil production, *Petroleum Review*, **52**(614), 33-34.

49. Cottrill, A. (1999) Deep rooted reasons for hope, *OilOnline* http://www.oilonline.com/news_spotlight_asianoil_deep090799.html, 27 October 1999.

50. Offshore Data Services (1999b) *World's deepwater basins poised to benefit from surge in oil prices*, press release 28 May 1999, http://www.offshore-data.com, 22 December 1999.

51. Anon (1999) EIA says OCS development lead time cut, *Oil and Gas Journal* **97**(12), http://os.pennwellnet.com, 28 October 1999.

52. George, D. (1997) Egypt becoming major gas province with Mediterranean reserves, *Offshore* **57**(9), http://os.pennwellnet.com, 28 October 1999.

53. These factors give rise to several definitions of reserves, including: 'proven reserves', determined on the basis of oil and gas which has been discovered and remains unused and probabilistic estimates of what proportion of this oil and gas in place can, with reasonable certainty, be recovered in the future under existing economic and operating conditions. 'Probable' and 'possible' reserves refer to additional reserves which may also be recoverable from the estimated total of oil and gas in place in the reservoir, with a probability of over 50% and under 50%, respectively.

54. BP Amoco (1999) *Statistical Review of World Energy 1999*, BP Amoco, London, http://www.bpamoco.com/worldenergy/review, 27 October 1999.

55. USGS (2000) *U.S. Geological Survey World Petroleum Assessment 2000*, U.S. Geological Survey, World Assessment Team http://greenwood.cr.usgs/energy/WorldEnergy/DDS-60/ESpt2.html 12 June 2000.

56. Nikitin, B. A. and Mirzoev, D. A. (1998) Complex scientific and technical trends of Russian Federation Arctic offshore hydrocarbon resource development, *Ocean and Coastal Management* **41**, 129-151.

57. Lavers, B. A. (1990) Some technical challenges for the upstream oil industry in the 1990s, in M. Ala, H. Hatamian, G. D. Hobson, M. S. King and I. Williams (eds), *Seventy-Five Years of Progress in Oil Field Science and Technology: Symposium, Imperial College of Science, Technology and Medicine, University of London, 12 July 1988*, Balkema, Rotterdam.

58. LeBlanc, L. (1993b) Re-engineering or streamlining: how major oil companies are changing, *Offshore International* (November), 25-26.

59. Corzine, R. (1994) N. Sea costs set to fall by up to 30%, *Financial Times* (11 March), 7.

60. LeBlanc, L. (1993) Technology closing in on deepwater solutions, *Offshore International* (October), 35.

61. George, D. (1994) 4D: The next seismic generation?, *Offshore* **54**(October), 21-22.

62. Bennett, J. (1995) The UK offshore survey industry: a rapidly changing scenario, *CSM* (April), 40-41.

63. Anon (1996b) Advances in offshore technology pave the way into world frontiers, *Oil and Gas Journal* **94**(20), 29-32.

64. Le Blanc, L. (1994) Operating budgets increasing at expense of capital outlays, *Offshore* **54**(February), 23-24.

65. Quinlan, M. (1994) Oil sector is keeping busy with new technology, *Petroleum Economist* (August/September), 18-19.

66. Decrane, A. C. (Jr). (1995) Petroleum: No. 1 source of abundant energy, *Oil and Gas Journal* **93**(50), 32-34.

67. Greenway, J. (1995) Revolutionary seismic vessel sees first action in North Sea, *Oil and Gas Journal* **93**(35), 72-76.

68. George, D. (1992) Marginal field production systems: new thinking on old technology, *Offshore/Oilman* (November), 47.

69. Buckley, N. (1992) Survey of oil and gas industry (6): North Sea mid-life crisis - Britain's offshore activity has passes its peak, *Financial Times* (3 November), iv.

70. Weener, R. (1996) Shell chooses floating units for central North Sea projects, *Oil and Gas Journal* **94**(34), 58-64.

71. Lawton-Davis, S. (1998) Tackling demanding offshore environments, *Petroleum Review* **52**(621), 35-36.

72. Anon (1993) Horizontal technology: moving the driller closer to the driver's seat, *Journal of Petroleum Technology* (July), 608-609,642-643.

73. Anon (1996c) The costs of cost-cutting, *Oil and Gas Journal* **94**(20), 27.

74. Knott, D. (1996) Drilling technology shows rapid change, *Oil and Gas Journal* **94**(12), 44.

75. Knott, D. (1996c) North Sea development action brisk - plays expand elsewhere off Europe, *Oil and Gas Journal* **94**(34), 45-57.

76. Crow, P. (1999) Watching government: abundant oil, *Oil and Gas Journal* **97**(51), http://ogj.pennwellnet.com, 4 February 2000.

77. Williams, J. L. (1999) Oil price history and analysis, *WTRG Economics*, http://www.wtrg.com/prices.htm 2 November 1999.

78. Snyder, R. E. (1999) New fields, new technology, *Offshore* **220**(7), http://www.worldoil.com/archive/archive_99-09/99-09_offshore.html 27 October 1999.

79. Crook, J. (1998) Ice protection on the Grand Banks, *Petroleum Review* **52**(620), 34-35.

80. Crook, J. (1999) Meeting the challenge of deepwater gas production, *Petroleum Review* **53**(624), 32-33

81. Pearson, R. (1995) Liverpool Bay development takes shape in Irish Sea, *Oil and Gas Journal* **93**(35), 60-66.

82. Williams, H. (1996) New field developments challenge contractor skills, *Petroleum Economist* **63**(7), 34-35.

83. Energy Information Administration (EIA) (n.d.) *Privatisation and Globalisation of Energy Markets*, U.S. Energy Information Administration, Washington D.C. http://www.eia.doe.gov/emeu/pgem, 9 December 1999.

84. Hindley, A. (1991) Libya searches for oil prospectors, *Middle East Economic Digest* **35**(34), 4.

85. Anon (1991) Oil's new world order: for a quarter of a century oil-producing countries have fought to keep western companies out; now they are wooing them back, *The Economist* **320**(7715), 67.

86. Anon (1993b) A shocking speculation about the price of oil, *The Economist* **328**(7829), 69.

87. Anon (1994b) Oil rolls over, *The Economist* **331**(7857), 65.

88. Ghanem, S. (1994) The paradox of capacity. *OPEC Review XVIII* (3), 335-351.

89. Kielmas, M. (1994). Hostile trends put producers on the spot, *Middle East Economic Digest* **38**(13), 7.

90. Hartshorn, J. E. (1993) *Oil Trade: Politics and Prospects*, Cambridge University Press, Cambridge.

91. Anon (1994) Aquzadeh predicts offshore oil and gas deals, *Middle East Economic Digest* **38**(14), 22.

92. Takin, M. (1996) Many new ventures in the Middle East focus on old oil, gas fields, *Oil and Gas Journal* **94**(22), 31-37.

93. Ismail, I. A. H. (1996) OPEC Middle East plans for rising world demand amid uncertainty, *Oil and Gas Journal* **94**(22), 38-42.

94. Wu, W. M. (1999) CNOOC sets strategies to reduce costs to $10/bbl by 2000: Reform is becoming more difficult, *Offshore* **59**(5), http://os.pennwellnet.com, 28 October 1999.

95. Locatelli, C. (1995) The reorganisation of the Russian hydrocarbon industry: an overview, *Energy Policy* **23**(9), 809-819.

96. Energy Information Administration (EIA) (1999h) *Russia*, U.S. Energy Information Administration, Washington D.C. http://www.eia.doe.gov/emeu/cabs/russia.html, 27 October 1999.

97. Dorian, J. P., Abbasovich, U. T., Tonkopy, M. S., Jumabekovich, O. A. and Daxiong, Q. (1999) Energy in central Asia and north-west China: major trends and opportunities for regional co-operation, *Energy Policy* **27**, 281-297.

98. Energy Information Administration (EIA) (1999j) *Energy in the Americas*, U.S. Energy Information Administration, Washington D.C. http://www.eia.doe.gov/emeu/cabs/theamericas.html, 27 October 1999.

99. DeLuca, M. (1998) Asian E&P, leasing on track despite troubled economies, *Offshore* **58**(11), http://os.pennwellnet.com, 28 October 1999.

100. Mack, T. (1991) Yankees (and other foreigners) come back, *Forbes* **147**(12), 83.

101. Anon (1994c) Onward and upward, *The Economist* **331**(7868), 11.

102. Kuo, C. S. (1997) *The Mineral Industry of Bangladesh*, http://minerals.usgs.gov/minerals/pubs/country 1999, 20 July 2000.

103. Anon (1996e) From major to minor, *The Economist* **339**(7966), 63-64.

104. Beckman, J. (1993) Cost clamp-down rumbling through European offshore sectors, *Offshore/Oilman* (August), 29-30l.

105. Murray, M. (1999) Cuba opens offshore oil to foreigners, *MSBC News* **12 October** http://www.msnbc.com/news/, 2 November 1999

106. US, Public Law 104-073, which authorises the imposition of U.S. sanctions against foreign companies investing in Iranian oil or gas projects of over $20 million (Energy Information Administration 1999l).

107. France's Total and Malaysia's Petronas stepped in place of Conoco

108. Energy Information Administration (EIA) (1999k) *United States of America*, U.S. Energy Information Administration, Washington D.C. http://www.eia.doe.gov/emeu/cabs/usa.html, 27 October 1999.

109. Energy Information Administration (EIA) (1999l) *World Energy "Areas to Watch"*, U.S. Energy Information Administration, Washington D.C. http://www.eia.doe.gov/emeu/cabs/hot.html, 27 October 1999.

110. Mossavar-Rahmani, B. (1998) Iran plays the great game to end its isolation, *The European* (15 June), 54-55.

111. Conn, C. and White, D. (1994) The Revolution in Upstream Oil and Gas, *The McKinsey Quarterly* (3), 71-87.

112. Kemp, A. G. and Stephen, L. (1999) Risk: reward sharing contracts in the oil industry: the effects of bonus: penalty schemes, *Energy Policy* **27**, 111-120.

113. Rose, B. (1999) Consolidation, changing core competencies alter offshore drilling responsibilities, *Oil and Gas Journal* **97**(50), http://ogj.pennwellnet.com 4 February 2000.

114. Plumb, R. L. (1994) Flexible facilities needed to fund maturing province, *Petroleum Economist* (August/September), 22-24.

115. Anon (1996) Chevron sells mature North Sea assets, *Petroleum Economist* **63**(9), 120.

116. Anon (1996d) BP plans to sell marginal oil fields in UK North Sea, *Oil and Gas Journal* **94**(13), 34.

117. Albeit predominantly among North American Institutions.

118. Mikdashi, Z. (1987) Oil funding and international financial markets, in J. Rees and P. Odell (eds.), *The International Oil Industry - An Interdisciplinary Perspective*, Macmillan Press, London, pp.88-106.

119. Energy Information Administration (EIA) (1999f) *North Sea*, U.S. Energy Information Administration, Washington D.C. http://www.eia.doe.gov/emeu/cabs/northsea.html, 27 October 1999.

120. Crow, P. (2000) Watching government: merger morass, *Oil and Gas Journal* **98**(5), http://ogj.pennwellnet.com, 4 February 2000.

121. Anon (1997) Interest grows in African oil and gas opportunities, *Oil and Gas Journal* **95**(19), http://os.pennwellnet.com, 28 October 1999.

122. Energy Information Administration (EIA) (1999) *Angola*, U.S. Energy Information Administration, Washington D.C. http://www.eia.doe.gov/emeu/cabs/angola.html, 27 October 1999.

123. Asamu, M. (1999) Nigeria's deepwater province taking off now that new government in place, *Oil and Gas Online*, **30 August**, http://www.oilandgasonline.com, 2 November 1999.

124. Energy Information Administration (EIA) (1999e) *Nigeria*, U.S. Energy Information Administration, Washington D.C. http://www.eia.doe.gov/emeu/cabs/nigeria.html, 27 October 1999.

125. Energy Information Administration (EIA) (1999c) *Colombia*, U.S. Energy Information Administration, Washington D.C. http://www.eia.doe.gov/emeu/cabs/colombia.html, 27 October 1999.

126. Schmidt, V. (1998) Latin prospects, interest growing despite oil price, Asian weakness. Atlantic, Caribbean margins looking up, *Offshore* **58**(9), http://os.pennwellnet.com, 28 October 1999.

127. Beanland, M. (1999) East Timor and the politics of oil, *Oil and Gas Online* **8 October**, http://www.oilandgasonline.com, 2 November 1999.

128. Anon (1998b) Iran to explore disputed Caspian area, *Oil and Gas Journal* **96**(51), http://os.pennwellnet.com, 28 October 1999.

129. Delmar, R. (1999b) Cambodia trying to end dispute with Thailand, *Offshore* **59**(8) 1 Aug, http://os.pennwellnet.com, 28 October 1999.

130. Anon (1998) Cautious progress on disputes affecting China seas exploration: Sovereignty issues continue to fester, *Offshore* **58**(10), http://os.pennwellnet.com, 28 October 1999.

131. Since 1997.

132. Priddle, R. (1999) The new millennium - new challenges for world energy, article for *Suddeutch Zeitung*, 27-28 November 1999, http://www.iea.org/new/speeches/priddle/1999/millen.htm, 25 July 2000.

133. Yamani, Sheikh A. Z. (1999) OPEC should take long-term approach to balancing oil supply-demand equation, *Oil and Gas Journal* **97**(38), http://os.pennwellnet.com, 28 October 1999.

134. Energy Information Administration (EIA) (1999b) *Caspian Sea Region*, U.S. Energy Information Administration, Washington D.C. http://www.eia.doe.gov/emeu/cabs/caspian.html, 27 October 1999.

135. Lester, C. (1996) Reforming the offshore oil and gas program: rediscovering the public's interests in the outer continental shelf lands, *Ocean and Coastal Management* **30**(1), 1-42.

136. Smith, E. R. A. N. and Garcia, S. R. (1995) Evolving California opinion on offshore oil development, *Ocean and Coastal Management* **26**(1), 41-56.

137. Side, J. (1997) The future of North sea oil industry abandonment in the light of the Brent Spar decision, *Marine Policy*, **21**(1), 45-52.

138. Pittard, A. (1997) Technology - field abandonment costs vary widely world-wide, *Oil and Gas Journal* **95**(11), http://os.pennwellnet.com, 28 October 1999.

139. McGinnis, M. V. (1998) An analysis of the role of ecological science in offshore continental shelf decommissioning policy, in O. R. Magoon, H. Converse, B. Baird and M. Henson (eds.), *Taking a Look at California's Ocean Resources: An Agenda for the Future*, ASCE, Reston, VA, pp.1384-1392.

140. Carr, M. H., Forrester, G. E. and McGinnis, M. V. (1998) Decommissioning of offshore oil and gas facilities: contrasts between southern California and the Gulf of Mexico and implications for ecological research, in O. R. Magoon, H. Converse, B. Baird and M. Henson (eds.), *Taking a Look at California's Ocean Resources: An Agenda for the Future*, ASCE, Reston, VA, p.1383.

141. Moritis, G. (1997) Industry tackles offshore decommissioning, *Oil and Gas Journal* **95**(49), http://os.pennwellnet.com, 28 October 1999.

142. 80% and approximately 450 platforms in the North Sea, costing of the order of $20 billion over the next 30 years.

143. Energy Information Administration (EIA) (2000c) *Monthly Energy Review*, U.S. Energy Information Administration, Washington. EIA http://www.eia.doe.gov /emeu/mer/prices.html 19 September 2000.

144. American Petroleum Institute (1999) *API Basic Petroleum Data Book*, American Petroleum Institute, Washington D.C.

145. Borgese, E. M., Ginsbury, N., and Morgan, J. R. (1993) Ocean Yearbook 10, University of Chicago Press, London.

CHAPTER 7. FISHERIES

J.L. SUÁREZ
G. GONZÁLEZ
S. FERIA

The end of the 20th century is the culmination of a period of sustained growth in catches for the fishing industry resulting from a model of industrial development which quickly gained ground after the end of the Second World War.

The fishing industry is currently going through a great crisis [1]: it is quite evident that natural resources are in decline and being exhausted. The biological crisis has triggered off another crisis amongst governing organisations that above all affects the credibility of public administrations which find themselves unable to put a halt to the down-turn in the fishing sector. As such, concepts such as public management of fishing stocks and even the legal regime that governs the rights to stock ownership are being thoroughly reviewed. All this is taking place against the backdrop of the ongoing process of economic and market globalisation which is having such a profound effect on fish products.

1. The State of Resources. The Generalisation of Overfishing

Over the past few decades the world's fisheries have been subject to a process of overexploitation resulting from gross overfishing, an absence of management actions and a lack of sustainable resource management. There are environmental problems as well as this lack of foresight. These have also contributed to the worsening of the present circumstances fisheries find themselves in, albeit to a lesser extent. As a result of all this alarm bells are ringing all round the world, pointing the finger at the unviability of the present system for exploiting sea resources.

Although the fact that certain fisheries are approaching a point of collapse is more and more widely recognised, nevertheless fishermen still carry on fishing clinging to the old belief that the oceans provide a limitless source of food, rather than making use of responsible management systems.

1.1. THE EVOLUTION OF CATCHES THROUGHOUT HISTORY

The great rate at which the size of catches has increased since the middle of the century has only slowed down over the past twenty-five years, with a fall from a growth of 6% during the fifties and sixties to 0.6% between 1995/96. [2,3,4,5,6] In spite of this, and despite the fact that the tendency is still for catches of seafish to stabilise throughout the world, every year a new record is set for marine species production, with the sum total of 87.1 million tonnes being reached in 1996. Catches are therefore approaching the 100 million tonne mark defined as the sustainable level for world fisheries in a 1971 FAO report (Figure. 1).

As we have seen, on the one hand figures for fish production in terms of volume continue to rise but, on the other, the slowing down in the rate of increase only serves to draw attention to the fisheries overexploitation prevalent throughout the world.

Fisheries scientists have employed a general model for fishing developments based on variations in landings over a period of time to explain the development process for any given fishery. This process is divided up into four stages ranging from the time a fishery is still undeveloped, through first a development and then a mature stage, to a senescent stage. [3,4,7] The relative rate of increase in yield, which can vary widely as the fishery nears its long-term maximum yield, is of special interest. It has been used to get a approximate calculation of the state of marine resources, both *in toto* and on a per ocean basis. The rate is zero during the pre-development stage, but increases sharply as the fishery begins to develop. It then begins to decrease during the fishery's constant growth stage, and goes back to zero when it reaches the point of maximum production. After this third stage, fishing capacity may still grow, which worsens depletion, and the relative rate of increase can become negative as overfishing goes on.

On the basis of these analyses, in 1994 it became clear that approximately 35% of the 200 major fishing species were in the senescent stage (ie: showing signs of decreasing yield), 25% were in the mature stage (ie: at a high level of exploitation) 40% were in the development stage and 0% were at a low level of exploitation. This means that 60% of the major world fishing resources are either in their mature or senescent stages and, given that only a few countries have established an efficient control over their fishing capacity, these species require management measures to be taken to either put a brake on the increase in fishing capacity or to allow affected stocks to recover. [3,4]

However, even though there seems to be some sort of agreement on estimates for fishing resources, there are contradictions with regard to estimating their potential size. These differences depend on whether estimates are made by the addition or subtraction of data on a per-ocean basis. Figures arrived at by the addition of per-ocean data have been rejected in recent times given that as more and more data is added, the smaller potential production seems to be. There are, therefore, three widely varying estimates for the potential production of the world's oceans:
1) If the total figures for all the world's oceans are considered, then resources reached a point of total exploitation in 1996 with there being no hope for any significant production increase at all.
2) With the addition of estimates on a per-ocean basis, the world's oceans are still considered to be at the development stage, which means a considerable production increase of up to 17.1 million tonnes could be expected.

3) If the total is arrived at by the addition of figures for different areas of the ocean, this would be at the development stage, with a likely increase of up to 42.1 million tonnes, for the main part through the implementation of management measures. [3,4]

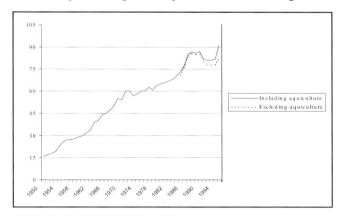

Figure 1. Sea-fishing production since 1950

Source: [3,4]

Nevertheless, these overall figures do not show how economic returns have evolved for these landings, which species are most affected by exploitation processes, or the areas that have been exploited most voraciously.

Increases in catches over the past forty years have resulted in a change in catch quality; whereas much smaller quantities of the most highly-valued species (cod, tuna, squid, etc.) have been landed due, to a great extent, to their high levels of overfishing, greater and greater amounts of other, less appreciated species (eg.: Peruvian Anchovy) are being landed. These are less profitable, however, which makes the whole viability of fishing questionable for certain (unindustrialized) fishing collectives.

In relation to large species groups, it should be recognised that production of pelagic species has continued to grow constantly. This has meant fishing for pelagic species has gained in relative importance, even though said production has fluctuated greatly due to natural variations in productivity (numbers of large pelagic species, such as tuna, swordfish, marlins and shark have fallen drastically. [9] On the other hand, the production of demersal species tended to increase until the mid-1970s, generally stabilizing since then. This, however, disguises the fact that there are regional differences in fisheries development and species that are overfished.

Data for landings of highly-migratory and straddling-stocks shows a significant rapid increase from 1970 onwards, with a decrease from the end of the eighties.

Overfishing is not the sole cause of the present state of the world's fisheries, however. There are other man-made and natural factors that have contributed to the present situation. Such is the case of wasteful fishing practice which results in the circumstantial catching of a large number of fish. Even more important, if possible, are the now generalized practices that increase the number of fish that are discarded. According to the FAO, the amount of fish discarded has reached 20 million tonnes, which equates to 25% of yearly production for marine species. [6] The discarding of fish would seem to be more prevalent in subsidised, industrialized fisheries rather than unindustrialized ones, where the figures are lower. These practices are encouraged by a number of reasons, all of them economic (where TACs are enforced and catches exceed them, fishermen discard the fish of least economic value; when income from sales does not compensate for the costs of going to sea)

It should also be borne in mind that certain fishing methods (trawling of the sea-bottom by drifters, dredging, etc.) are not only responsible for fish mortality; they can also destroy certain marine habitats, such as beds of sea grass, coral reefs and other species, which eventually affects the ecosystem's food chain. The food chain can also be broken by the disappearance of certain species as a result of overfishing. [6,9]

Another problem which gravely affects sea species is pollution (principally from land sources: fertiliser, pesticides, sediment deposits, dumping of sewage, other types of dumping, etc.). This affects the whole of the oceans, but especially bays, estuaries and semi-enclosed seas, such as the Mediterranean. The reduction in fishing resources can also be attributed to natural phenomena such as "*El Niño*". [10]

1.2. GEOGRAPHICAL DISTRIBUTION OF FISHING POTENTIAL

The general overexploitation of fisheries began after the Second World War. By the sixties economic returns were already peaking for traditional and high-value species but, from that time on, an ever-greater process of sea-fish over-exploitation in traditional fishing areas began, which meant new areas had to be sought out. Traditional fishing areas such as the Atlantic, Mediterranean and the Black Sea were thus joined by certain areas of the Pacific, and the Indian Ocean, where volumes for catches are still on the increase, eventually became the last new area of exploitation.[2,3,4,10]

With regard to the present state of sea stocks, total marine catches in the majority of Atlantic fishing areas and some of those in the Pacific would seem to have their reached their maximum yield a few years back. As such, these areas are not likely to register any considerable increase in catches. Even though the Atlantic Ocean (1980) was fully exploited a decade before the Pacific (1990), no future increase in production is expected for either of them. [3,4]

It should be specified that by area the north-western Atlantic, the south-eastern Atlantic and the central-eastern Atlantic were the first to reach their maximum production levels some ten or twenty years ago, and there is now a general trend towards a smaller overall total of catches. In the north-eastern Atlantic, the south-western Atlantic and the central-western Atlantic, the central-eastern Pacific, the north-eastern Pacific and the Mediterranean and Black Seas, catches would seem to be stable, or have even gone down slightly, after reaching their maximum potential some years ago. The general trend towards the flattening-out or decline in catches found in these areas coincides with the observation that it is precisely in these areas that the greatest number of totally-exploited sea species are to be found, along with species that are over-fished, exhausted, or recovering after a period of exhaustion (Figure 2).

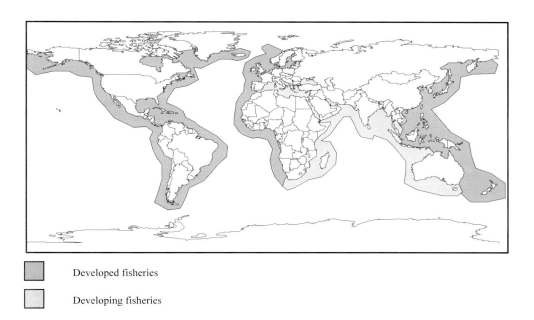

Developed fisheries

Developing fisheries

Figure 2. State of the resources by oceans

The eastern and western Indian Ocean, and the central-eastern and north-eastern Pacific are the main regions where overall figures for catches are still rising and where, theoretically, there is still room for more increase. Generally-speaking, these areas are home to fewer species that are totally-exploited, over-exploited, exhausted or recovering, and resources, which are still under-exploited or only moderately exploited, are still quite numerous. Nevertheless, these areas also contain a greater number of species whose level of exploitation is either unknown or unsure, which means estimates for production and stock levels are less reliable.

2. New Fishing Powers: The Boom in the Market, A Slump for Fishers.

In the past, fishing powers were traditionally economically-developed countries. At the present time, it is the highly-populated countries that are the leading fishing powers.

The economics of fishing have not been unaffected by trends towards a global market and the consequent heightened levels of competition in the sector and increase in demand for products. This situation, which on the surface seems favourable, is offset by other factors such as stock exhaustion, an over-large world fishing fleet, and a fall in employment in the fishing sector. The present world fishing industry scenario is the result of the way these opposing factors have come together.

2.1. FISHING IN THE POST-INDUSTRIAL ERA

Over the last ten or twenty years under-developed countries have taken the lead role in the world fishing sector as fishing constitutes an important source of foreign income with which to pay off their foreign debt. At the same time, the gap in volume of catches has widened between the countries that top the world ranking and those that follow (Table 1).

Attention must be drawn to the spectacular progress in China's contribution to world fishing production, with an increase of nearly 18 million tonnes over the last twenty years, which has resulted in a wide gap being opened up between China and the other countries immediately behind her. China is considered to be the current leading supplier of crustaceans to the international market. Special attention should also be given to Peru which, over the same period of time (1975 to 1995) has become the world's leading producer of animal fodder.

Paradoxically, although Japan's share in catch volume has gone down since 1988, given recent difficulties regarding access to fishing grounds, the levels of imports that have been reached and an ever-greater urge to control processing and trading have converted Japan into the international market's greatest mainstay.

Table 1. Changes in the classification of major producing countries*

1975	Tonnes	1995	Tonnes
1. Japan	10,524,204	1. China	24,433,321
2. USSR	9,935,606	2. Peru	8,943,208
3. China	6,880,000	3. Chile	7,590,947
4. Peru	3,447,490	4. Japan	6,757,570
5. USA	2,742,703	5. USA	5,634,419
6. Norway	2,550,438	6. India	4,903,659
7. India	2,328,000	7. Russian Federation	4,373,827
8. Korean Republic	2,133,371	8. Indonesia	4,118,000
9. Denmark	1,767,039	9. Thailand	3,501,772
10. Thailand	1,552,984	10. Norway	2,807,549

*Includes aquiculture products
Source: [3,4,11]

As far as developments in production according to economy-type (developed, transitional or under-developed) are concerned, it can be said that although US and European [24] production have continued at their natural rate – stable or on the increase- Japan, on the other hand, has continued on the downward trend started in 1988. Fish production has gradually decreased in transitional economies except in the Russian Federation, where production has increased slightly. At the other extreme are the under-developed countries, where there is still a large ongoing increase in production. It is fundamentally countries such as China, India, Bangladesh, Morocco, Indonesia and the Philippines that have driven the increase registered in Low-Income Food-Deficient Countries of this type. These countries are all large fish producers and, at one and the same time, make up 73% of the world's Low-Income Food-Deficient Countries. In some cases there has been a notable fall in production (Figure 3).

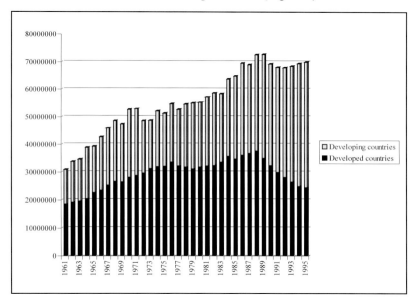

Figure 3. World fishing production per economic group

Source: [3]

In recent times the enormous growth of the world fishing fleet capacity resulting from the trend towards industrial fishing has caused a number of problems which are becoming progressively more serious (Figure 4).

A world-wide analysis of developments in tonnage for decked fishing vessels reveals a growth rate of 2.9% above the average, reaching 26 million tons in 1992. Two thirds of the total number of vessels (approximately 3.5 million) are small boats without decks. Nevertheless, more than 5,000 of the total number of fishing boats have a GRT of more than 500 tonnes. In twenty years the growth rate for the total world fishing fleet has doubled the growth rate of catches. This situation is worse still if we bear in mind the increased power of fishing vessels due to technological developments. [12]

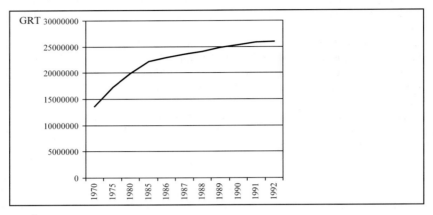

Source: [8]

Figure 4. Development of world fishing fleet tonnage

In economic terms, the over-capacity of the fishing fleet has meant that fishing has become economically unviable, with a loss of 15,000 million dollars world-wide every year. [12] The state of the fishing grounds, exhausted by overfishing, and continually taken advantage of by government subsidies, only makes the situation worse.

There are 12.5 million fishermen at the present time (10 million of whom belong to non-industrial, small or family concerns), with over 3 million vessels and 200 million people depending directly on the fishing sector. [7] If countries continue to support large scale industrial fishing development, there will be serious results for the smaller fishing businesses as there will be a drastic reduction in the number of local fishermen. This will in turn bring about the progressive disappearance of communities with age-old traditions and a rich cultural heritage. This process will go hand in hand with high social costs which will become progressively more difficult to contend with. What is even more worrying is the fact that the increase in the employment rate in fishing is slow compared to the sharp rate of population increase.

2.2. SUPPLY AND DEMAND OF FISH PRODUCTS

The internationalisation and deregulation of markets has resulted in a fall in the importance of local products and local fishers to the benefit of service companies, especially in the distribution sector.

The relationship between producer and distributor is becoming unbalanced in Europe due to an ongoing process of mergers amongst the big distribution chains, which means they are gaining an advantage over the producers thanks to their deeper understanding of consumer habits. As such they control strategic information about the demand for food products. [13] In short, the survival of the fishing industry no longer depends exclusively on the supply of resources but on its ability to adapt to new trends in international markets, and the development of knowledge. [14]

In this context, international fishing trade channels have adapted to the requirements implicit in the sharp increase in demand. As such, the volume of products exported has tripled over a period of twenty years (1976-1996) driven by developments in aquaculture, and now makes up 40% of total production. [5,6]

The majority of exports come from non-economically developed countries and are almost wholly imported into developed countries. These countries, on the other hand, import about seven times the amount imported by non-developed countries. Japan (with 30% of world production), the USA and European countries are all cases in point, making up 75% of external trade in terms of value. [5,6] The future development of global domestic demand will depend on the following factors: population growth and changes in fish consumption, income levels and fish prices. If we look at the two big areas of consumption, food and non-food consumption, estimates suggest the former will reach a figure of 110-120 million tonnes in 2010 compared with the 75-80 million tonnes for 1994-95 [3] Fish meal, the non-food staple *par excellence*, will remain stable (figure 5).

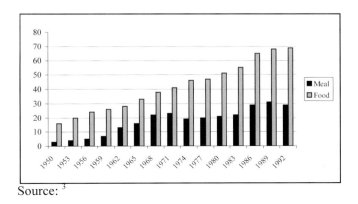

Source: [3]

Figure 5 Use of fish worldwide (millions of tonnes, live weight)

A stagnation in the way demand develops is foreseen, with mean annual per capita consumption of fish for the year 2010 standing at 13.5-14 kilograms, a figure that has not altered since 1993-1995. A rise in fish prices is expected, however, as a result of unsatisfied demand. One of the effects present trends will have is a reduction in domestic consumption in non-economically developed countries, notwithstanding their extreme dependence on the consumption of fish products, on which their diets are based.

Aquaculture has again been looked to as an alternative way of satisfying the sharp increase in demand for fish products for human consumption. At the present time, most of the world aquaculture fish production is to be found in under-developed countries, especially China, which leads the way with 70% of world production.

In spite of this, and in spite of the rapid increase in aquaculture products over the past years, this may not be enough to combat excess demand and compensate for the present decline in marine fisheries. Marine aquaculture also entails a number of drawbacks, with a deterioration in the marine ecosystem and a lack of adequate fishing areas. As such, the future looks brighter for fresh-water aquaculture, as prices are more within reach of low-income consumers and it is not as hostile towards the environment; fish nurseries and farming activities are integrated so as to recycle waste produced by the former. It even helps to maintain water quality. [10]

3. Fishing and the Governance of the Oceans

On the brink of the new millenium, with fishing having developed as never before, there is a wide-ranging consensus throughout the world, both among states and international organisations, that there is a need to reinforce and support fisheries management. This general agreement also recognises the fact that fisheries management over the past four or five decades has failed to find solutions to the problems affecting the industry today: the over-exploitation of stocks, and the threat to millions of fishermen and their respective communities that the present high level of over-capacity implies.

A number of changes have been suggested during the last few years of the present century which have a vision of fishing management that goes beyond a strictly bio-economic concept. The integration of fishing into the management of fishing areas, a change in the property rights regime for resources and the opening up of management to other, non-State, bodies are some of the trends that point to the future of fisheries management.

3.1. THE CRISIS OF THE BIO-ECONOMIC MODEL AND THE NATIONALIZATION OF RESOURCES

The scientific management of fish is a practice which was introduced on the back of the modernisation process that affected resource management. This in turn requires scientific and political bodies to be created in order to take charge of its development. [15]

From a scientific point-of-view, biology has been fundamental in creating models for fish species and their development, and these models, together with economics, have created a management pattern which has been a basis for the way the various ocean fisheries have been managed for the past hundred years or so.

Scientific management is based on the supposition that there is a lack of resources and a need for exploitation to be optimised. The concept which expresses the point of balance between a determined fishing effort and the ability for stock to recover is known as the Maximum Sustainable Yield (MSY), which is the maximum number of catches a stock can sustain without risking depletion. Both this characteristic point in stock evolution and maximum economic profit can be determined by applying bio-economic models.

The failure of fishing under a regime of free access and free competition has only recently been recognised. The reasons given are either related to the property rights regime concerned (common, state or free property), with emphasis placed on how it is impossible to conserve stocks in the absence of a regime of private ownership, or based on deficiencies in MSY calculation. Sustainability means that catches should not surpass a level that might put the survival of stocks at risk. Although this notion is not difficult to understand, there is, nonetheless, no single figure that represents the sustainable maximum; instead of a single figure there are a great number of possibilities, which depend on a variety of factors [16]: make-up and quantities of stock; short and long-term environmental conditions; the social and economic evolution of catches. Each of these points represents a *sustainable maximum,* and the calculation of each depends upon the accuracy of data and the reliability of techniques used for gathering statistics. The fish management model that has been used from the beginning of the century is for the most part based on biological evaluation, and does not include social, economic or environmental factors.

At the same time that the bio-economic model started to dominate fisheries management, a legal process was begun aimed at widening sovereignty rights of coastal States over their offshore waters. The first claims to large areas of the ocean of up to 200 nautical miles began to be made after the end of the Second World War, mostly in Central and South America. This process came to a head in the nineteen-seventies during the negotiation stage of the United Nation's Third Conference on the Law of the Seas.

These legal claims take on a new significance compared to previous processes of marine expansion; the objective is no longer solely the exclusion of third-parties, but, it is argued, there is also a need for the protection and conservation of stocks.

The process to establish exclusive economic zones (EEZ) and fishing zones (EFZ) prevailed throughout the international community at the end of the seventies, although it was the under-developed countries who took on the leading role, by associating the expansion of their area of jurisdiction with their post-colonial development aims and political emancipation.

The creation of EEZs and EFZs on a worldwide scale brought about a redistribution of fishing resources throughout the world. Ninety percent of catches are now made in waters under national jurisdiction. [5,6] From a management point-of-view, the new jurisdictional rights had an effect on institutional mechanisms; international regional organizations disappeared to be replaced by other bodies. It was especially difficult for under-developed countries to manage large areas of ocean, which is why the FAO created a special aid programme through its Fisheries Committee in 1979 to help developing countries face up to the challenge. [12]

3.2. THE HIGH SEAS: AN INTERNATIONAL CHALLENGE

As far as fish supplies are concerned, about 10% of stocks are made up of migratory species or straddling stocks that live outside the bounds of national jurisdiction, ie: on the high seas. [8] The nationalization of fishing resources brought about a gradual increase in pressure on stocks that inhabit the high seas, especially those found in areas bordering on exclusive economic zones and fishing zones; fishing here is in free waters, but on the very edge of areas of national jurisdiction. This has generated a new conflict concerned with taking actions regarding stocks that move freely to and fro between waters that fall under national jurisdiction and others that are not. Given the excess fishing capacity that exists in the world and the difficulties that gaining access to EEZs and EFZs entails, a large part by far of the high-tonnage fishing fleet is to be found on the high seas. Amongst the most conflict generating questions to be faced by the international community and the international organizations responsible for the fishing sector are: the regulation of straddling and highly migratory stocks and the protection of marine mammals, the use of non-selective fishing-gear (drift-nets), reflagging and the increasing use of open registers by fishing vessels, and the role to be played by regional organizations in managing fisheries on the high seas.

The management and conservation of living resources on the high seas as regulated by the UNCLOS (arts. 116-120) is a compromise between the right to free fishing that is enjoyed by all countries, and the duty to take measures to exploit these resources in a sustainable manner, as well as an obligation for states to cooperate with one another. The Convention text does not include any preferential rights for coastal states. Nevertheless, inadequacies in high sea fishing regulations and pressure from certain states led the United Nations to call a Conference on Straddling and Highly-Migratory stocks (United Nations General Assembly, Resolution 47/192 of 22nd December, 1992). The agreement reached at this conference (the Agreement for the Implementation of the Provisions of the United Nations Convention of the Law of the Sea of 10th December, 1982, Relating to the Conservation and Management of Straddling Fish Stocks and Highly Migratory Fish Stocks, 1995) gives coastal states greater management powers over resources outside their EEZs, giving rise to what jurists call the "creeping jurisdiction" [17] fomented by countries with long coastal strips and stocks that move back and forth between their EEZs and the high seas, prime examples of which are Argentina, Chile, Iceland and Canada.

Migratory and anadromous species as well as marine mammals are also gravely threatened on the high seas due to the use of non-selective gear, especially driftnets, which were the object of a UN Resolution on Large Scale Pelagic Driftnet Fishing (1989), in which a moratorium on the use of this type of gear was requested from 1992 onwards.

Although there have been many initiatives throughout the nineties aimed at regulating fishing on the high seas, during the same period of time there has been an increase in the number of high-sea vessels that have changed the flag they sail under in order to elude any commitments that States that have been party to international agreements are bound to enforce. This problem was subsequently included in Agenda 21 and the

Cancun Declaration. Although the number of vessels changing flags went down from the beginning of the nineties, a new strategy was nonetheless observed: the use of open registers or flags of convenience which allowed the vessels to act under the flags of States that have little commitment to conserving stocks on the high seas.[12] According to FAO calculations, there are about 1,000 registered vessels of 500 tons or more that undertake this practice, which accounts for about 20% of the total number of vessels in this segment[8] (Table 2).

Table 2. International initiatives on fishing on the high-seas

	Agreement/Declaration
1992	Cancun Declaration on Responsible Fishing (FAO/Mexico)
1992	UNCED: Chapter17, Agenda 21, Protection of the Ocean (UN)
1993	Agreement to Promote Compliance with International Conservation and Management Measures by Fishing Vessels on the High Seas (FAO)
1995	Code of Conduct for Responsible Fisheries (FAO)
1995	Agreement for the Implemetation of the Provisions of the UN Convention of the Law of the Sea Relating to the Conservation and Management of Straddling Fish Stocks and Highly Migratory Fish Stocks
1995	Rome Consensus on World Fisheries (FAO/UNCED)
1995	Kyoto Declaration and Action Plan on Sustainable Contribution of Fisheries to Food Security
1996	Resolution Adopted by the 95th Interparliamentary Conference

Reproduced from [20] Source: [21,22]

3.3. NEW APPROACHES TO FISHERIES MANAGEMENT: DEREGULATION, DECENTRALIZATION AND PRIVATIZATION

The FAO has identified good fisheries management on a world-wide scale as one that comprises the following elements: "a strategy that expressly leads to ecological, economic and social sustainability; effective fishing and research organisations…; a co-operative, organised and well-trained fishing sector; adequate legislation and judicial organisations…, and having effective links with regional international bodies".[5] Developing countries are still having great difficulty creating an institutional, administrative and scientific framework of this type.

On the other hand, even though developed countries have a long history of and experience in management practice, the crisis and decline of the fishing sector have brought with them a crisis in institutional bodies. This has resulted in dwindling government legitimacy in their respective fishing policies due to discord between hyper-developed legislative apparatus and a sector that is clearly in a phase of regression. This situation is clearly apparent in the North Atlantic. [18,19]

In modern times, the governance system for fishing has been based on three institutional orders (Apostle *et al*): the State, the community and the market. The changes that fishing management has undergone can be explained by and analysed as changes in the relationship between these three elements. The State is characterised by its hierarchical system, bureaucratic structures and relations based on authority. Fishing communities have provided the work force, knowledge and technical skills, as well as the organisational network. At the community level, even today personal relationships, an archaic system of payments and a feeling of identifying with the community still prevail. The market is founded on cost-effectiveness and reasoning. Competition is part of its very essence and, unlike the State, the dominating relationships are founded on decentralised exchanges rather than a governmental principle of authority.

Alternative management systems to the State have been proposed based either on giving more sway to the laws of the market, or on greater participation by the various collectives connected with fishing. The alternative systems tend towards three objectives, schematically-speaking: less regulation, decentralisation and privatisation. Deregulation implies "less State"; decentralisation - also known as regionalisation- is aimed at greater participation in governance (co-management); through formulas such as Individual Transferable Quotas (ITOs), privatisation is supposed to correct failings in market workings resulting from the common-property nature of fishing resources which make fishing non-cost-effective.

Co-management would entail an agreement between governments and end-users or people who actively participate in fishing. This is an institutional process of integration and redistribution of powers and responsibilities. The co-management system is seen as a solution to the grave problem of overfishing, as it entails greater commitment on the part of user and participating groups to undertake some kind of sustainable exploitation. A lot has been written about co-management in scientific papers, although it has not been applied to a great extent in practice; it is, nevertheless, regarded as desirable, and it seems that its implementation may become more widespread in the future. Joint funding by the fishing sector and the State has not been ruled out. [5,6]

As far as a reform of resource ownership is concerned, the proposed ITQ system would entail the most far-reaching modification, as if it were applied wholesale, fishing resources, which are in the main property of the public domain, would be turned into private resources. Supporters of this system believe it would be the solution to the biological crisis in fisheries known as "the tragedy of common property", by analogy with the assertion that Garrett Hardin made in his well-known article on the conservation of such property entitled "The Tragedy of the Commons" (1968), in which he puts forward privatisation as a means of avoiding the non-existence of an answerable owner. As such, this system consists of making fish quotas a market commodity like any other, up for purchase, sale, surrender or hire; in short, a private piece of property.[23]

The challenge that awaits us next century will be to put the proposed concepts and most important guidelines into practice, especially those put forward over the past ten years: sustainable development, responsible fishing, precautionary foresight and integrated fishing management for coastal area administration, environmental management, and policies designed to promote social well-being, while at the same time relying on ample participation and commitment from not only the fishermen themselves, but everyone in related sectors; in short, a complex and dynamic solution, capable of providing answers to societies' new and difficult requirements in the new millenium.

References

1. Goodwin, J.R. (1990) *Crisis in the World's Fisheries. People, Problems and Policies*, Stanford University Press, Stanford.

2. Chaussade, J & Corlay J.P. (1990) *Atlas des Pêches et des Cultures Marines. France, Europe, Monde*. Éditions Ouest- France – Le Marin, s.l.

3. FAO (1997) *El Estado Mundial de la Pesca y la Acuicultura 1996*, FAO, Rome.

4. FAO (1997) *Estadísticas de Pesca. Capturas y Desembarques 1995*, vol. 80, FAO, Rome.

5. FAO (1999) *El Estado Mundial de la Pesca y la Acuicultura 1998*, FAO, Rome.

6. FAO (1999) *Estadísticas de Pesca, Productos 1997* vol. 85, FAO, Rome.

7. García, S.M. et al (1999) Towards Sustainable Fisheries: A Strategy for FAO and the World Bank, *Ocean and Coastal Management* 42, 369- 398.

8. FAO (1994) *Examen de la Situación de las Especies Altamente Migratorias y las Poblaciones Transzonales. Documento Técnico de Pesca* 337, FAO, Rome.

9. WWF (1998) *Peces Marinos en su Hábitat Natural, Informe de WWF sobre el Estado de las Especies 1996*, WWF, Madrid.

10. Weber, P.(1995) *Pérdidas Netas. Pesca, Empleo y Medio Ambiente Marino. Cuadernos Worldwatch*, Bakeaz, Bilbao.

11. FAO (1977) *Capturas y Desembarcos 1976. Anuario Estadístico de Pesca 42*, FAO, Rome.

12. Swan, J.; Satia, B.P.(1999) *Contribution of the Committee on Fisheries to Global Fisheries Governance 1977- 1997. Fisheries Circular* 938, FAO, Rome.

13. Friis, P. (1996) The European Fishing Industry: Desregulación and the Market, , in Crean, K. & Symes, D. (ed.) *Fisheries Management in Crisis*. Fishing News Books/ Blackell Science, Oxford, pp. 175- 186.

14. Friis, P.& Vedsman, T. (1998) From Resource to Knowledge- based Production. The Case of the Danish Fishing Industry. *European Urban and Regional Studies* vol. 5, nº 4, 343- 354.

15. Apostle, R. et al (1998) *Community, State, and Market on the North Atlantic Rim*. University of Toronto Press, Toronto.

16. Kesteven, G.L.(1997) MSY revisied. A Realistic Approach to Fisheries Management and Administration. *Marine Policy* vol. 21, n° 1, 73- 82.

17. Casado Raigon, R. (1994) *La Pesca en Alta Mar*. Junta de Andalucía, Sevilla.

18. Crean, K. & Symes, D. (1996) *Fisheries Management in Crisis*. Fishing News Books/ Blackell Science, Oxford.

19. Symes, D. (1998) *Northern Waters: Management Issues and Practice*. Fishing News Books/ Blackel Science, Oxford.

20. Suárez de Vivero, J.L. (1998) Political Geography and Politics in the North Atlantic, in Symes, D. *Northern Waters: Management Issues and Practice*. Fishing News Books/ Blackell Science, Oxford, pp. 52- 63.

21. Marashi, S.H (1996a). *Summary Information on the Role of International Fishery and otherBodies with Regard to Conservation and Management of Living Resources on the High Seas. Fisheries Circular* 908, FAO, Rome.

22. Marashi, S.H. (1996b) *The role of FAO Regional Fishery Bodies in the Conservation and Management of Fisheries Fisheries Circular* 916, FAO, Rome.

23. Eythorsson, E. (1996) Theory and practice of ITQs in Iceland. Privatization of Common Fishing Rights. *Marine Policy* vol.20, n° 3, 269- 281.

24. Europe, including the European republics of the ex USSR, is the main regional producer

CHAPTER 8. THE USE OF THE SEA FOR RECREATION AND TOURISM

A Management Challenge for the 21st Century

M.B. ORAMS

1. Introduction

We are fortunate to be alive at an unique time in history. The beginning of the third millennium is a natural point at which we should pause, look back and consider many things. The increasing influence that humans and our activities have had on the health and functioning of our planet is one of those things. When considering this, one of the issues of significance is the important role that oceans, coasts and harbours have played in the history of human societies. We should also consider the influence that humans and their activities have had on those marine environs. While the marine environment has certainly been central to human activities throughout history, it is only in the latter part of the twentieth century that its use has become all encompassing. Only 50 years ago many parts of the marine world were unexplored and inaccessible to almost all, now however, there are few locations and few species on our marine planet that have escaped the effects of humans. Probably the most recent and most dramatically increasing use of our seas is our use of them for recreation. In particular the rapid rise of the tourism industry has had a marked impact on the marine world, there are now few locations that humans do not visit for recreational purposes. The marine world has become a playground with few boundaries.

The beginning of the third millennium is also a natural point to look forward. To consider where we are going as a global community. It seems extremely unlikely that the recent rapid growth of tourism will abate. In fact the World Tourism Organisation predicts that there will be 1.6 billion international travellers by the year 2020, [1] three times the number there were in 1995. In addition, it must be remembered that many more millions of people travel for recreational purposes domestically. Tourism is, therefore, going to continue to have a major influence on life on our planet. Over the next one hundred years our curiosity for new and unexplored environs, combined with increased technological capability will see us travel to many new locations previously "protected" from direct human influence.

Some consider space travel to be the next recreational/tourism frontier [2] while others argue that polar tourism is the next growth area. [3] There is much evidence, however, that the next frontier for recreational development and influence is already well underway and likely to grow exponentially over the early part of this, the third millennium. This frontier is the sea. Marine environments have been largely protected from human recreational use for the great majority of history. However, indications from the last decade of the 20th century are that marine tourism is the fastest growing segment of the global tourism industry. [4] Furthermore, marine ecosystems, as before with terrestrial environments, are experiencing significant detrimental impacts as a result of this increasing visitation.

This chapter examines the issue of marine recreation and tourism and considers its growth and influence. It adopts a "macro" view, largely because the turn of the millennium seems to engender such a perspective. As a consequence, the issue of conservation becomes central to the discussion. For if one is to consider the state of the planet over the next hundred or next thousand years, the ideals of conservation and sustainability are inescapable.

2. The Importance of the Sea

There is no doubt that the future health of our planet and all things that live on it is totally dependent on the sea. It is widely understood that the health of our world's oceans is critical to the health of our planet.

'The living ocean drives planetary chemistry, governs climate and weather, and otherwise provides the cornerstone of the life-support system for all creatures on our planet, from deep sea starfish to desert sagebrush.

That's why the ocean matters. If the sea is sick, we'll feel it. If it dies, we die. Our future and the state of the oceans are one'. [5]

It is important therefore, when looking at the recreational use of marine resources, to consider the influence of those activities on the health and viability of marine ecosystems. Thus, I believe that the basis for analysing and managing marine recreation and tourism activities must be carried out for the purpose of ensuring the sustainability of the resource upon which not only the recreation is dependent, but upon which the health of all living things is dependent. This may seem to be somewhat of an over-reaction. The sea is vast and by far the great majority of marine-based recreation occurs in but a small portion of that vastness. How then, can recreational activities in the sea threaten the survival of the planet? This question is addressed in a reflection from the most well known of ocean explorers, Jacques Cousteau:

'But soon I had to face the evidence: the blue waters of the open sea appeared to be, most of the time, a discouraging desert. Like deserts on land, it was far from dead, but the live ingredient, plankton, was thinly spread, like haze, barely visible and monotonous. Then, exceptionally, areas turned into meeting places; close to shores and reefs, around floating weeds or wrecks, fish would gather and make a spectacular display of vitality and beauty ...

The 'oasis theory' was to help me to understand that the ocean, huge as it may be when measured at human scale, is a very thin layer of water covering most of our planet - a very small world in fact - extremely fragile and at our mercy'. [6]

The oceans of our planet are, therefore, not the vast endless resource that many humans still perceive them to be. Despite the fact that nearly 70 per cent of our planet is covered by ocean, only small portions of this area form the basis for most forms of marine life. The lesson from the last millennium with regard to our use of natural resources is clear. The view that the supply of forests, animals and minerals is so vast that removing as many of them as we like or need would have little impact was selfish, short sighted and wrong. We have also learned (I hope) that the discard of waste from our activities can negatively impact the ability of all life including our own - to survive. Unfortunately, however, there is still a distinct disparity in our attitude toward terrestrial resources compared with marine resources. The great majority of the developed world accepts that the uncontrolled cutting of forests and the dumping of toxins on land is fraught with danger. However, uncontrolled fishing and the dumping of waste at sea remains commonplace. We must apply the lessons we have learned from our use of the land to our use of the sea and avoid making the same mistakes. Our ability to survive into the next millennium may depend on it.

In my view it is therefore imperative that the principles of conservation and sustainability become central to all human "use" of marine resources. The great opportunity and challenge is that the rapidly growing marine tourism industry can make a significant contribution to that end. My hope is that through visiting and enjoying the marine environment through recreation, that many more people will come to view the oceans as worthy of protection. Perhaps in the same way as recreation and tourism are now used as a legitimate justification for the protection of land based resources, (as when land based resources are protected as parks), we will see more marine conservation advocates and marine parks. This hope may be somewhat naive, for the rapid and widespread development of marine based recreation and tourism suggests that it may merely be another form of exploitation of marine resources rather than an agent for marine conservation. However, there are a number of examples where recreation and tourism has produced positive results for things marine, it can and has happened. This gives us hope, for if it can be done once, it can be done again and again. This is the challenge for the new millennium.

3. The Importance of Marine Settings for Recreation and Tourism

The use of coastal areas and the sea for recreation has existed for probably as long as humans have. The sea has a strong attraction for people, not surprising given its importance as a source of food and transport. This importance is reflected in the fact that the great majority of the world's population resides along the coast. [7] Coastal and marine recreation and tourism is, quite simply, huge business that forms a significant component of the wider recreation and tourism industry. For many island and coastal nations it is the primary focus of their tourism industries. [8] A number of important changes in the latter part of the 20th century have contributed greatly to the importance of the marine environment as a host for human recreational activities.

3.1 THE RAPID GROWTH OF MARINE RECREATION AND TOURISM

It is difficult to separate marine recreation from general recreation data. There appears to be, however, a consensus in the literature that coastal and marine recreation and tourism is growing at a faster rate than the general tourism sector. [7] Recent data from the World Tourism Organisation [9] confirms that international travel and tourism continues to grow with 1999 recording around four per cent growth in international visitor arrivals, with many regions experiencing double digit growth. International arrivals reached 657 million in 1999 representing expenditure of US$455 billion. International tourism is, therefore, big business experiencing widespread growth. Marine tourism forms a significant component of that growth. This reflects not only increasing opportunities for marine recreation but also a "generally increased level of interest in anything to do with marine environments" [10] and an increase in the number and type of vessels that allow people to access the sea. A limited number of studies suggest that the growth of marine recreation and tourism has been relatively recent. For example, research on marine tourism businesses in New Zealand revealed that over 60 per cent of operators (400) had been in business less than five years. [11]

An important indicator of the relative importance of the sea as a tourism attraction is shown by a recent study on the value of beaches in the United States. Houston [12] found that:

'Beaches are key to U.S. tourism, since they are the leading tourist destination, with historical sites and parks being second most popular, and other destination choices minor by comparison. Coastal states receive about 85 per cent of U.S. tourist-related revenues, largely because of the tremendous popularity of beaches. For example, a single beach, Miami Beach, has more annual visits than Yellowstone, the Grand Canyon, and Yosemite National Parks combined'.

West [13] agrees but also points out that beaches in close proximity to urban areas are by far the most popular.

'The demand for beach and bathing facilities has largely paralleled the demographic developments ... Urban beaches are increasingly seen as the single most important recreational outlet for a large segment of the urban population'. [13]

The popularity of beaches (or regions of beaches) as settings for recreation is also reflected in popular culture, as represented in music, art, movies, television and writing. Examples include Surfer's Paradise (Queensland, Australia), Copacabana (Brazil), Waikiki (Hawaii), The Riviera (France), San Sabastian (Spain), Venice (California), Acapulco (Mexico), The Golden Mile (Durban, South Africa) and Uluwatu (Bali, Indonesia). Each of these areas hosts well in excess of a million visitors each year. This pattern of intensive use is repeated around the world at virtually every beach located close to an urban area.

While beaches are without doubt the most popular locations for marine recreation a second important location is islands. Of course these islands also include beaches, however, they provide a base for many marine activities and have proved to be immensely popular settings for tourism and associated development including resorts, hotels, restaurants and activity providers. Once again, many of these islands (or groups of islands) have become famous locations for marine tourism, examples include Hawaii, Tahiti, Fiji, Bali, Catalina, San Juan's, Key West, Martinique, Aruba, Jamaica, Bermuda, Majorca, Mikinos, Cyprus, Seychelles, Palau, Maldives, Canaries, and the Galapagos. For many of these islands marine tourism is the mainstay of the local economies. For example, the Seychelles, a small island nation in the northern Indian Ocean, derives approximately 70 per cent of its foreign exchange earnings from tourism, [14] and this tourism is entirely ocean based". [15]

The popularity of marine recreational activities is undoubtedly influenced by the strong positive image that small islands, beaches, coasts and the sea have. For many people the term "relaxation" evokes images of waters gently lapping against sandy beaches. These images have tremendous power and have been used successfully to influence people's decision making regarding their use of leisure time. The mental image of the "3-S's", sun, sand and sea is automatic for many when they think of holidays. The fact that there are now more locations, more activities and more opportunities to experience the "3-S's" contributes to the demand for marine recreation and tourism. It appears that a widespread latent demand exists for marine recreation. However, the question as to why this demand exists is a difficult one to answer. Part of the answer is provided by Jones [16] who claimed:

'This increase in interest is fuelled by a better educated public, a public that is rapidly developing an almost insatiable curiosity about the wonders of the sea'.

Other environmental settings, mountains for example, offer a similarly diverse range of recreation opportunities as well as beautiful natural settings. However, the world-wide use of coastal environments for recreation far exceeds any other. Why is this the case? The answer lies in the complex number of influences on demand. Fabbri [17] points out

'... that quite often in market-oriented economies, the demand tends to be a response to the offer. The more pressing the offer, the higher the demand, and there is no doubt that the offer of tourism and/or vacation from coastal areas exceeds by far, both in intensity and variety, offers from any other place'.

Fabbri's point is well made. It is important to recognise that demand for marine recreation and tourism is not a simple function of an inherently desired setting or activity. Rather it is an outcome of a complex relationship between opportunity, image, perceived benefit, cost and history. A major influence over the demand for marine recreation activities has been the invention of new technology and through this, the creation of new activities and new locations. Thus, "the offer" and more specifically the marketing of that "offer" has created demand for the activity or location.

The invention and marketing of activities such as surfing, wind-surfing, water skiing, SCUBA diving and para-sailing have created a demand for these activities. It is difficult to find any marine recreation activity that is not experiencing rapid growth. This growth is indicative of the massive interest in and demand for marine based recreation. Kenchington [18] states that:

> 'Studies in Australia and elsewhere have demonstrated that fishing is the most popular participatory recreational activity. The image of leisure pleasurably anticipated is a day off to go fishing, or retiring to go fishing'.

Whilst this may be true for a proportion of the population, particularly older males, marine environments have a diverse appeal for many recreational activities and there are a number of other activities that have experienced spectacular growth in participation. A good example of this is the growth of the SCUBA diving industry.

Since its invention in the 1950s SCUBA has spawned an entire new marine tourism industry. Television programs, such as "Jacques Cousteau's Undersea World", helped to stimulate an interest in exploring and enjoying the underwater environment. Thus, in a pattern often repeated in marine recreation, the invention of new technology and publicity associated with its use created a "demand" for the activity.

A number of commentators claim that SCUBA is amongst the fastest growing sports in the world [19,20] and whilst this claim seems to be made by almost all new sports there is no doubt that SCUBA has become an immensely popular activity. Davis and Tisdell, [21] who review estimates of numbers of SCUBA divers in Australia, claim that around 100,000 people learn to SCUBA dive each year in Australia. They also point out that, in addition to certified divers, so-called "resort dives", where non-certified divers are taken out diving under supervision, are growing rapidly. Wilks [22] estimated that there were around one million recreational dives undertaken annually off the Queensland (Australia) coast alone. Data from North America also reveals the popularity of SCUBA diving, West [13] claims that there are between four and five million certified SCUBA divers in the United States. The publication of many dive magazines and dive videos, the memberships of dive clubs, the operation of "live-aboard" dive vessels, the creation of dive oriented resorts, the establishment of many dive shops and dive tour operators attest to the growth of SCUBA as a significant marine tourism activity.

SCUBA is but one example of a new invention that has spawned a popular marine pastime. The influence of technology has been instrumental in the increasing levels of use of the sea for recreation and tourism. In the past the majority of our planet's marine environment has been "protected" from recreation and tourism because of its inaccessibility, safety concerns and the relatively high cost of recreating in the sea. However, over the past three decades a significant number of new inventions have made the marine environment more accessible both in real and economic terms. Examples include, electronic satellite based navigation and emergency location aids like global positioning systems (GPS) and electronic position indicator radio beacons (EPIRBs), massive aluminium high speed catamarans, personal water craft (Jet-skis) wind-surfers, kite-surfers and submarines.

In addition, interest in the marine environment has grown, television nature shows, such as those pioneered by Jacques Cousteau, magazines and films have exposed millions of people to a world that was once unknown. The result has been that the sea is now not only interesting, but also accessible. The resultant increase in demand for marine activities has added to a tradition of sea, sand and sun holidays. As a consequence, millions of people now visit marine environs for recreational pursuits. The potential impact of this increase in demand is beginning to be recognised.

'As the 21st century comes into focus, tourism is being revealed as a major sociocultural force with a potential to destroy, protect, or otherwise dramatically reconfigure coastal and marine ecosystems and societies'. [8]

The causes of this increased amount and diversity of marine recreation and tourism is threefold. First, the world's population continues to grow at a rapid rate. There are now twice as many people on the planet as there were 20 years ago, we now number over six billion. Our population will continue to increase and is predicted to reach eight billion by the year 2118. [23] Because the majority of this population will reside close to the coast the use of the marine environment for recreation and other purposes will show a corresponding increase. [24] Second, the rapid rise in mass tourism has resulted in more people travelling away from their places of residence for recreational purposes. A significant (but as yet unquantified) proportion of this travel is to coastal areas. Of particular interest is travel to areas of the world that were previously undeveloped and unexposed to high levels of human use. The rise of tourism has contributed significantly to the geographical spread of human recreational activities. The third factor that has been of enormous influence with regard to the marine environment has been the invention and mass production of materials and vehicles that have improved access to and safety in marine settings. Many hundreds of machines and equipment now permit safe and relatively easy access to the sea, the result is an increasingly diverse range of activities in an increasing number of settings.

When the development of marine recreation is traced, the most obvious features that emerge are, first, that it has become increasingly popular and second, related to that popularity, it has become increasingly diverse. Today there are more ways and means of accessing the marine world for recreation than ever before. It appears highly likely that there will continue to be an increase in this diversity as we invent more ways to go under it, get in it and on it. This complex array of activities, some complimentary and many not, create many management challenges as agencies responsible for marine resources try to reduce conflicts, decrease risks for participants and minimise damage to natural resources.

3.2 THE DISTRIBUTION OF MARINE TOURISM ACTIVITIES

Within this trend of increasing diversity and increasing popularity several important patterns can be identified. First, as one would expect, greater use tends to occur close to areas of human concentration, namely cities. Despite the emphasis on analysing impacts of travel and tourism in the literature, the majority of recreation occurs close to people's place of residence, this is also true with regard to marine recreation. Thus, it must always be remembered that the great majority of marine recreation is a result of regular "day trip" visits from local residents. This point is emphasised in an observation by Miller: [7]

'It should be kept in mind that six out of ten people around the world reside within 60 kilometres of the coastline and two-thirds of the world's cities with populations greater than 2.5 million are located by tidal estuaries. The population of the coastal zone is projected to double within the next 20-30 years'.

Because much marine recreation occurs in close proximity to urban areas, the environments upon which the recreation is based are subject to increased pressure. Marine environments closer to cities receive large amounts of urban "runoff" and other discharges resulting from human activity (sewage, storm-water etc.). In addition, they are often subject to dredging, foreshore alteration and reclamation and dumping of waste products. They are more frequently fished, have larger numbers of vessels and navigation aids. The effect of this higher level of use is that these marine areas are more vulnerable to additional pressure such as that produced by recreational activities.

A second main trend that is easily identified with regard to the spatial distribution of marine recreation is that there is an inverse relationship between distance from shore and intensity of use. A wider variety of activities and a greater level of use are associated with near shore environments. This pattern is, of course, entirely logical because humans are terrestrial based animals who find it difficult to survive in a liquid medium. It is important to recognise the dilemma this causes. Greater levels of human use occur close to shore and close to cities, however, it is these environments that are the most critical to the health and long term survival of the oceans and all that live within in it. [5] Marine recreation predominantly occurs, therefore, in those ecosystems that are most vulnerable to disturbance.

One of the important functions that marine recreation experiences provide for people is that of an escape from urban living - those experiences encapsulated within the expression "getting away from it all". For many, the opportunity to recreate in a natural setting, uninfluenced by human activity, is extremely important. Scherl [25] argues that there are significant psychological benefits that arise from wilderness experiences. There is no doubt that, for many, wilderness experiences are provided by marine based tourism, for it is the sea that provides one of the few truly wild areas left on our planet. However, the opportunity for peace, solitude and "wildness" is decreasing, even on the sea. Aldo Leopold's comments of half a century ago are even more relevant today:

'Wilderness is a resource which can shrink but not grow. Invasions can be arrested or modified in a manner to keep an area usable for tourism, or for science, or for wildlife, but the creation of new wilderness in the full sense of the word is impossible ... One of the fastest - shrinking categories of wilderness is coastlines. Cottages and tourist roads have all but annihilated wild coasts ... No single kind of wilderness is more intimately interwoven with history, and none nearer the point of complete disappearance'. [26]

The issue for management is clear - wilderness experiences are desired by a segment of the marine tourism market, however, opportunities for such experiences are decreasing.

The supply of marine recreation opportunities is closely linked with the issue of access. New inventions are creating new activities and allowing access to previously unused areas. Thus, it could be argued that the supply of opportunities for marine recreation is increasing. There are certainly many marine activities available now which were not available 30 years ago. However, an important issue with regard to the supply of opportunities for marine recreation is that of environmental quality. Most marine recreational activities are dependent on the quality of the resource, for example, fishing cannot occur if there are no fish! Or, an even more extreme scenario would see a complete loss of marine recreation opportunities if an area is so polluted it is harmful to human health. Unfortunately, this is a reality for many harbour and beach areas close to large cities. Thus, the supply of marine activities while increasingly diverse, is constrained by environmental quality. The impacts of recreation and other human activities on marine environments inevitably affects our ability to utilise those environments for recreation. The issue conservation is therefore critical to the future of marine recreation and tourism.

4. Problems and Challenges for the 21st Century

Despite the significant efforts being made to mitigate the impacts of marine tourism through coastal zone management, marine parks and protected areas and fisheries management, the reality of ever increasing numbers and fixed resources remains. Our interest in the marine environment seems unlikely to abate. Increasing interest in and use of the sea for recreation has been a long-term trend. Furthermore, it is certain that our invention of new ways to access the sea and utilise it for recreational purposes will continue to increase. The result of these two basic trends is simple, marine based recreation will continue to grow in popularity into the 21st century. Because of the increasing demand, the supply of marine tourism opportunities will become a critical issue.

Supply of marine tourism opportunities is constrained by several important factors. First, marine resources are limited. Whilst the oceans are vast, the locations sought for marine tourism are relatively small. A significant factor in this will become environmental quality. Marine tourists do not want to conduct their recreational activities in polluted areas, many of the attractions they seek are not present in polluted locations and their own health is threatened. Thus, management of marine resources in order to maintain, or improve environmental quality will become the major challenge in the next century.

A further issue with regard to marine recreation will become cost. A basic economic principle - high demand and scarce supply - will continue to force the cost of marine recreational activities up. Many cultures regard free access to and use of marine resources as a basic human right. However, it is already obvious that access to high quality areas and popular activities, in some areas, is only afforded by the wealthy.

An additional challenge derived from increasing demand and limited supply is the conflict between incompatible activities. This is common in popular areas now and it will escalate over the coming decades. Underwater photographers who value large fish species alive will compete with spear fishers who wish to hunt them. Jet skiers who wish to wave ride and wave jump will endanger surfers who wish to use the same "resource". Wildlife watchers will conflict with water skiers who wish to use sheltered bays for their sport. Indeed, all of these conflicts are common today and they will be an increasing challenge for marine resource managers in the future.

It is important to note however, that most degradation of marine resources is not the result of human recreational activities. The damage caused by the pollution of our coastal environs from human activities on land and from commercial use of our oceans for fishing, the dumping of waste, dredging and so on far outweighs the influence of recreation and tourism. Consequently, the future of marine recreation and tourism is inextricably linked with all other human activities that affect the sea.

One of the most significant challenges, and perhaps, one of the greatest opportunities for marine recreation and tourism is to ensure that the demand for the use of marine resources for recreational purposes is given due consideration in marine resource utilisation decision making.

While the future for marine resources is considered by many to be bleak, there are a number of cases that show positive results. As has occurred on land, the uniting of large numbers of people who wish to use natural environments for recreation and who value high quality natural environments has produced positive change for nature. The marine conservation movement, while not as well established as terrestrial equivalents, appears to be gathering momentum. Save the oceans campaigns, beach clean-ups, marine mammal and sea bird rehabilitation centres and many other efforts have shown that humans who are interested in things marine can make a positive difference through their efforts.

One of the most encouraging signs of positive change is the increasing number of marine parks and marine protected areas that are being established world-wide. When the decisions of humans over the past millennium are reviewed, close to the top of the list of "good decisions" will be the establishment and protection of large areas of high natural quality as national parks and other protected areas. These areas not only provide habitat for many of the planet's species, they also significantly contribute to the health and functioning of our planet and all things that live as a part of it.

As our use of marine resources for recreation and for other needs continues to grow, marine protected areas will also become critical to the health of our seas. They will also provide important locations for marine recreation activities and most importantly, as their terrestrial counterparts have done, provide locations where a marine conservation ethic can be nurtured in those who visit them. [27] The role of marine resource managers, environmental groups, management agencies and marine tourism operators is critical. If we are to exist as a species for the next thousand years we must utilise every opportunity to change the way people behave towards things marine. If people are not inspired and changed as a result of their experiences recreating on, in and under the sea, then the industry is simply exploitative and will ultimately become destructive.

As we move into this, the third millennium, the use of our oceans needs to be carefully considered. Recreation and tourism has an important role to play in the use of those resources, the industry could become an agent for positive change - a contributor to the healthy functioning of our marine environment, or it could be yet another significant cause of the demise of the quality of life on our planet. Given the size and diversity of the industry, it is likely to be both. The way that it is managed will be the difference.

4.1 POTENTIAL SOLUTIONS AND AREAS FOR RESEARCH

Complex challenges seldom have simple solutions and consequently the classic resource management dilemma of conserving natural attractions whilst allowing for their use will never be solved completely. Rather, the solution is more like a "battle" where resource managers must continually work to minimise costs and maximise benefits by developing strategies from the range of techniques available to meet the unique challenges provided by individual situations.

In fighting that battle, answers to a number of important questions will provide additional weapons for resource managers. These questions include:

- What motivates people to pursue marine based recreational activities?
- What are the characteristics of these marine tourists?
- What techniques are most effective in controlling human behaviour in marine settings?
- How can the impacts of recreational activities on marine resources be minimised?
- How can the benefits of marine recreation and tourism activities be maximised?
- What techniques are most effective in reducing conflict between competing uses?
- What management regimes/decision making approaches are most effective?

The role of marine scientists is crucial in providing answers to these questions and thus adding to the "menu" of management strategies available to marine resource managers. It is ironic that almost all of the challenges faced by the marine environment are the result of human activities, including recreation, however, the great majority of research that occurs on our oceans remains in the biological and physical sciences. It is critical for our marine environment that marine scientists incorporate and encourage the social sciences in the marine science community. Increasing our understanding of humans, what they do and why, is fundamental to finding answers to the challenges posed by marine recreation and other human uses of marine resources. In particular, a focus on applied research - that which is focussed on providing solutions to problems rather than simply outlining the extent of the problems is paramount.

The challenges faced in managing marine recreation and tourism are huge and there will probably always be more tales of failure rather than cases of success. However, because there are success stories, where the management of the activity or area has resulted in positive change for local communities and for marine ecosystems, there is hope. If it can be done in one area, it can be done in others. Continued investigation, critical thinking, learning from experiences and communicating those experiences are ways forward to a better future.

References

1. World Tourism Organisation (1997) *Tourism 2020 Vision.* World Tourism Organisation, Madrid.

2. Brownhill, M. (1998) When the sky's not the limit. *New Zealand Herald,* 8 August.

3. Hall, C.M. and Johnstone, M.E. (eds) (1995*) Polar Tourism: Tourism in the Arctic and Antarctic Regions.* John Wiley and Sons, Chicester.

4. Orams, M.B. (1999) *Marine Tourism: Development, Impacts and Management.* Routledge, London.

5. Earle, S.A. (1995) Sea change. *Ocean Realm* (June), 30-34.

6. Cousteau, J.Y. (1985) *The Ocean World.* Harry Abrams Inc, New York.

7. Miller, M.L. (1990) Tourism in the coastal zone: Portents, problems, and possibilities. In *Proceedings of the 1990 Congress on Coastal and Marine Tourism Vol* 1, eds M.L. Miller and J. Auyong. National Coastal Resources Research Institute, Corvallis, Oregon, 1-8.

8. Miller, M.L. and Auyong, J. (1991) Coastal zone tourism. A potent force affecting environment and society. *Marine Policy* (March), 75-99.

9. World Tourism Organisation (2000) *News from the World Tourism Organisation. Asia/Pacific Comes Back to Drive World Tourism.* World Tourism Organisation Web Site, http://www.worldtourism.org/pressrel/00 01 25.html, 25 January.

10. Shackley, M. (1996) *Wildlife Tourism.* International Thomson Business Press, London.

11. McKegg, S., Probert, K., Baird, K. and Bell, J. (1998) Marine tourism in New Zealand; A profile. In *Proceedings of the 1996 World Congress on Coastal and Marine Tourism (19-22 June 1996, Honolulu, Hawaii),* eds M.L. Miller and J. Auyong, Washington Sea Grant Program and School of Marine Affairs, University of Washington, Seattle, Washington and Oregon Sea Grant College Program, Oregon State University, Corvallis, Oregon, 154-159.

12. Houston, JR. (1998) The economic value of U.S. beaches. In *Proceedings of the 1996 World Congress on Coastal and Marine Tourism (19-22 June 1996, Honolulu, Hawaii),* eds M.L. Miller and J. Auyong, Washington Sea Grant Program and School of Marine Affairs, University of Washington, Seattle, Washington and Oregon Sea Grant College Program, Oregon State University, Corvallis, Oregon, 329-332.

13. West, N. (1990) Marine tourism in North America. In *Recreational Uses of Coastal Areas,* ed P. Fabbri. Kluwer Academic, Dordrecht, 257-274.

14. Gabbay, R. (1986*) Tourism in the Indian Ocean Island States of Mauritius, Seychelles, Maldives and Comoros.* The University of Western Australia and National Centre for Development Studies, Islands Australia Project.

15. Sathiendrakumar, R. and Tisdell, C.A. (1990) Marine areas as tourist attractions in the southern Indian Ocean. In *Proceedings of the 1990 Congress on Coastal and Marine Tourism Vol* 1, eds M.L. Miller and J. Auyong. National Coastal Resources Research Institute, Corvallis, Oregon, 78-88.

16. Jones, B.L. (1993) The emerging undersea leisure industry. *Sea Technology* (February), 38-42.

17. Fabbri, P. (ed) (1990*) Recreational Uses of Coastal Areas.* Kluwer Academic, Dordrecht.

18. Kenchington, R.A. (1990) Tourism in coastal and marine environments: A recreational perspective. In *Proceedings of the 1990 Congress on Coastal and Marine Tourism Vol* 1, eds M.L. Miller and J. Auyong. National Coastal Resources Research Institute, Corvallis, Oregon, 23-34.

19. Tabata, R.S. (1990) Dive travel - a case study in educational tourism: policy implications for resource management and tourism development. Paper presented at *The Global Classroom Symposium.* Christchurch, New Zealand (August), 19-22.

20. Dignam, D. (1990) Scuba gaining among mainstream travellers. *Tour and Travel News,* (26 March).

21. Davis, D. and Tisdell, C. (1995) Recreational scuba-diving and carrying capacity in MPAs. *Ocean and Coastal Management 26(1),19-40.*

22. Wilks, J. (1993) Calculating diver numbers: critical information for scuba safety and marketing programs. *SPUMS Journal* 23(1), 11-14.

23. Wright, J.W. (1990) *The Universal Almanac.* Andrews and McMeel, New York.

24. Griffin, R. (1992) Threatened coastlines. *CQ Researcher* 2(5), 97-120.

25. Scherl, L.M. (1987) Our need for wilderness: a psychological point of view. *Habitat Australia 15(4),* 32-35.

26. Leopold, A. (1949) *A Sand County Almanac.* Oxford University Press, Oxford.

27. Orams, M.B. (1993) Towards a marine conservation ethic: Our marine protected areas can lead the way. *Trends* 30(2), 4-7.

CHAPTER 9. MANAGING MARINE WASTE DISPOSAL

R.C. BALLINGER

1. Introduction

"The sea washes away all the ills of men."
Euripedes

The earliest maritime civilisations recognised the relative ease of disposing of wastes at sea. However, it was the rapid industrialisation and urbanisation of the nineteenth century and associated concerns over sanitation and health issues, which saw the development of sewage collection systems on a vast scale [1], and which established the offshore disposal route as a *bone fide* option for modern society. Although this view has been contested recently and the offshore route for many types and methods of disposal has been gradually eliminated, marine waste disposal, associated with the increasing littoralisation of the world's population, still remains a significant issue. Persistent problems associated with land-based sources of pollution, along with other issues including the over-exploitation of coastal resources, continue to degrade the global marine environment. [2] At the start of a new Millennium it is timely to review marine waste disposal and assess the factors which may influence its application and impact on the marine environment and future generations.

This chapter provides an overview of marine waste disposal practice and management, focusing on the evolution and relationship between perception and practice of marine disposal, particularly during the last half century. The dimensions of the marine waste issue are explored first: a brief description of the types and characteristics of marine waste is provided along with a discussion of the geography of marine waste including population and environmental impact issues. In the following section key developments associated with the evolution of general environmental management, waste management and pollution control are explored, including the extent to which the Precautionary Principle has been adopted. The final discussion provides an analysis of current, emerging and future issues, which will dominate discussions and development of marine waste management well into the current century.

2. The Waste Issue

2.1 WASTE CHARACTERISATION

Although only three basic approaches to ocean waste disposal have been recognised, namely containment (e.g. disposal in containers), decomposition (e.g. sewage treatment) and dilution (e.g. ocean dumping), [3] marine waste itself encompasses a plethora of substances from a wide range of sectors and activities. Figure 1 provides a characterisation of marine wastes highlighting aspects relevant to waste management. The waste source, described in terms of the sector or activity generating the waste, is particularly important to identify. With each sector frequently forming a separate administrative and management sub-system, strong sectoral controls are a vital component of general waste and pollution management. For example, port waste reception facilities are not only essential to port waste and environmental management, but also determine the quality of the surrounding marine environment. In the Wider Caribbean region this is particularly relevant: here ship-generated waste accounts for approximately 80% of solid wastes in coastal areas. [4]

In providing a relevant description of marine waste for management purposes, it is important to specify the components or the 'fingerprint' of the waste output, as many wastes, such as sewage, are complex mixtures of different chemical and biological substances. Each of these influences the longevity and severity of the resultant impacts of the waste output on the receiving environment as a result of differing toxicities and levels of persistence. Extensive scientific effort has been targeted at understanding the pollution characteristics as well as the environmental and public health implications of these components. This has led to specific pollution controls for particular pollutants or groups of pollutants. Related legislation, such as the European framework directive on the discharge of certain dangerous substances into the aquatic environment (76/464/EEC) and its daughter directives, have had a profound influence on offshore waste disposal practice.

Finally, it is essential to define the category of entry of the wastes to the marine environment (Figure 1). Offshore waste disposal practice has been largely regulated through control systems targeting these entry points. For example, much early international legislation and associated management effort relating to offshore waste disposal, such as that associated with the global London Dumping Convention (1992) and 'regional' sea legislation, focused on offshore dumping. Currently, global programmes, notably the recent Global Programme of Action to protect the Marine Environment against Land-based Activities, focus on other entry points as a result of increased concern over less tangible non-point, coastal and land-based catchment sources (Section 4).

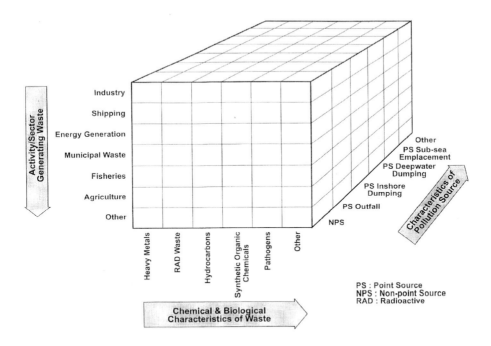

Figure 1. Waste Categorisation

2.2 GEOGRAPHY OF MARINE WASTE

The global geography of marine waste disposal is determined largely by population distribution, population growth and levels of development. With over a third of the world's population residing within 100km of the coast and sixty of the world's largest cities being coastal,[5] the pressures on the coastal, including the nearshore global environment, are significant and rising.[6] Development-related impacts, including those linked to waste and pollution, affect about half of the world's coast. [7,8] Littoralisation of populations is a phenomenon of both More Developed Countries (MDC) and Less Developed Countries (LDC): it is estimated that 50% of the North America population is coastal [9] and that 500 million people are concentrated in urban coastal areas in Asia.[10]

Although population growth rates have been falling steadily throughout most of the developing world since the 1980s, the massive population base still translates into much additional waste resulting from even moderate growth rates. The problem is compounded by urbanisation: many urban areas in Latin America have doubled their population at least every fifteen years. [11] Urbanisation poses significant local issues for waste disposal in LDCs. It frequently results in sewage management shifting from reclamation to disposal, with little or no treatment, and from resource use to resource waste. [12] For example, Calcutta dumps 400 million metric tons of raw sewage and other wastes into the Hooghly Estuary causing significant biological impacts. [13] In MDCs, however, higher standards of living result in higher quantities and more complex waste products. Barrow [14] estimates that urban waste production amounts to between 500 to 8000 tonnes per day in developed cities. However, increased affluence and more environmentally aware constituencies result in significant capital expenditure being devoted to waste management: Jin and Kite-Powell [15] report that New York City spent over US $160 per ton of sewage sludge in 1994 and currently are spending US $48.15 per ton for ash management. Consequently, neo-Malthusian perspectives on population waste impacts, though useful, are limited: linkages are complex and include socio-economic, technical and political factors.

3. The Context for Marine Waste Management: General Issues and Trends

The development of environmental management theory and practice along with the parallel evolution of general waste management and pollution control principles and practice have had a profound impact on offshore waste disposal. This section explores the relative influence of these and explores the perceptions and issues, which have shaped current offshore waste management practice.

3.1. EARLY TRENDS

Although the modern concept of solid general waste management came about through health and sanitation issues in the late nineteenth century most marine waste concerns in the early to mid twentieth century revolved around safety and other strategic issues rather than aesthetic, health or environmental matters. For example, regulations controlling the discharge of oily wastes from ships were only introduced during WWI once it was realised that harbour fires caused by oily ship discharges were disrupting the war effort in western Europe and on the east coast of America. [16] Even as late as 1958 Convention on the Continental Shelf established that disused offshore installations should be removed from the marine environment to prevent hazards to shipping and future legitimate uses of the sea rather than referring to any specific environmental concerns *per se*.

3.2. AWARENESS AND INSTITUTIONALISATION OF THE ENVIRONMENT

During the late 1960s and early 1970s there was an intense general debate on the relationship between population growth and resource use, fired by Hardin's essay on the *Tragedy of the Commons* [17] and the Club of Rome's report. [18] Within such discussions the vulnerability of finite environments, including the marine environment, became important elements. Well-publicised pollution incidents including the *Torrey Canyon* in 1967 demonstrated the interconnectivity of the environment and the transboundary nature of pollution. These provided the catalyst for international meetings to develop policy, regulation and an 'international infrastructure' to deal with such events. The United Nations Conference on the Human Environment (1972) was one of the earliest of these, focusing on the condition of the global, including the marine environment. This was swiftly followed by the establishment of the United Nations Environment Programme (UNEP) and the Regional Seas Programme [19] within which marine pollution and waste management became key elements. [20]

Since this early period environmental management has evolved dramatically. It has moved away from simplistic, reactionary approaches to more mature proactive, precautionary and participatory ones, emphasising stewardship and multidisciplinary, holistic views of environmental problems and solutions. The concept of sustainable development has emerged, initially promoted by the World Conservation Strategy [21] and the Brundtland Report, *Our Common Future* [22] and more recently by the work of the World Commission on Environment and Development [23] and the UN Conference on Environment and Development (UNCED) process. Despite the concept's inherent ambiguity and difficulties in operationalisation, organisations at all levels of governance have signed up to it. [24] Some of the related principles, notably the Polluter Pays and the Precautionary Principle, have major implications for marine waste management (Section 3.5). In the last decade *Agenda 21*, Chapter 17 also focused attention on the need for state involvement in the prevention, reduction and control of degradation of the marine environmental as part of a global programme of action for sustainable development.

3.3. GENERAL WASTE AND POLLUTION CONTROL TRENDS

During the last half century considerable effort has been devoted to refining definitions of waste and waste management as a result of appreciation of the wider issues associated with waste disposal and resource use. These include those associated with land availability for land-based disposal, pollution impacts, cost implications of waste treatment and the energy potential of waste products. [25] With heightened awareness about environmental matters and resource efficiency, there has been a gradual shift from waste reclamation and disposal towards waste prevention and integrated waste management, including life cycle analysis and cradle-to-grave technologies. [26,27] A preferred waste hierarchy has been suggested, which defines the following methods in priority order:

- waste reduction/minimisation
- waste recycling, recovery and reuse
- waste treatment and disposal

3.3.1. Waste reduction and minimisation
Reduction of waste remains a major issue, particularly in MDCs where consumerist life styles result in high quantities of garbage per capita. [28] A recent report revealed that the total output of wastes and pollutants in Austria, Germany, Japan, the Netherlands, and the USA increased by 28 percent between 1975 and 2000 despite increasing efficiency in natural resource use. [29] Ironically, end-of-pipe improvements sometimes lead to increased waste production at other points in the waste stream and consequent problems for waste management. For example, additional volumes of sewage sludge from higher level treatment of municipal wastes, along with the phasing out of offshore sewage

sludge dumping (Section 3.5), resulted in a significant waste disposal problem for many countries at the end of the 1990s. This was particularly true for those, such as the UK, where there was only a limited range of alternative options, notably landfill, incineration or 'recycling'. [30]

3.3.2. Waste recycling, recovery and reuse

Waste recycling is frequently deterred by high-energy requirements and is dependent on specific waste stream characteristics. Consequently, its use is limited even in MDCs. For example, only 5% of Japanese and 10% of US plastics, and 50% and 21% of municipal wastes in each of these countries were recycled in the late 1990s. [31] The frequent mixing of industrial and domestic waste streams, particularly in urban and industrial areas with old sewerage systems, has precluded the total re-use of the sewage sludge waste on land in agriculture and forestry, with dredging spoil becoming one of the last types of waste disposed offshore and a contentious issue around many hub-ports in MDCs, [32] attention has focused on alternative uses of this material, including foreshore and nearshore nourishment, construction, landfill, reclamation, habitat reclamation and derelict land rehabilitation. However, dredging material is frequently contaminated and its treatment remains expensive: for example, costs between £60 and £120 per tonne were quoted for the treatment of wastes in Hamburg in the mid-nineties. [33] Despite this, specialist techniques continue to be developed so that, in some countries, such as the Netherlands, high percentage recoveries (80%) of reusable materials are achieved. [34] In addition, there are some promising signs in the chemical industry, where healthy profits can be gained from recycling wastes, such as titanium dioxide, which were previously disposed of at sea.

3.3.3. Waste treatment and disposal

In the context of marine waste disposal, significant attention remains focused on the bottom end of the hierarchy, particularly on waste treatment, disposal routes and points of entry to the environment. Improvements in waste treatment have frequently been driven by concerns over the deleterious effects of waste disposal on the environment and by the need for specific water quality objectives. For example, recently the European *Urban Waste-Water Directive* (91/271/EEC) stipulated minimum levels of treatment required for sewage wastes from coastal settlements in response to concerns over nutrient and other pollution inputs into coastal waters. However it is the legislative programmes, which control water quality directly, which have had the most impact on waste management practice, particularly in MDCs. Many have arisen through concerns over specific pollutant impacts on coastal amenity and use. The European *Bathing Waters* (76/160/EEC and COM (2000) 860) and *Shellfish Waters* Directives (79/923/EEC) are particularly noteworthy, establishing uniform water quality objectives for bathing and shellfish water across Europe. These directives developed amidst heightened concern about the health risks of swimming or harvesting shellfish in areas contaminated by urban waste-water, provoked by effective media campaigns by pressure groups, such as Surfers Against Sewage and the Marine Conservation Society in the UK. Although these directives have been fraught with problems, [35] they have been a significant impetus for sewage treatment improvements in Europe over the last few decades.

3.3.4 The solution to pollution is dilution

The notion that the 'solution to pollution is dilution' has frequently resulted in discussion over the diluting and assimilative capacity of the offshore environment. Many early guidelines, offshore permitting systems and related monitoring programmes relating to offshore dumping (GESAMP and the US EPA) were based on crude estimations of the capacity of offshore waters to assimilate wastes without significant adverse effects to the environment as well as the identification of legitimate environmental uses of the nearshore environment, including shell-fisheries and offshore recreation. However, the early development of coastal water quality legislation was fraught with doubt over the validity of such an approach. Within Europe, the UK was always in a minority of one, with respect to its view on maximising the assimilative capacity of coastal waters to determine discharge consents. [36] As a result of Britain's intransigence, a special dispensation was made to the UK for the framework *Directive on the discharge of certain dangerous substances into the aquatic environment* (76/464/EEC). This allowed the UK to set its own discharge standards based on local assimilative capacities, whereas the remaining eight Member States, preferred and were obliged to set Uniform Emission Standards. More recently, however, the varying assimilative capacities of the European offshore environment have been more widely recognised. The European *Urban Wastewater Directive* (91/271/EEC), for example, has instructed Member States to define high dispersion and sensitive areas. The former, less sensitive areas with good water exchange, require less stringent treatment standards whereas the latter, characterised by poor water exchange and the possible build up of nutrients, necessitate tertiary treatment.

3.3.5 Points of entry to environment

The recent introduction and application of the cross-media, integrated approach to pollution control should be mentioned. This seeks to minimise total pollutant inputs into the environment across all sectors and has been the preferred approach for the control of pollution from hazardous operations within many MDCs over the last decade. However, for offshore waste disposal, most attention has been devoted to targeting waste disposal sites away from sensitive environments and activities as well as to choosing sites maximising dilution and dispersive capacities. Prior to recent bans on ocean dumping, selection of offshore dumpsites was based on these principles. Initial reactions to concerns about the validity of offshore dumping commonly included movement of disposal sites further offshore, where there were thought to be larger assimilative capacities and fewer sensitive human uses. For example, the US Deep-water Municipal Sludge Sites were seen as the solution to the mid-Atlantic site (off Delaware Bay) and the 12 Mile sewage sludge dump sites in New York Bight. [37,38] Similarly, long sea and deep ocean outfalls replaced short sea outfalls to maximise the greater dilution and mixing capacities of the offshore receiving waters. A deep ocean outfall off Hawaii, for instance, replaced a municipal outfall into Kanoehe Bay in the late 1970s to reverse coral reef decline associated with the disposal of secondary sewage effluent into the poorly flushed waters of the local lagoon. [39] Even current debates on municipal wastewater disposal frequently revolve around discussions over points of entry. In the UK for example the relative merits of long-sea outfalls versus high land-based treatment of municipal wastewater have recently been considered by a House of Commons Environment Select Committee. [40] However, with increasing concern over beach and coastal water quality in the UK, many companies have opted for both higher treatment levels and longer outfalls.

3.4. INSTITUTIONALISATION OF WASTE MANAGEMENT

3.4.1. Global and regional institutions

Since the 1970s a large number of global and regional institutions have developed to address the offshore waste issue. These include UN programmes, processes and intergovernmental activities as well as major international conventions, such as the London Dumping Convention (1972), MARPOL (1973/78) and the Paris and Oslo Conventions for the north east Atlantic (Figure 2). In addition to the comprehensive environmental legislative framework and perspective provided by UNCLOS and UNCED for the global ocean, [41] these conventions, evolving in response to changing perceptions and scientific evidence, have provided the backbone of international controls on offshore waste disposal. They have been particularly useful in identifying practical steps to prevent, reduce and control land-based marine pollution, providing guidance, relevant knowledge and experience for national and regional levels of administration. [42] Under the Waste Assessment Framework (Annex 2) of the London Convention for example, there is guidance on waste prevention audits, waste management options, waste characterisation, dump site selection and permit conditions as well as on assessment and monitoring of potential effects. The conventions have played a key role in encouraging international and other donors to prioritise marine pollution projects and have emphasised the need for a national commitment to marine waste management, both in terms of actions and funding. [43] The role of these institutions is considered further in the following section (Section 3.5) issues surrounding their current and future functions are then outlined in Section 4.

3.4.2 Sectoral institutionalisation of waste management

Over the last few decades, it has become clear that there is a need to move away from an entirely sectoral approach to environmental management and protection to one, which influences the policy, and practice of other key marine and coastal sectors. [44] Only when environmental and waste management has become institutionalised within these constituent sectors can holistic and sustainable waste management be achieved. Despite the lack of involvement of many sectors in waste management, some, such as the cruise and ports industries with major roles in offshore waste management, have begun to address the issue in some depth. Various management tools are being used and developed by these sectors, including Environmental Management Systems, Total Quality Management and Self-reporting. [45,46] For example, the Caribbean cruise industry, which produces three quarters of all ship-generated waste in the wider Caribbean region, [47] has been encouraged to adopt improved waste management practices as a result of stricter international and US legislation, greater public scrutiny and growing awareness of environmental issues. Many cruise lines now voluntarily adhere to comprehensive waste management strategies emphasising source reduction and recycling and several cruise lines have voluntarily adopted 'zero discharge' rule for solid waste into the sea. [48] Similarly, many ports have instigated their own waste management schemes and environmental auditing. Wooldridge [49] provides a useful synthesis of the efforts of the European port sector.

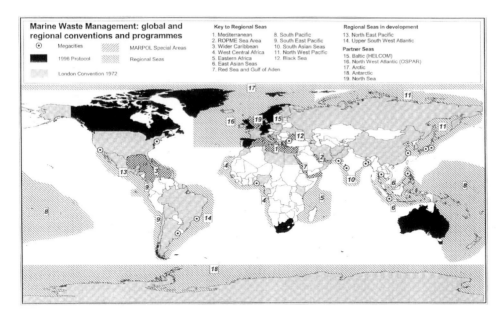

Figure 2. Marine Waste Management Conventions and Programmes

3.5. THE PRECAUTIONARY APPROACH AND MARINE WASTE DISPOSAL

Heightened public awareness and a catalogue of incidents, including algal blooms off Long Island, US and medical waste wash-ups along the New Jersey shore in 1987, [50] resulted in mass public rejection of most offshore waste disposal routes in the 1980s. Initially a move further offshore allayed fears. However, many revised offshore disposal routes were short-lived. Despite extensive monitoring, [51] there were substantial uncertainties surrounding the nature and significance of environmental damages associated with offshore disposal, particularly for marine dumping and incineration. [52,53,54] There was increasing concern over the diffusability of wastes and transboundary transfer of contaminants. Regarding waste inputs in deep-sea environments there were also mounting fears over the comparative difficulties of monitoring impacts and the irretrievability of the wastes, particularly for radioactive and other hazardous waste disposal. [55]

Some authors suggest that economic factors and the need to allay mounting public opposition were the most significant influences on offshore waste disposal practice. [56] The high profile media campaigns of Greenpeace and other environmental NGOs of the late twentieth century are particularly noteworthy. Reference should be made to Peet [57] for an informed discussion of their impact on the evolution of the London Dumping Convention for example. Through persistent, sophisticated and often emotive campaigning, driven by a belief that ocean dumping and incineration are 'primary disincentives to clean technologies and clean production methodologies', [58] they have been a major catalyst to the abandonment of many offshore waste disposal options. Such campaigns, though sometimes scientifically uninformed, have been able to mobilise wide-scale opposition and influence both government and industry decision-making. Examples include the ban on US ocean incineration in the mid 1980s which resulted from a strong region-wide coalition of small coastal interest and environmental groups [59] and the more recent victory of Greenpeace in relation to the disposal of the *Brent Spar*.

Amidst mounting public concern and increasing uncertainty regarding the implications of offshore disposal, the 1980s and early 1990s saw the first significant abandonment of ocean waste disposal. International, regional and national legislation and agreements were reformed ensuring that the Precautionary Principle became enshrined in post-modern marine waste management practice. Resolutions of the London Convention in the 1990s included the phasing out of industrial waste disposal (Resolution LC.49 (16)) and the placing of moratoria on incineration (Resolution LC.50 (16)) and radioactive waste disposal at sea (Resolution LC.51 (16)). Similarly, at a regional scale, the North Sea Inter-ministerial conferences phased out industrial and sewage offshore waste dumping as well as offshore incineration in the early 1990s. Several 'veteran' conventions dealing with offshore waste issues were also renamed and reformulated. Notable examples include the renaming of the London Dumping Convention to become the London Convention and the establishment of OSPARCOM, the convention for the Northeast Atlantic, which replaced the Oslo and Paris Conventions, addressing offshore dumping and land-based sources, respectively.

The 1996 Protocol of the London Convention is one of the most significant milestones in international regulations, enhancing the application of the precautionary approach and the polluter pays principle. The 'Reverse list' in Annex 1 replaced the former black and grey lists of the parent convention so that there is now a blanket prohibition (except in emergencies) of most materials rather than just restrictions on the dumping of selected toxic and persistent substances. Only a short list of relatively unharmful, inert and non-persistent wastes, which could possibly be considered for offshore dumping, has been included. The new Protocol has also closed a possible loophole relating to the sub-seafloor disposal of waste by redefining offshore waste disposal to include 'any storage of wastes or other matter in the seabed and the subsoil'. However, only a small number of signatories to the convention have agreed to the new Protocol since its inception in 1996 (Figure 2).

4. Discussion: Current, Emerging and Future Marine Waste Issues

4.1. CURRENT ISSUES

4.1.1. Pollutant sources
Concern revolves around the quantification of waste inputs into the marine environment especially for non-point sources and particularly for LDCs where monitoring is less developed. Over the last decade land-based and particularly non-point sources of marine pollution have taken on a greater relative importance [60,61,62,63] with recent estimates

indicating that between 75 – 85% of all marine pollution inputs come from land-based sources. [64] Nutrients and bacteria from urban and agricultural runoff and from municipal wastewater discharges pose significant problems: eutrophication is now considered a major priority at both regional and global scales. [65] Although most municipal wastewater discharges in MDCs receive secondary treatment prior to disposal, nearshore and catchment disposal via Combined Sewer Overflows (CSO) along with excessive nutrient loading from fertiliser use continue pose potential dangers to human health, food supplies and sensitive coastal ecosystems. [66] In LDCs problems arise not only through the sheer volume of sewage wastes, but also the inadequacy and poor operating conditions of sewage treatment facilities. Increased fertiliser consumption in many developing countries associated with the expansion of modern agricultural practices is also causing significant deterioration in water quality.

Considerable effort is being made to attempt to control these pollution sources at all levels of decision-making. At a global level, the *Global Programme of Action (GPA) for the Protection of the Marine Environment from Land-Based Activities* 1995 is noteworthy. This has developed a clearing-house mechanism for national and regional decision-makers, providing access to current sources of information, practical experience and scientific and technical expertise. As many problems are cross-state issues, use of existing regional sea networks is being sought. However, for seas, such as the Black Sea, where the drainage network extends far inland through several land-locked sovereign states, the political and technical challenges are immense. Even where there are existing and well developed non-point source pollution programmes, such as in the US, eutrophication remains an issue for sensitive coastal waters. Under the reauthorised US Coastal Zone Management Act lead agencies are required to identify critical coastal areas and land uses that individually or cumulatively may cause or contribute to coastal water quality degradation.

As noted above (Section 3.3) offshore dumping of dredging spoil remains one of the last remaining waste products dumped offshore. However, this practice is increasingly coming under scrutiny as concern over the disposal of potentially contaminated dredged material reaches critical levels. [67] The US Environment Protection Agency has recently indicated that approximately 60 million cubic yards of the sediment dredged annually from US waterways is disposed offshore at designated sites. [68] However, public pressure to stop this activity continues to escalate. The economic consequences of these controversies can be far-reaching. For example, it has been suggested that some maritime commercial activity left affected US ports, particularly the Port of New York / New Jersey, in the late 1980s and early 1990s as a result of inaction in maintenance dredging because of environmental concerns. [69,70]

The decommissioning of redundant structures associated with the offshore hydrocarbon industry is no less contentious (Section 3.5). Over a thousand structures have been decommissioned, including many small structures in US waters in the Gulf of Mexico and some 30 small structures and sub-sea installations in the shallower waters of the Southern North Sea sector. [71] However, installations on the continental shelves of over fifty countries are scheduled for imminent decommissioning, including those installed during the heyday of hydrocarbon exploitation in the North Sea. [72] Although the activity is regulated by a formidable array of international, regional and national legislation, much of this, including the London Convention 2000 Guidelines, is recent and untested. Advanced technological expertise is required to ensure that environmental, economic, health and safety issues are suitably addressed. This is particularly so in the harsh and relatively deep waters of parts of the North Sea where some of the largest structures in the world will have to removed. To address this and other related issues, a North Sea Decommissioning Group, including representatives of the oil industry, operators and governments, has been established recently to develop technology and encourage co-operation and collaboration between interested parties. [73]

Other critical issues include marine litter, ballast water and the release of hormone-mimicking pollutants. Marine debris pollution affects even the most remote coastal locations, impairs many species of marine organisms [74] and causes massive losses in tourism and fishing revenue. It was estimated that US $1.5 million is spent annually in New Jersey removing debris from beaches and coastal waters. [75] A recent UK study reported a similarly large beach-cleaning bill for Local Authorities. [76] Although considerable efforts are targeted at pollution removal, litter sourcing and upstream waste reduction remain under-investigated. There are also many issues associated with the need to modify the operating systems and economics of other marine and coastal sectors to rectify the debris problem. There is considerable potential to improve garbage discharges from vessels, which are inadequate for a number of reasons, including the inadequacy of port reception facilities and the increasingly tight schedules under which many vessels operate. [77]

The transfer of non-native marine species within ballast water is not a new concern. The introduction of the infamous European zebra mussel *Dreissena polymorpha* into the North American Great Lakes system and its subsequent spread into 40% of US internal waterways has required over US$5 billion in expenditure on control measures since 1989. [78] With estimates that about 10 billion tonnes of water is transferred globally each year and over 3000 different species are carried in ships' ballast tanks at any one time, [79] the associated potential transfer of harmful and non-indigenous aquatic organisms poses a considerable threat. In response to a call from UNCED to the International Maritime Organisation (IMO) and other bodies to take action to address this issue in 1992, the IMO has adopted voluntary guidelines (IMO Resolution A.868 (20), 1997), recommending management and control measures to minimise the transfer of harmful aquatic organisms and pathogens. In addition, the Marine Environment Protection Committee of IMO has recently drafted a new global instrument [80] and the Global Ballast Water Management Programme (GloBallast), a US$10.2 million initiative under the Global Environment Facility, is helping IMO assist developing countries to implement suitable schemes. At a national and sub-national level, many countries and sub-national jurisdictions, including Australia, Canada, New Zealand, the USA, various individual States within the USA, as well as several individual ports around the world, including Vancouver, have unilaterally developed or are developing related legislation. [81]

4.1.2. Economic issues
Although considerable progress has been made advancing technology and improving water quality, economic issues still thwart many aspects of marine waste management. The lack of resources and investment to implement legislation is a major impediment. All too often limited resources hamper infrastructure building and institutional strengthening at sub-regional and national levels and improved education and public awareness at base-levels. [82] For example, the *Global Investigation of Pollution of the Marine Environment*, whilst providing for a suitable monitoring programme, has no resources for infrastructure-building and institutional strengthening. [83] Similarly, port waste reception facilities require substantial investments to strengthen the implementation of MARPOL 73/78. In some areas, such as the Caribbean, a 'Special Area' under Annex V of MARPOL 73/38, the problem is critical.

Problems frequently still lie in balancing economic and environmental concerns. These are compounded by the relatively naïve state of environmental economics and the concomitant difficulties in placing economic values on the invaluable 'intangibles' of the marine environment, such as clean water and a healthy coastal ecosystem. Consequently, the true costs of waste management decisions are still to be determined. However, for political reasons and as a result of simplified enforcement of legislation, increasing funds may be frequently dedicated to improving wastewater treatment and disposal even though there may be not be any definite 'added value' for the marine environment and human health. Decision-making associated with the Los Angeles disposal of municipal wastewater illustrates this clearly. Here, Garber and Gunnerson [84] maintain that environmental legislation has been based more on the financial equity than environmental efficiency.

4.1.3. Public awareness issues
The last few decades have witnessed an unprecedented increase in public environmental awareness in MDCs, where populations share a strong concern about the impacts of water pollution on human health, environmental quality and recreational water quality. [85] This can influence marine waste management decision-making as constituent populations opt for NIMBY approaches rather than the Best Practicable Environmental Option (BPEO). As shown earlier (Section 3.5), the public, via lobbying, campaigns and exercising their democratic voting rights, may relatively easily mobilise political processes and influence public policy and government decisions. This is nowhere more graphically illustrated than in the development of nuclear waste disposal policy. [86] However, public perception and the associated judgements of risk, may not necessarily be accurate and may be harnessed by media campaigns and the sensationalisation of pollution events. Moore, [87] for instance, has described the influence of a range of pollution incidents, including the algal booms off Long Island in 1987 and the sewage wash-ups of Massachusetts in the 1980s on the ODA amendments of 1988. [88] Considerable attention, therefore, needs to be devoted to this issue, which lies at the heart of marine waste management decision-making.

4.2. EMERGING AND FUTURE ISSUES

4.2.1 Emerging issues

In addition to the above issues, there are a number of emerging issues, which need addressing. There is the urgent need to improve the scientific and knowledge base for waste management decision-making. The lack of high quality data and harmonisation of data sets for ascertaining trends poses problems for understanding global processes. [89] These are compounded by the increasing uncertainty surrounding global environmental change, particularly climate change and sea level rise. Associated with this are further uncertainties regarding future oceanic and atmospheric circulation patterns [90] and the role of the ocean in balancing emissions and absorption of greenhouse gases. Although scientific understanding and technological expertise have always limited the development of new approaches to offshore waste disposal practice, there is now a much greater awareness of the implications of our limited knowledge and understanding. [91] Problems remain as out-dated pollution standards, bearing little resemblance to scientific risk, take considerable time, resources and political will to be updated. Although there are promising developments in this area, including the up-to-date specific assessment frameworks under development for each substance on the London Convention 'Reverse list'. [92] Apart from being able to define 'safe' waste limits, there is still a need to continually redefine waste itself, especially as new forms, such as decommissioned platforms, need consideration. Recently, there has been a lack of consensus regarding the definition of 'industrial waste: London Convention members have been taxed with whether or not the ocean disposal of carbon dioxide from fossil-fuelled power plants is a form of such waste.

Another emerging issue facing waste management is the relative contribution of offshore waste disposal in relation to other inputs. As noted above (Section 4.1), concern currently focuses on non-point sources, which are much more difficult to control, and on natural inputs. Recent studies of beach and coastal water quality have revealed considerable inputs of pathogens from natural sources and have shown that where there are very high levels of sewage treatment such sources may cause beaches still to fail to comply with environmental quality criteria. [93] It is also becoming increasingly evident that impacts of offshore waste disposal must be viewed in context of those associated with other marine and coastal uses and activities, particularly urbanisation, fishing and deforestation. In terms of management effort, overfishing, resource degradation and reduction in biodiversity, are priority issues. As many of these are induced by a complex web of direct and indirect impacts of human activities, understanding these inter-relationships, including the relative impact of waste disposal, is a key area of work.

There are also some important emerging institutional and political issues. Despite the positive contribution of the international conventions (Section 3), critics maintain that there is frequently insufficient attention given to their implementation at regional and particularly national levels. [94] The most sceptical even maintain that their very presence has given legitimacy to offshore dumping in the past. [95] All too often, the focus is on monitoring frameworks with limited or no resources for implementation (Section 4.1). The relationship between some of the legal frameworks, notably between UNCLOS and the London Convention, is also complex and as a result problematical.[96] In addition, many regional seas programmes, though relatively successful in gaining regional co-operation and in promoting environmental quality and protection, have been hampered by the limited involvement of major sectors, notably tourism, fisheries and shipping.[97] The lack of focus and expertise of some of the international institutions and even advisory bodies such as GESAMP on socio-economic aspects of waste and pollution is also of concern.[98] However, it is frequently contended that rather than the contents of any particular international convention, it is the philosophy and practice of the United States of America (US) which provides the lead for other nations in offshore waste management.[99] Certainly, many early provisions concerning ocean dumping found in international law were championed by this and other leading users of the ocean for waste disposal.[100] However, this has been much less so recently. Efforts to phase out sewage sludge from the London Convention's 'Reverse List' were rejected by, amongst others, the US even though it had agreed a prohibition on ocean disposal of sewage sludge on its own territory within the US Ocean Dumping Act.

4.2.2 Future issues
Looking to the future, the most significant issue for offshore and indeed waste disposal is the alarming trend in global population growth coupled with the widening economic gap between LDCs and MDCs.[101] With the largest projected rise of population in littoral areas waste management is likely to remain high on the agenda of coastal states. It is estimated that by 2025 75% of the world's population, or 6.3 billion people, will live within coastal areas.[102,103] In MDCs, with approximately half to three-quarters of resource inputs returned annually to the environment as waste,[104] population and economic growth will drive the waste management process. For coastlines of some LDCs, where the offshore environment is already a tempting resource for waste disposal usage, population increase is set to rise most. Along the Western and Central African coast population increase will contribute to the formation of a continuous chain of coastal cities, with potentially significant waste problems, along the 1000km of the Gulf of Guinea.[105] The increasing divergence of income between LDCs and MDCs, exemplified by the doubling of the gap between the income of the world's richest and poorest 20% of the population between 1960 and 1995,[106] could preclude the LDCs from technological advances. This ever-widening gap could even destabilise the global political environment in which waste management decision-making is made.[107]

Apart from these global trends, wastes containing cocktails of new, synthetic pollutants with unknown and synergistic and antagonistic effects will continue to test waste management technology. Further challenges may also arise dealing with the legacy of former seabed activities, notably the covert disposal of munitions, chemical and radioactive wastes. Gourlay[108] estimates, for instance, that there have been at least fifty nuclear weapons lost at sea and McIntyre[109] singles out the dumping by the former USSR in the Kara and Barents Sea as being of particular concern. The nuclear problem, however, is not only a legacy from the past. It has been estimated that decommissioning of obsolete UK nuclear power stations, most of which are coastal, will generate a bill in excess of £30 billion.[110] Barrow[111] also notes that there were at least 356 nuclear-power generation reactors in over thirty countries, operating or under construction in 1989. Finally, the potential of future waste disposal technologies, particularly abyssal floor emplacement and carbon dioxide waste disposal, though still largely the domain of academic speculation and study [112,113,114,115,116,117] must not be overlooked. With oceans of over 3000m depth covering half of the Earth's surface [118] there are suggestions that abyssal sites may provide reduced risk for future waste management. Although Jin and Kite-Powell [119] estimate that the potential savings for New York could be up to $60 million a year, the uncertainty associated with the external costs of such schemes is currently so large to preclude their use. [120]

5. Conclusions: Final Comments

Over the last few decades there has been a remarkable shift in opinion and practice relating to offshore waste dumping. In 1985 the UK Royal Commission on Environmental Pollution recommended that marine disposal of waste, particularly sewage, was the 'Best Practicable Environmental Option'. [121] Just over ten years later, the 1996 Protocol of the London Convention established its 'Reverse List': a blanket prohibition of all materials, apart from a short list of substances, to be dumped offshore. Certainly, offshore waste disposal is now either highly regulated or absolutely prohibited in many parts of the world. [122] With the rise of sectoral waste management and pollution control efforts and the use of more refined science and technology, significant progress has been made over the last quarter century. Such efforts have been initiated through mounting concern over the deterioration of the marine environment and the uncertain, but potentially irreversible environmental damages associated with offshore waste disposal.

The largest challenges for offshore waste management remain the interdependence and inter-relatedness of ocean and coastal activities and the associated complexity of economic, cultural, technological and environmental interactions, which shape waste policy decisions. These make it a necessary to move from the sectoral approach, which pervades current waste management to a more integrated and holistic approach (Figure 3). Although educational, economic and institutional structures and theories are insufficiently advanced to deal with this, [123] it is clear though that legal instruments and institutions must try to take account of these interactions. [124] There is limited evidence to suggest that this is already taking place. For example, the latest North Sea reports of Inter-ministerial Conferences have shown a much more holistic view of the situation, not only covering marine pollution issues, but also fisheries impacts and requirements for integrated coastal management. [125] At local levels there are a few notable examples of dedicated and holistic waste programmes, such as that for the Puget Sound, and related programmes linking catchment and coastal management, such as the Chesapeake Bay Program.

Finally, although current focus revolves around the quantification and management of waste inputs from land-based and non-point sources, looking further into the future, the use of the offshore and particularly the deep sea environment for dedicated waste disposal is not an impossibility. Indeed, the ocean's cleaning capacity is both large and of considerable economic value. [126] The public is also becoming increasingly aware of the environmental, social and economic costs of all waste options, including land-based alternatives. So, with global warming and the growth of the human population remaining as two very large and significant imponderables, [127] it is vital that continued attention and debate is devoted to the offshore option to waste disposal and its management.

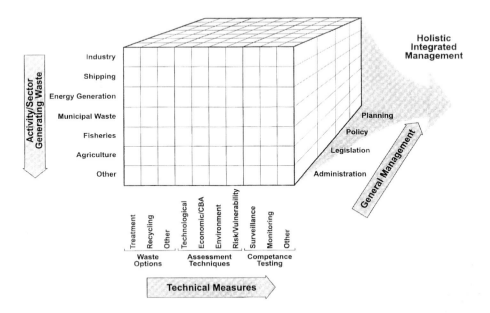

Figure 3. The Holistic Approach to Waste Management

References

1. Nihoul, C.C. (1991) Dumping at sea, *Ocean and Shoreline Management*, **16**, 313 – 326.

2. Joint Group of Experts on the Scientific Aspects of Marine Pollution (GESAMP) (1998) *Report of the 28th session*, Geneva, Switzerland, 20 – 24 April 1998, Rep. Std. no. **66**, 1998.

3. Duedall, I.W., Ketchum, B.H., Kilho Park, P. and Kester, D.R. (1983) Global inputs, characteristics and fates of ocean-dumped industrial and sewage wastes: an overview, In: Duedall, IW, Ketchum, BH, Kilho Park, P and Kester, DR (1983) *Wastes in the ocean*, Volume 1: Industrial and sewage wastes in the ocean, John Wiley, New York.

4. McCreary, S., Twiss, R., Warren, B., White, C., Huse, S., Gardels, K. and Roques, D. (1992) Land use change and impacts on the San Francisco Estuary: a regional assessment with national policy implications, *Coastal Management*, **20**, 219 – 253.

5. Cohen, J.E., Small, C., Mellinger, A., Gallup, J. and Sachs, J. (1997) Estimates of coastal populations, *Science*, **278**, 5341, 1211 – 1212

6. Kullenberg, G. (1999) Approaches to addressing the problems of pollution of the marine environment: an overview, *Ocean and Coastal Management*, **42**, 999 – 1018.

7. World Resources Institute (1995) *World Resources Institute brief: coastlines at risk: an index of potential development-related threats to coastal ecosystems*, Washington DC.

8. Hinrichsen, D. (1998) *Coastal waters of the world: trends, threats and strategies*, Island Press, Washington DC.

9. United Nations Environment Programme (1999) *Geo-2000: UNEP's Millennium report on the environment*, Nairobi: UNEP.

10. World Resources Institute, ICLARM, WCMC and United Nations Environment Programme (1998) *Reefs at risk: a map-based indicator of threats to the world's coral reefs*, World Resources Institute, Washington DC, US.

11. Gilbert, A. (1990) *Latin America*, Routledge Introductions to Development Series, London: Routledge.

12. Barrow, C.J. (1999) *Environmental management: principles and practice,* Routledge Environmental Management Series, London.

13. Hinrichsen, D. (1998) O*p. Cit.*

14. Barrow, C.J. (1999) *Op. Cit.*

15. Jin D. and Kite-Powell, H.L. (1998) Cost assessment for abyssal seafloor waste isolation, *Journal of Marine Systems*, **14**, 289 – 303.

16. Anon (1919) *Journal of Commerce and shipping telegraph*, 26th August.

17. Hardin, G. (1968) The tragedy of the commons, *Science*, **162 (3859)**, 1243 – 1248.

18. Meadows, D.H., Meadows, D.L., Randers, J. and Behrens, W.W. III (1972) *The limits to growth* (a report to the Club of Rome's project on the predicament of mankind), Universal Books, New York.

19. Hulm, P. (1983) *A strategy for the sea: the Regional Seas Programme; past, present and future,* UNEP, Geneva.

20. Keckes, S. (1991) The Regional Sea Programme. Integrating environment and development in the next phase. Paper presented at the *Pacem in Maribus XIX*, Lisbon, Portugal, 18 - 21, Nov, 91

21. IUCN, UNEP and WWF (1980) *World Conservation Strategy: living resource conservation for sustainable development*. International Union for Conservation of Nature and Natural Resources, Gland.

22. World Commission on Environment and Development (1987) *Our Common Future* (Brundtland Report), Oxford University Press, Oxford.

23. IUCN, UNEP and WWF (1991) *Caring for the Earth*, Earthscan, London.

24. Bishop, K. D. and Flynn, A. C. (1999) The National Assembly for Wales and the Promotion of Sustainable Development: Implications for Collaborative Government and Governance, *Public Policy and Administration*, **14(2)**, 62-76.

25. Tammemagi, H. (1999) *The waste crisis: landfills, incinerators and the search for a sustainable future*, Oxford University Press, Oxford and New York.

26. Young, J. (1990) *Post Environmentalism*, Belhaven, London.

27. Bradshaw, A.D., Southwood, R. and Warner, F. (Eds) (1992) *The treatment and handling of wastes*, Chapman and Hall, London.

28. Tammemagi, H. (1999) *Op. Cit.*

29. World Resources Institute (2000) *The weight of nations: material outflows from industrial economies*, World Resources Institute, Washington DC, US

30. House of Commons Select Committee on the Environment, Transport and Regions (1998) Second Report, House of Commons, UK, Ref. no.: 171/98.

31. Tammemagi, H. (1999) *Op. Cit.*

32. Copeland, C. (1999) *Ocean Dumping Act: a summary of the law*, Congressional Research Service, Report for Congress, RS20028, January 22, 1999, National Council for Science and the Environment, http://www.cnie.org/nle.

33. Anon (1995) *Seminar on disposal of dredged material*, HR Wallingford, UK.

34. *Ibid.*

35. House of Commons Select Committee on the Environment, Transport and Regions (1998) *Op. Cit.*

36. Haigh, N. (1984) *EEC Environmental Policy and Britain*, Environmental Data Serves, London.

37. Santoro, E.D. (1987) Status report – phase out of ocean dumping of sewage sludge in the New York Bight Apex, *Marine Pollution Bulletin*, **18 (6)**, 278 – 280.

38. Swanson, R.L. (1993) The incongruity of policies regulating New York City's sewage sludge: lessons for coastal management, *Coastal Management*, **21**, 299 – 312.

39. Clark, J.R. (1998) *Coastal seas: the conservation challenge*, Blackwell Science, Oxford.

40. House of Commons Select Committee on the Environment, Transport and Regions (1998) *Op. Cit.*

41. Kullenberg, G. (1999) *Op. Cit.*

42. Claussen, E.B. (1998) *Critical coastlines: outline action to implement the Global Programme of Action for the Protection of the Marine Environment from land-based activities*, Global Programme of Action for the Protection of the Marine Environment.

43. *Ibid.*

44. Group of Experts on the Scientific Aspects of Marine Pollution (GESAMP) (1991) Global strategies for marine environmental protection, *UNEP Regional Seas Reports and Studies*, Nairobi, Oceans and Coastal Areas Programme Activity Centre, UNEP.

45. Letson. D., Suman, D. and Shivlani, D. (1998) Pollution prevention in the coastal zone: an exploratory essay with case studies, *Coastal Management*, **26**; 157 – 175.

46. Wooldridge, C.F. Tselentis B.S. and Whitehead, D. (1998) Environmental management of port operations - the ports sector's response to the European dimension. In: *Maritime Engineering and Ports* (Eds. G Sciutto and CA Brebbia) pp 227-242.WIT Press.

47. International Maritime Organisation (1994) *The Caribbean – a very special area: Wider Caribbean initiative for ship-generated waste*, Port of Spain, Trinidad: Media and Editorial Projects.

48. *Ibid.*

49. Wooldridge C.F. (2000) Quality assurance in European port operations, *BIMCO Bulletin*, **95 (1)** 55-58.

50. Moore, S.J. (1992) Troubles in the high seas: a new era in the regulation of US ocean dumping, *Environmental law*, **22**, 913 – 951.

51. Oostdam, B.L. (1983) Sewage sludge dumping in the Mid-Atlantic Bight in the 1970s: short, intermediate and long-term effects, In: Duedall, IW, Ketchum, BH, Kilho Park, P and Kester, DR (1983) *Wastes in the ocean*, Volume 1: Industrial and sewage wastes in the ocean, John Wiley, New York, 313 – 336.

52. Oslo and Paris Commission (OSPARCOM) (1993) *North Sea Quality Status Report 1993*, Oslo and Paris Commissions, London.

53. Copeland, C. (1999) *Op. Cit.*

54. International Maritime Organisation (1991) *The London Dumping Convention: the first decade and beyond*, IMO, London.

55. Curtis, C. (1993) The London Convention and radioactive waste dumping at sea: a global treaty regime in transition., In: *Proc. Radioactivity and Environmental Security in the Oceans: new research and policy priorities in the Arctic and North Atlantic*, Woods Hole Oceanographic Institution, Woods Hole, MA.

56. Kite-Powell, H.L., Hoagland, P. and Jin, D. (1998) Policy, law and public opposition: the prospects for abyssal ocean waste disposal in the United States, *Journal of Marine Systems*, **14**, 377 – 396.

57. Peet, G. (1994) The role of environmental non-governmental organisations at the Marine Environment Protection Committee of the International Maritime Organisation and at the London Dumping Convention, *Ocean and Coastal management*, **22**, 3 – 8.

58. Parmentier, R. (1998) *Twenty years protecting the ocean from waste dumping*, Greenpeace International Report.

59. Bailey, C. and Faupel, C.E. (1989) Out of sight is not out of mind: public opposition to ocean incineration, *Coastal Management*, **17**, 89 – 102.

60. McCreary, S., Twiss, R., Warren, B., White, C., Huse, S., Gardels, K. and Roques, D. (1992) *Op. Cit.*

61. Kullenberg, G. (1999) *Op. Cit.*

62. Joint Group of Experts on the Scientific Aspects of Marine Pollution (GESAMP) (1998) *Op. Cit.*

63. McIntyre, A.D. (1995) The current state of the oceans, *Marine Pollution Bulletin*, **25**, 28 – 31.

64. Parmentier, R. (1998) *Op. Cit.*

65. McIntyre, A.D. (1995) *Op. Cit.*

66. Kite-Powell, H.L., Hoagland, P. and Jin, D. (1998) *Op. Cit.*

67. Valent, P.J., Palowitch, A.W. and Young, D.K. (1998) Engineering concepts for the placement of waste on the abyssal seafloor, *Journal of Marine Systems*, **14**, 273 – 288.

68. US Environmental Protection Agency (1998) *Perspectives on marine environmental quality today*, US EPA Webpage: http://www.epa.gov.

69. Haggerty, B. (1993) Dredge-lock, *Workboat Magazine*, Nov – Dec, 28 – 31.

70. Munson, D. (1996) Rising mud, *NE – MW Economic Review*, Northeast-Midwest Institute, US, Nov / Dec., 4 – 9.

71. UK Offshore Operators Association (UKOOA) (2001) *Decommissioning* webpage; http://www.ukooa.co.uk

72. *Ibid.*

73. *Ibid.*

74. US Environmental Protection Agency (1998) *Op. Cit.*

75. *Ibid.*

76. Hall, K. (2000) *Impacts of marine debris and oil, Economic and Social Costs to Coastal Communities*, KIMO, Shetland.

77. Ball, I. (1999) Port Waste Reception Facilities in UK Ports, *Marine Policy,* **23**, 307-327.

78. International Maritime Organisation (2001) *http://www.imo.org/index.htm*

79. *Ibid.*

80. Marine Environment Protection Committee (2000) *Report of meeting, 45th session*: 2-6 October 2000, London.

81. International Maritime Organisation (2001) *Op. Cit.*

82. Parmentier, R. (1998) *Op. Cit*

83. Kullenberg, G. (1999) *Op. Cit.*

84. Garber, W.F. and Gunnerson, C.G. (1995) Science, technology, and willingness to pay in ocean disposal of wastes: America's misplaced search for equity. An introductory case study of Los Angeles, *Water Science and Technology*, 32 (2), 9 – 17.

85. Bloom, D.E. (1995) International public opinion on the environment, *Science*, **269**, 354 – 458

86. Slovic, P., Flynn J.H. and Layman, M. (1991) Perceived risk, trust and the politics of nuclear waste, *Science*, **254**, 1603 – 1607.

87. Moore, S.J. (1992) *Op. Cit.*

88. Kite-Powell, H.L., Hoagland, P. and Jin, D. (1998) *Op. Cit.*

89. Joint Group of Experts on the Scientific Aspects of Marine Pollution (1998) *Op. Cit.*

90. Broecker, W.S. (1997) Thermohaline circulation, the Achillles Heel of our climate system: will man-made Co2 upset the current balance? *Science*, **278**, 1582 – 1588

91. McIntyre, A.D. (1995) *Op. Cit.*

92. Kite-Powell, H.L., Hoagland, P. and Jin, D. (1998) *Op. Cit.*

93. Kay, D. (2000) Presentation at the Irish Sea Conference, *An Environmental Review of the Irish Sea*, October, 2000, Isle of Man.

94. Kullenberg, G. (1999) *Op. Cit.*

95. Parmentier, R. (1998) *Op. Cit*

96. Kite-Powell, H.L., Hoagland, P. and Jin, D. (1998) *Op. Cit.*

97. Kullenberg, G. (1999) *Op. Cit.*

98. *Ibid.*

99. Garber, W.F. and Gunnerson, C.G. (1995) *Op. Cit.*

100. Kite-Powell, H.L., Hoagland, P. and Jin, D. (1998) *Op. Cit.*

101. United Nations Environment Programme (1998) *World population prospects, 1950 – 2050* (the 1996 revision), United Nations, New York, US.

102. Hinrichsen, D. (1996) Coasts in crisis, *Issues in Science and Technology*, **12 (4)**, 39 – 47.

103. United Nations Environment Programme (1999) *Op. Cit.*

104. World Resources Institute (2000) *Op. Cit.*

105. World Bank (1995) *Towards environmentally sustainable development in Africa*, World Bank, Washington DC, US.

106. United Nations Environment Programme (1998) *Op. Cit.*

107. United Nations Environment Programme (1999) *Op. Cit.*

108. Gourlay, K.A. (1992) *World of Waste*, Zed Books, London.

109. McIntyre, A.D. (1995) *Op. Cit.*

110. Barrow, C.J. (1999) *Op. Cit.*

111. *Ibid.*

112. Angel, M.V. (1990) The deep ocean option for waste disposal, *Underwater Technology*, **16 (1)**, 15 – 23.

113. Angel, M.V. (1992) The deep ocean-option for the disposal of sewage sludge, *Environmentalist*, **8(1)**, 19 – 26.

114. Hollister, C.D. (1992) Potential use of the deep seafloor for waste disposal, use and misuse of the seafloor, *In Hsu, K.J.. and Thiede, J. (Eds.)*, John Wiley, Chicester, 127 – 130.

115. Jin, D. (1994) Multimedia waste disposal optimisation under uncertainty with an ocean option, *Maritime Resource Economics*, **9**, 119 – 139.

116. Jin D. and Kite-Powell, H.L. (1998) *Op. Cit.*

117. Valent, P.J., Palowitch, A.W. and Young, D.K. (1998) *Op. Cit.*

118. Angel, M.V. (1990) *Op. Cit.*

119. Jin, D. and Kite-Powell, H.L. (1998) *Op. Cit.*

120. Jin, D. (1994) *Op. Cit.*

121. Royal Commission on Environmental Pollution (1985) *Eleventh Report: Managing waste: the duty of care*, HMSO, London, 214 pp

122. Jin, D. and Kite-Powell, H.L. (1998) *Op. Cit.*

123. Kullenberg, G. (1999) *Op. Cit.*

124. Group of Experts on the Scientific Aspects of Marine Pollution (1991) *Op. Cit.*

125. McIntyre, A.D. (1995) *Op. Cit.*

126. Kullenberg, G. (1999) *Op. Cit.*

127. McIntyre, A.D. (1995) *Op. Cit.*

CHAPTER 10. MARINE CONSERVATION AND RESOURCE MANAGEMENT

S. PULLEN

1. Introduction

In 1960, Rachel Carson wrote "Although man's record as a steward of the natural resources of the earth has been a discouraging one, there has long been a certain comfort in the belief that the sea, at least was inviolate, beyond man's ability to change and to despoil. But this belief, unfortunately, has proved to be naïve". [1] Over three decades later, Norse [2] identified five major threats to marine biodiversity, including pollution of the seas, over exploitation of living things, physical alteration of the environment, the introduction of alien species, and finally, increased ultraviolet radiation and alterations of climatic conditions.

While Carson was referring principally to the widespread occurrence of radioactive contamination in the world's oceans it is now clear that this world-wide distribution of persistent pollutants is also true for many highly toxic chemicals. For example, polychlorinated biphenyls, DDT and its derivatives, and tributyltin (TBT) and the dibutyl (DBT) and monobutyl tin (MBT) breakdown products are found world-wide in the muscle and fatty tissues of marine invertebrates, fish and wildlife at the top of marine food chains, as well as humans. [3]

The exploitation of marine wildlife has led to extinctions globally and locally, for example the Atlantic gray whales and the great auk became extinct following massive overexploitation while the northern right whale is extinct in British waters for the same reasons. Poor management of commercial fish stocks has led to 70% of the world's important fish stocks being depleted or slowly recovering, overexploited or fully exploited. [4]

In 1982, two eminent fisheries biologists – Tony Pitcher and Paul Hart - in a standard text book on fisheries ecology wrote: "The current picture is rather bleak. Over the past 30 years protein resources have been squandered by failure to manage stocks properly. Until ten years ago there was perhaps some excuse for ignoring the advice of biologists because classical fishery management models were often misleading and inadequate. ….It is a sad reflection on our ability to manage resources that the past decade has seen no amelioration of overfishing. A long list of stocks endangered in 1970 would be longer still in 1980, the only removals being those stocks that have collapsed."

Pitcher and Hart [5] go on to list herring, cod, hake, sardine, anchovy, pilchard, tuna, mackerel and many flatfish amongst the species where many stocks are in a worse state in the early 80s than ever before. By 2000, little had changed! The list has simply grown. At the end of the 20th century, the International Council for the Exploration of the Sea (ICES) consider the cod stocks in the North Sea, Eastern Channel, West of Scotland, Irish Sea, Western Channel and Celtic Sea to be "outside safe biological limits". Because of the plight of the cod stocks in UK waters, the European Commission reduced the Irish Sea cod quotas for 2000 by 80%. The fishermen, however, have not been able to find sufficient fish to meet even these lowered quotas. A cod recovery programme has been introduced but there is now concern that climatic changes resulting in warming of sea temperatures and changes in the abundance and distribution of plankton and / or pollution impacts on cod eggs and larvae may mean that recovery of these stocks is not possible. A report published in 2000 by the 15 Governments of the North East Atlantic reports that two-thirds of the commercial fish stocks of the NE Atlantic are being fished at unsustainable levels. [6]

One of the most horrifying examples of destruction of habitats experienced on this planet is the loss of the intertidal wetlands. Habitats such as coastal lagoons, saltmarsh and mudflats have been systematically destroyed for coastal development, agricultural use and tourism. In the UK it is estimated that 25% of the mudflats have been lost between the early 1950s and 1980s, while 75% of the saltmarsh in England and Wales has been lost. [7] Furthermore, the Hadley Centre- a leading UK research body - predicts that globally 50% of the world's coastal wetlands will be lost by the 2080s due to coastal erosion and rising sea levels (in [8]). As well as the UK's coastal wetlands being vital "kitchens" for internationally important populations of coastal wildfowl and waders, these habitats are also the spawning and nursery areas for many commercial fish species, sinks for nutrient run-off and natural coastal defences – dissipating the energy of the wind and the waves.

We still know relatively little about the oceans. In the past 30 years one of the largest fish in the sea – the megamouth shark – was seen for the first time and has still only been recorded a handful of times. Similarly, the 7m long oarfish has only been recorded occasionally, including one record from the North Sea. A new species of beaked whale was recorded during the 1980s. The second largest fish in the sea – the basking shark – frequents British waters in the summer months but where these animals go between September and May is not known. Sightings of these gentle giants indicate that populations have declined. Twenty-five years ago we were only beginning to get an inkling of what mysteries the hydrothermal vent systems held – new species and new phylums. Communities where the building blocks of life are not carbon or oxygen but sulphur! Imagine the surprise of scientists in a deep sea submersible diving over the young geological formations of the spreading earth's crust – on finding a Nike shoe amongst the basalt rocks or plastic bags of rubbish at 1,900m depth in the Sea of Cortez.[9] Nowhere, it seems remains pristine!

It is widely recognised that a series of events pushed marine environmental issues onto political agendas in the late 1980s. For example, the 1988 mass mortality of seals around the North Sea, was followed by dolphin deaths in the Mediterranean and turtle deaths in the USA. Each event created widespread publicity. "Toxic" ships travelled the world's oceans laden with wastes which all too frequently "disappeared". This was coupled with an increasing awareness of climatic changes, coastal pollution and over-exploitation of fish stocks. Due to its complexity and inaccessibility, it is frequently difficult for those not directly involved in marine and coastal activities to appreciate the importance of the environment and its wildlife and habitats to the changes that have been experienced. However, awareness now – at the turn of the Millennium - is probably higher than ever before, not only because of the major 'events' of the last 10 to 15 years, but also because technology has enabled better access, and communicating messages about the oceans has improved considerably.

Along with the increased awareness and concern has come an encouraging proliferation of international agreements and conventions, EU directives and national legislation relevant to marine conservation. Box 1 provides an indication of some of the international and regional commitments that have been secured - but practically, has much been achieved? Recent assessments of the health of the seas, for example, the NE Atlantic Quality Status Report [6] would suggest that despite some successes not enough has been done.

2. Out of Sight - Out of Mind

A major difficulty in protecting and sustainably managing the marine and coastal environment has traditionally been the lack of knowledge about the wildlife, habitats and natural processes accompanied by little awareness of the impacts of people's activities. It has, however, been recognised for sometime that there is no single solution to the plight of the oceans – it is pointless managing commercial fish stocks properly if we continue to poison the seas or destroy the spawning and nursery grounds. Similarly, it is useless protecting highly diverse or highly productive sites if a transient oil tanker runs aground spilling thousands of tonnes of crude oil over tens or hundreds of kilometres of sea and coastline ignoring the boundaries of marine protected areas.

An integrated approach to the management of the oceans, seas and coasts is fundamental to the protection of biodiversity as well as to the sustainable management of the resources. Such an "ecosystem approach" requires not only the management and protection of the species and habitats, but also the processes that generate and sustain the habitats and species and the multitude of activities that take place – research, exploitation of living and non-living resources, transport, waste disposal, offshore development, recreation and others. It also recognises the need to involve and accommodate human communities in the management of the oceans.

While awareness of the need to introduce an ecosystem approach has improved, the delivery remains poorly understood. It should involve three goals - the maintenance of biodiversity and ecological processes, the sustainable and equitable use of marine resources, and the restoration of marine and coastal ecosystems where their functioning has been impaired. [10]

> **Box 1. Examples of international and regional agreements with an impact on marine conservation and management developed during the 1990s.**
>
> ❖ 1992: The UN Conference on Environment and Development (UNCED) concluded with an Agenda for the 21st Century, including a chapter on Oceans, and a Convention on Biological Diversity.
>
> ❖ 1992: Environment Ministers of the Northeast Atlantic states agreed a new regional seas convention – the "OSPAR" Convention.
>
> ❖ 1992: The European Community adopted a new directive on the Protection of Habitats and Species, which complemented the 1979 Birds Directive and was aimed at both terrestrial and marine species and habitats.
>
> ❖ 1995: A Global Programme of Action was agreed on Land-based Activities which degrade the marine environment
>
> ❖ 1995: Discussions were concluded on a UN Agreement on highly migratory fish stocks and a FAO Code of Conduct on sustainable fisheries.
>
> ❖ 1997: The UK signed the UN Convention on the Law of the Sea (UNCLOS).
>
> ❖ 1998: North East Atlantic Environment Ministers agreed the "Sintra Statement" along with a series of strategies for the protection and management of the NE Atlantic and a new Annex on the protection of biodiversity and ecosystems.
>
> ❖ 1987, 1990, 1993,1995 and 1997: North Sea environment ministers met, often with their agriculture and fisheries counterparts present, and agreed measures aimed at restoring and protecting the wildlife and habitats of the North Sea.

The delivery of an ecosystem approach, requires managing human use of the marine environment and will out of necessity involve resource management, nature conservation and pollution prevention. A number of management tools can be utilised including, for example, assessment tools such as strategic environmental and social assessment, risk assessment and environmental impact assessment. Other management tools, grouped as delivery mechanisms, include protected areas, fish stock recovery mechanisms, identification of and mitigation measures for "high risk" areas, pollution prevention and reduction practices and restoration measures where systems have already been degraded. [11]

3. Marine Resource Management - Sustainable Development!

3.1. FISHERIES

Commercial fish stocks world wide are threatened by over-exploitation – the UN Food and Agriculture Organisation (FAO) recognise that 70% of the world's commercial fisheries are depleted, exploited at biological limits, overexploited or slowly recovering. The threats to fish stocks and fisheries, experienced all around the world, has resulted in sustainable management of fish stocks being high on the agendas of most coastal nations. The collapse of the North Sea herring and mackerel stocks in the 1970s, due to overfishing, led to the closure of those fisheries in the 1980s - and decline in stocks of these species was still reflected by North Sea fishing suspensions in 1998. In the early 1990s the cod fisheries off the coasts of Halifax and Newfoundland, Canada, closed with huge economic consequences - 30,000 Canadians were put out of work. Now in 2000, the UK is experiencing a virtual closure of the Irish Sea cod fishery and cod stocks in the North Sea are close to collapse. [12]

Focusing on highly migratory fish stocks or those which straddle national boundaries, in 1995 the UN negotiated an agreement aimed at sustainably managing these stocks – the Implementation Agreement on Straddling Fish Stocks and Highly Migratory Fish Stocks. However, while these serious threats to the future of high seas fisheries have been recognised, few measures to fully implement this new international agreement have been introduced. [13]

The need for improved management of fish resources, incorporating precautionary measures, is paramount to secure the livelihoods of coastal fishing communities and the long-term future of marine wildlife populations dependent on commercial fish species as a source of food. 'No-take' or 'Fishing Free' zones, a measure that has been used with some notable successes in New Zealand, South Africa and Australia, can support existing management measures and be utilised to help facilitate recovery plans by safeguarding critical areas for commercial species and the ecosystems upon which they depend. The extent and location of no-take zones is debatable, but data from existing ones - supporting evidence of increased marketable yields of fish stocks in the long term - is sufficient to warrant a network of pilot projects in temperate waters. [14] It is important to recognise however, that No Take Zones are not a panacea and will not work in isolation of other fisheries management measures. Of particular importance is the need to manage fisheries on an ecological basis and to involve the fishers in the management of the fishery. So-called "zonal" management is the preferred management option currently being promoted by fishermen and conservation bodies alike. [14]

The incidental mortalities or bycatch of not only marine mammals but also seabirds, sharks, turtles and other non-target fish species, is a particular problem which can often be area specific depending on the resident or migratory species present. Levels of harbour porpoise bycatch in bottom set gill nets in the Celtic Sea can be six times that suggested to be sustainable. Drift nets (gill nets set at or near the surface) are responsible for significant casualties of marine wildlife, including rissos and bottlenose dolphins, and pilot, sperm, minke and fin whales. Precautionary measures are required to secure the future for cetaceans that are particularly vulnerable to being accidentally caught in fishing nets. It is essential to reduce the bycatch of small cetaceans such as harbour porpoises and dolphins in set gill nets and drift nets, and sharks in both net and line fisheries. A number of British sharks and rays have been over-exploited by directed commercial fisheries and bycatch due to poor management of these species. While some of these catches are landed, many including most blue sharks are discarded - sometimes after finning - because of the low market value of their meat. The ecological impact of removing millions of these top predators from the oceans is unknown, but it is likely to be highly significant.

Increasing evidence suggests that there is also a significant problem with the bycatch of marine turtles. The leatherback turtle, which can grow to 2.8 metres in length, is a regular visitor to UK waters between July and October after a journey across the Atlantic when it follows and feeds on luminescent jellyfish. While there is little data available on the bycatch at sea of these animals, those found washed up on the coasts of the British Isles and displaying marks consistent with fishing gear are a reminder of casualties at sea.

Better control of fishing effort, the impact of fishing gear on the seabed and on the discarding of 'non-target' fish and wildlife and 'over-quota' fish, is required. The use of surveillance systems could not only assist monitoring and enforcement but also help increase the safety of fishing vessels. Indeed, it could benefit all shipping in UK waters.

Fundamental to the future management of all commercial fisheries will be the recognition of and introduction of two concepts – the ecosystem approach and a precautionary approach. Current management methods focus too heavily on individual fish stocks and short-term reactionary measures. In future, greater emphasis is needed on the interactions between different trophic levels, the impacts of other anthropogenic activity (such as the loss of nursery and spawning grounds and sub-lethal impacts of pollutants), and on variables beyond human control.

Consumer pressure is another important tool in achieving sustainable use of the world's resources. The Marine Stewardship Council, a non-profit, non-governmental body, is creating market-led economic incentives for sustainable fishing. It is establishing a broad set of principles for sustainable fishing and will aim to set standards for individual fisheries. Fisheries meeting these standards will be eligible for certification by independent, accredited certifying firms. Products from certified fisheries will eventually be marked with an on-pack logo which will allow consumers to select fish products coming from a sustainable source. By 2000 3 fisheries had been certified as sustainable – the Alaskan salmon, the Western Australian rock lobster and the Thames herring, with more expected in the near future. [15]

3.2. COASTAL RESOURCES – CONSERVING COASTLINES

A multitude of human activities are concentrated in the coastal zone – coastal flood defences, residential, commercial and industrial developments, barrages, port development and recreation facilities to name but a few. Ten years ago, few recognised the importance of integrated coastal management as a tool for managing the UK's coastal zone, but in 1994 a Coastal Forum (for England) was set up to help develop a national policy. Since then, two new documents - *Policy Guidelines for the Coast* and *Coastal Zone Management: Towards Best Practice* - have been published by the Department of the Environment (as was). Similar forums have also been set up in Scotland, Wales and Northern Ireland.

Despite some progress in the past 10 years, however, much more effort is needed to manage coastal resources and to conserve and increase the resilience of British estuaries and coastlines. On a number of estuaries 50 per cent or more of the intertidal areas have been lost to land-claim and development. These highly productive habitats act not only as the 'restaurant' of marine and coastal ecosystems but also as our natural coastal defences against the power and energy of the seas. Loss of these defences, coupled with industrial and urban development in coastal lowlands and on eroding coasts, has resulted in the construction of mile upon mile of 'hard' sea defences.

The threats posed by rising sea levels in response to eustatic changes, global climate change and hard defences means the intertidal lands will be squeezed even further. Populations of coastal waders and wildfowl, intertidal fish spawning and nursery areas and highly productive marine habitats will also be squeezed - in some areas possibly out of existence. Measures must be taken now to improve coastal and estuarine management, to reduce conflicts resulting from our multi-use demands and to ensure that 50% of coastal wetlands are not lost in the next century as sea levels rise.

New coastal development should not be considered acceptable in protected sites or on eroding coastlines. A strategic approach at a national level, backed by legislation, is necessary to manage all human activities and demands in the coastal zone and ensure sustainable use. Furthermore, innovative programmes for the restoration and recreation of coastal and near-shore sub-tidal habitats must be developed - where possible returning land claimed from the seas and estuaries to the former habitats. We must also seek to recreate lost or shrinking habitats elsewhere, including migrating freshwater wetlands inland to allow space for the recreation of brackish and saltwater wetlands in coastal locations.

3.3. OFFSHORE RESOURCES

Ten years ago it was unlikely that the offshore oil and gas industry or many other industrial sectors would have included environmental groups in its consultations, or indeed accommodated the views of the environmental movement. But now a wide range of industry bodies is consulting environmental groups, and meetings with government ministers and civil servants take place regularly. Although environmental groups feel that their influence varies considerably and their advice is, too frequently, not accepted!

With respect to developments offshore, the need for a more systematic and strategic assessment of the threats to marine wildlife and habitats remains necessary even after comprehensive surveys of the environment and ecology of an area have been undertaken. Areas that are simply too sensitive for exploitation must also be identified - something that has been done in the USA at Stellwagon Bank and the Florida Keys National Marine Sanctuaries and in Australia for the Great Barrier and Ningaloo Reefs. In areas where development is considered to be compatible with the sensitivity of an area, appropriate controls need to be introduced to ensure that the impact of the development is kept to a minimum. Strategic environmental assessment (SEA) and environmental impact assessment (EIA) are fundamental tools for the management of marine resources and will not only benefit offshore oil & gas developments [16] but also a range of other offshore activities such as aggregate extraction and wind farms.

4. Marine Nature Conservation - Protecting Marine Wildlife and Habitats

4.1. PROTECTING BIODIVERSITY

In 1999, the UK Government published Maritime Biodiversity Action Plans for a number of marine species and habitats, to contribute to the fulfilment of UK obligations under the Convention on Biological Diversity (CBD). [17] Not surprisingly, little progress has yet been made on implementing these Action Plans, but work is in progress. The list of marine species and habitats for which Action Plans has been produced is short in comparison with the one for terrestrial habitats and species. However, as both the impacts and the solutions tend to be similar for a variety of marine wildlife, it has been possible to group species – so for example, there is a single action plan for whales, dolphins and porpoises. Many of the marine and coastal BAPs are somewhat similar in nature, highlighting the vital need for marine and coastal biodiversity objectives to be integrated into other sectoral policies and management. For example, a local shoreline management plan addressing flood defence should include biodiversity objectives for a variety of coastal habitats and wildlife dependent on the coastal wetlands.

In providing protection for marine wildlife, consideration must be given to the broader measures necessary to protect animals that range or migrate over large distances and spend time in other parts of the world. In some cases we do not know where these other places may be: for example, the basking shark is seen in British waters during the summer but we do not know where it goes for the rest of the year, or where or when it breeds. This lack of information can create real difficulties when effort is required to protect the largest fish found in British waters, particularly as numbers appear to have declined.

There are numerous examples of highly mobile or migratory wildlife around our coastline including whales and dolphins, sharks, other large fish, seabirds, coastal waders and wildfowl. Broader measures across the whole of their range will be necessary for these species. The legal mechanisms exist via the Bern and Bonn Conventions, which should be used more proactively to provide better protection for marine wildlife across their range.

The UK Government also has nature conservation responsibilities in dependent territories. These territories in the Caribbean, the south Atlantic, the Indian Ocean and the south Pacific support a great diversity of marine wildlife. Protecting and maintaining this wildlife is not only important to meet international biodiversity commitments, but is also crucial to ecological and economic sustainability for cultural and social well-being. The ratification and implementation of international conventions aimed at protecting and maintaining natural resources in these territories is dependent on technical and financial resources for baseline research, data management systems, development of legislation, training and enhancing capacity. Further resources will be necessary in the coming decade to ensure that the UK's commitments to dependent territories are met.

4.2. MARINE SITE PROTECTION

One of the most traditional management tools for nature conservation is the protected area. In 1988, the UK had only one marine protected area - the waters surrounding the island of Lundy in the Bristol Channel. Now, three marine nature reserves (MNRs) have been designated – Lundy (Bristol Channel), Skomer (Pembrokeshire, Wales) and Strangford Lough (Northern Ireland) and in 2000 over 50 sites had been proposed to the European Commission for protection under the EU Habitats Directive as marine Special Areas for Conservation (SACs). SAC sites, some of which already have management plans being prepared, will protect a range of marine habitats. These include large shallow inlets and bays such as Plymouth Sound, Morecambe Bay and Loch Maddy; major estuaries including the Solway Firth and Essex estuaries; and the rocky shores and reefs of the Berwickshire and Northumberland coasts and the Pembrokeshire islands. Along with marine Special Protection Areas (SPAs) to be designated under the Birds Directive, marine SACs will form part of a network of protected areas throughout Europe called "Natura 2000".

For some time it was believed that the identification of Special Areas of Conservation (SAC) was restricted to sites within the 12 nautical miles territorial waters and that only sites of European importance could be encompassed. While sites identified under an EC Directive must be of European wide importance it is now clear that qualifying sites further offshore should also be identified. Following a High Court ruling in November 1999 (Case No: CO/1336/1999) the provisions of the Habitats Directive will now be applied to all the waters beyond 12nm offshore over which the UK has competence. Currently, most of the candidate marine SAC sites are estuarine or coastal, however offshore sites are anticipated to supplement the list of sites presented to the European Commission in 2000. In particular, the UK Government needs to identify reef and sand bank sites beyond 12nm which should be protected under the Habitats Directive.

The establishment of Natura 2000 - a network of protected areas across the European Union - will be a major achievement. However, marine habitats and species receive too little attention under the EU Birds and Habitats Directives. In particular, the list of marine habitats and species to which protection should be afforded is extremely limited. While it is possible that the species and habitats annexes to the EU Habitats Directive may be reviewed at some, as yet undetermined future date, the adoption in 1998 and subsequent ratification in 2000 by the UK Government of a new Annex to the OSPAR (NE Atlantic) Convention offers a new opportunity to protect important deeper water or offshore habitats and for marine species. Annex V addresses the protection of biodiversity and ecosystems and could be used to provide better protection for marine wildlife such as sharks, deepwater fish and marine invertebrates, for which little or no protection currently exists.

New environmental legislation will be required to deliver the levels of nature conservation required under international obligations such as the Convention on Biodiversity and the new Annex V to the OSPAR Convention. A network of coastal and marine sites, selected on the basis of scientific quality, should represent the full range of diversity found in UK waters and encompass sites of international and national importance, sites in deep water beyond territorial waters as well as sites within the territorial waters limit. A new regime for marine protected areas in the UK should also recognise the importance of voluntary, community-based marine conservation initiatives. This network will not only require new legislation but also adequate resourcing. And should not be considered a massive 'no-go' area but would rather restrict only those activities incompatible with the conservation objectives of each site. In this way, activities compatible with the conservation objectives can also be managed in a way that can generate local income and maintain local livelihoods.

5. Marine Pollution – Prevention and Cure!

5.1. LAND BASED MARINE POLLUTION

It is estimated that between 70 and 80 per cent of pollution reaching the oceans results from human activity on land. It reaches the sea via rivers, direct run-off, deposits from the atmosphere and direct dumping. Pollution of the oceans has been a focus of many international meetings in the past decade. In 1990, it was agreed to stop incineration of industrial waste in the North Sea from the end of 1991. By the end of 1998, following considerable pressure from the public and North Sea governments, the UK ended dumping of sewage sludge at sea. Inputs of a number of industrial and agricultural chemicals such as pentachlorophenol, heavy metals and some pesticides, to the North Sea via rivers and the air have been halved between 1985 and 1995. In the past 10 years, much has been done to reduce the input of some sources of pollution and a few chemicals - but a large number of man-made chemicals are still released into the environment.

Many pollutants, particularly persistent organic pollutants or POPs, are eventually carried to the sea where they build up in the food chain. Scientists have identified a number of persistent and degradable compounds that are sub-lethal and disrupt the reproductive, nervous, immune or developmental systems. The health effects of these chemicals (known as hormone or endocrine disruptors) on humans and wildlife are potentially enormous. Although there is a great deal of scientific uncertainty regarding their precise impact, the implications of their sub-lethal effects on marine biodiversity and sustainable use of marine fisheries are very serious. Some of these endocrine disrupting compounds are detectable in marine wildlife, the water or sediments, despite their use having been largely onshore or in offshore developments and shipping. Impacts have also been recorded among some marine species including beluga whales, common seals, polar bears, Caspian terns, turtles, dog whelks and herring.

Action is needed internationally, since many of these compounds are manufactured and/or utilised world wide and since those that are persistent are spread round the globe via air and water currents. An international "POPs" Convention has been negotiated and in 1998 in Sintra, Portugal, further reductions in emissions into the North East Atlantic were agreed. The aim now, of the 15 Contracting Parties to the OSPAR Convention, is to reduce discharges, emissions and losses of hazardous substances, and move towards a target of stopping inputs within 25 years. This ambitious target is to be met by 2020 – to begin with a "priority" list of chemicals or chemical groups has been identified.

At a national level it is necessary to implement a precautionary approach and introduce strategies to reduce exposures. Toxicity tests must be improved and compounds already in use should be re-tested for endocrine disrupting effects. Further research is necessary and should be adequately funded, prioritised and co-ordinated. All sectors releasing man-made chemicals into the environment should also be required to report releases and place such information on publicly-accessible registers.

Man-made chemicals are not the only pollutants to have an impact on marine wildlife, although they do have the highest potential for significant long-term consequences. Nutrients, litter, sewage and radioactive substances also add to the pollution load of our coastal waters and the waters of the North East Atlantic. A dedicated programme of work is required to ensure the implementation of measures that will reduce to background or near-background all inputs of pollution from land-based sources.

5.2. SHIPPING – REDUCING THE RISK FROM OFFSHORE ACTIVITIES

Each year numerous small incidents occur around the coast and three of the world's largest spills from oil tankers have occurred in British waters - the *Torrey Canyon* (1967), the *Braer* (1993) and the *Sea Empress* (1996). The coastal communities of south west England, the Shetland Islands (Scotland) and Pembrokeshire (Wales) need little reminder of how devastating a major oil spill can be for local economies and livelihoods, as well as for the environment and wildlife. During the 1990s, Atlantic Europe experienced three of the largest oil tanker spills in the world – the Aegean Sea, in north-west Spain in 1992, the *Braer* and *Sea Empress*, as well as numerous smaller spills. In the UK, following the *Braer* oil spill in the Shetland Islands, a public inquiry into pollution from merchant shipping resulted in 103 recommendations. Further recommendations were made following the *Sea Empress* disaster in 1996. Yet more measures are likely to result from the International Maritime Organisation (IMO) and the European Commission following the break-up of the Erika, resulting in the estimated death of 300,000 seabirds, off the coast of Brittany in December 1999. Many of the recommendations following the *Braer* and *Sea Empress* oil spills have been accepted and implementation undertaken.

Of particular note is the fact that the North Sea and a large area to the west of the British Isles is now recognised internationally as a 'Special Area'. While this will not have any impact on pollution from shipping accidents, it does mean that no legal discharge from ships of oily wastes from tank washings or engine rooms are permitted. Monitoring and enforcement measures have also been strengthened to facilitate implementation of the Special Area provisions. [18]

Accidents, regrettably, are inevitable. One management tool to help addresses the risks associated with shipping activity is the Particularly Sensitive Sea Area (PSSA). Yet, in the 1990s two particularly sensitive seas areas (PSSAs) have been identified – the Great Barrier Reef and the Cuban Sabana-Camaguey Archipelago. While in 2000 no PSSAs had been formally proposed for European waters, following risk assessment case studies a number of areas are being promoted for consideration as PSSAs by environmental groups. These include the waters off the west coast of Wales and the waters of the Minches off the west coast of Scotland. [19] Meanwhile, the UK Government has recently commissioned a risk analysis of UK waters and the coastline with the purpose of identifying those areas that are most sensitive and most vulnerable to shipping activity. [20] Identification of such areas as Marine Environmental High Risk Areas (MEHRAs) or Particularly Sensitive Sea Areas (PSSAs) should be supported, along with the introduction of measures to reduce the risks and improve the safety of shipping around the UK.

Further measures are required to address and eliminate chronic discharges from ships and from offshore oil and gas installations. A package of measures should include improved monitoring, enforcement of existing regulations and more openness and transparency, with public reporting of chemical and waste emissions from ships and offshore installations.

With respect to other forms of shipping pollution, global agreement has been reached on the need to completely ban the use of organotins in antifouling paints following increasing evidence of pollution of marine food chains and fish harvested for human consumption by tributyl tin and its breakdown products. [3] A partial ban was introduced in the late 1980s on vessels under 25m in length, however, it is now recognised that while this measure significantly reduced TBT contamination around marinas and areas of high concentrations of smaller vessels it was insufficient and a complete ban on all vessels is required. Such a global ban will, however, not become fully effective until 2008 and that assuming that the legal instrument to bring into effect such a ban is adopted in 2002 and can be brought into force in 2003.

6. Awareness, Human Resources and Partnerships

It is now almost four decades since the plight of the marine environment began to receive serious recognition and in the past decade, considerable progress has been made in introducing major global treaties and commitments, yet we are continuing to witness the collapse of fish stocks and dramatic declines in many marine wildlife populations, while coastal livelihoods struggle for survival. The traditional threats facing the marine environment, including over-fishing, coastal development, marine pollution, etc, are well recognised by scientists and environmental experts, but in order to resolve these problems it is necessary to address the lack of general knowledge and awareness within governments, industrial sectors and the general public. In addition, the inadequate allocation of financial resources and lack of capacity and expertise also require urgent attention.

A major difficulty in protecting and sustainably managing the marine and coastal environment has traditionally been the lack of knowledge and awareness of both the importance and diversity of the oceans and the impact humans have on marine wildlife and habitats. Awareness initiatives such as World Oceans Day, celebrated annually on 8 June, provide an opportunity for everyone involved in ocean issues to put aside their differences of opinion and join together in raising awareness. In addition, events such as WWF's 'Beach Days', The Wildlife Trusts SeaQuest Project, the Marine Conservation Society's "Adopt a Beach" programme and 'Marine Week' in south-west England each summer help to raise awareness of the importance of the marine environment and the need for conservation.

It is also increasingly important to ensure that adequate information and experience flow between different levels of decision-making and management. International and national policy should be based on 'on the ground' experience of the local communities who use the marine environment and, where possible, good scientific information. At the same time, local initiatives can benefit from the impetus provided by development of international marine policies. The connections between the different levels of policy-making and management will be fundamental to implementing marine conservation and management in the coming decade.

Lack of capacity and resources at the local level and even national level remains one of the biggest problems facing the world's oceans and ocean management today. While there remain some loopholes in international law especially in the area of enforcement, if all the international commitments agreed since the Rio Earth Summit in 1992 were to be fully implemented a huge amount will have been achieved to help secure the future of the world's oceans and seas, marine wildlife and the local communities dependent on them. It is understandable that some underdeveloped countries refuse to take on further political commitments which they know will be meaningless while lacking the necessary capacity and resources to implement them. In some cases the underdeveloped countries are unable to even be present in the international marine arena due to lack of funds, expertise or capacity. Less understandable, however, is the developed countries continued inadequate attention to issues such as international aid, trade, national debt, provision of resources, capacity building, education and awareness and transfer of technology and expertise. All of these have a bearing on the management and conservation of marine resources. Lack of attention to integration of science, policy and practice also remains a fundamental difficulty.

Many international conservation organisations are now attempting to address these issues as well as the traditional marine threats. The importance of developing capacity and expertise locally is recognised, in both developed and underdeveloped countries, but to really address these issues all the stakeholders need to be accepted and involved in order to be able to co-operate and work together with international bodies who are trying to address ocean management. This, however, poses difficulties! While the involvement of non-governmental organisations and civil society in ocean management has certainly improved since the Rio Earth Summit, 1992, a number of restrictions remain to be addressed, including capacity, access, and recognition of non-governmental organisations.

Capacity remains an issue within the non-governmental / civil society sector. There are relatively few environmental non-governmental organisations focusing on marine conservation and management at a global level. Even when they do, a number of the international policy and regulation frameworks require the non-governmental organisations to be true "observers" there is, therefore, no opportunity for real involvement of the non-governmental organisation. Furthermore, at many international frameworks, only truly international organisations have "observer" status, thus the views of national and local bodies are not well reflected. Some countries still do not recognise the roles or rights of non-governmental organisations. Therefore there is still not a "level-playing field" at local, national or international levels for the involvement on non-governmental organisations or civil society. In many countries, however, both in the developed and the underdeveloped world, it is the non-governmental organisations and civil society that are frequently attempting to facilitate the implementation of marine conservation and resource management through the practical application of marine programmes.

Partnerships are increasingly recognised to be fundamental to achieving protection of the marine environment and sustainably managing ocean resources. Significant progress is possible if local communities, environmental interests, industry, academic bodies, local and national government find ways of working together to bring about change. And because financial and human resources will always be in limited supply, far more will be achieved if collaborative partnerships are developed. Voluntary marine conservation areas such as those at St Mary's Island in Northumberland, Kimmeridge in Dorset and the Helford River in Cornwall are good examples of how local voluntary initiatives driven by local community interests can help to achieve marine conservation and marine management. For many years, conservation organisations have co-operated with national governments in underdeveloped countries to deliver nature conservation and management of the environment. Increasingly, partnerships between the corporate sector and voluntary bodies are also being developed. The Marine Stewardship Council, an independent body responsible for certification of sustainable fisheries, resulted from a joint initiative between Unilever and WWF. Partnerships are likely to be the only way in which sufficient commitment and resources can be drawn together to achieve the integrated or ecosystem approach to the management of the oceans, seas and coasts, that is now recognised to be fundamental to the protection of biodiversity as well as for the sustainable management of the resources. Partnerships are also the way forward for involving and accommodating human communities in the management of the oceans.

This chapter attempts a broad overview of the threats facing the marine environment, marine wildlife, and the coastal communities dependent on marine resources for livelihoods. At the beginning of the new millennium, global awareness and understanding of the importance of the world's oceans and seas is probably higher than at any point in history. International and national political commitments and agreements to redress the mistakes and failures of the past have been given at unprecedented rates, yet still too little is being done.

All the issues outlined in this chapter, from overexploitation of living and mineral resources and our continued use of the seas as a disposal site for chemicals and waste products, to the delivery of strategic environmental and social assessment, planning and management; adequate resourcing of recovery and management programmes; and the delivery of novel partnerships must be addressed if an ecosystem approach to the management of the marine environment is to be achieved. The future of the world's oceans and indeed the future of this planet depends on it.

Acknowledgements: I would like to thank all the colleagues at WWF who have provided ideas and advice with respect to the development of marine resource management and conservation policy.

References

1. Carson, R., 1961 edition. The Sea Around Us.

2. Norse, 1993. Global Marine Biological Diversity – A Strategy for Building Conservation into Decision-Making.

3. Linley Adams, G. The accumulation and impact of organotins on marine mammals, seabirds and fish for human consumption.

4. FAO, 1995. The State of World Fisheries and Aquaculture, Rome: FAO.

5. Pitcher, T.J., & Hart, P.J.B., 1982. Fisheries Ecology.

6. OSPAR, 2000. Quality Status Report for the NE Atlantic

7. Environment Agency, 1999. The State of the Environment of England & Wales: Coasts. Environment Agency, Bristol.

8. WWF, 1998. Keeping the Seas At Bay.

9. Van Dover, C.L., 1996. The Octopus's Garden. Hydrothermal Vents and Other Mysteries of the Deep Sea.

10. WWF / IUCN, 1998. Creating a Sea Change. The WWF / IUCN Marine Policy.

11. WWF, in prep. 2001, A Vision for the North Sea – Making Ecosystem Management Operational.

12. MAFF (Ministry of Agriculture, Fisheries and Food), Press Release: North Sea Cod in Crisis, 9 November 2000.

13. Porter, G., 1998. Too Much Fishing Fleet, Too Few Fish –A Proposal for Eliminating Global Fishing Overcapacity.

14. WWF, 2000a. Marine Update 41: WWF's Oceans Recovery Campaign – ORCA.

15. Marine Stewardship Council, 2000. www.msc.org

16. Joint Links Oil & Gas Environmental Consortium, 1997

17. UK Biodiversity Group, 1999. Tranche 2 Action Plans Volume V – maritime species and habitats. Published by English Nature.

18. The Wildlife Trusts / WWF, 1999. Marine Update 38: Ships of Shame or vessels of virtue?

19. WWF, February 2000b. Shipping Briefings:
Briefing 1: Should PSSAs be implemented in the UK?
Briefing 3: North West Scotland as a Particularly Sensitive Sea Area?
Briefing 4: The Need for Management of Oil Tanker Risks Off West Wales?

20. DETR, 2000. Identification of Marine Environmental High Risk Areas (MEHRAs) in the UK. Safetec UK Ltd.

CHAPTER 11. THE NORTH ATLANTIC

LEWIS M. ALEXANDER

The North Atlantic is one of the world's most utilised ocean bodies, most strategic militarily and commercially, and one of the most marine-resource rich. The North Pacific is probably its closest rival in these respects. The North Atlantic has a rich history of political and economic development both among its littoral states and in comparison with other ocean regions around the world. With the break-up of the Soviet Union and the growing economic integration of eastern and central Europe, the geopolitical structure of the North Atlantic is changing somewhat; however, its status as an economic force is still great.

As defined here, the North Atlantic area runs from the Labrador Strait between Greenland and Canada to include the Canadian and U.S. east coasts to the Straits of Florida between Florida and Cuba. The southern limit follows approximately the 23-1/2 degree North Latitude line eastward to Africa, intersecting the coast near the border between Mauritania and Western Sahara. From here the eastern boundary extends north along the coast of Morocco, across the Strait of Gibraltar, and follows the coastline of Western Europe, including the North and Baltic Seas, and terminates in Iceland. As so described, the total water area of the North Atlantic is approximately 11,200,000 square nautical miles (14,800,000 square statute miles; 37,000,000 square kilometres).

This defined sector is bordered by 21 independent states, six non-independent countries, and one entity in dispute. Of the non-independent countries, Denmark's Greenland and the Faroe Islands both have home rule; Portugal's Azores and Madeira are autonomous regions, while France's St. Pierre and Miquelon, off south-western Newfoundland, is a "territorial collectivity". [1] In 1995, a majority of the voters in Bermuda rejected a British offer of independence; hence, the island remains a dependent territory of the United Kingdom.

The entity in dispute -- Western Sahara -- has been claimed by Morocco since the Spanish withdrew their control in 1975. Morocco's claim has been challenged by the Polisario, a largely indigenous group seeking independence for the area.

Since at least the 16th century, the North Atlantic has been the world's primary water body in terms of economic, strategic, and geopolitical concerns. With the end of the Cold War, the economic developments in Europe (including Russia) and the virtual end of European colonialism world-wide, the relative importance of regional bodies has shifted somewhat from the North Atlantic to the North Pacific area. The volume of trade between U. S/Canada and the countries of Eastern Asia (Japan, China, Korea, etc.), the commercial fisheries development in the North Pacific, and the rising strength of China, as a maritime power, suggests that the status of the North Atlantic as a prime ocean body will be -- and perhaps already has been -- eclipsed as the world's most important ocean area.

1. Geographic Site

The North Atlantic is an oval-shaped water body with three major connections to the Arctic Ocean:
- Davis Strait between Canada and Greenland,
- Denmark Strait separating Greenland from Iceland, and
- the Norwegian Sea between Iceland and Norway. The greatest distance across the North Atlantic is between Florida and Morocco -- 3,886 nautical miles (7,200 kilometres) (km). [2] Marginal seas include the North and the Baltic, the Bay of Biscay (bordering both France and Spain), and the Labrador Sea between Canada and Greenland, leading to the Davis Strait.

There are several major international straits associated with the North Atlantic. One is the English Channel/Strait of Dover, connecting the Atlantic with the North Sea. To the east is the Skaggerak/Kattegat between the North Sea and the Baltic. Further south, the Strait of Gibraltar connects the Atlantic with the Mediterranean Sea; it is one of the only two waterways joining the Mediterranean with the open ocean, the other being the Suez Canal. In the Western Hemisphere, the principal strait connecting water areas is the Straits of Florida between the North Atlantic and the Gulf of Mexico.

On the west side of the Ocean, significant island groups include: The Bahamas, which straddle the 23-1/2-degree N. Lat. line; Bermuda, located 502 miles (940 km) off the U.S. coast; and France's St. Pierre and Miquelon. On the east side there are, to the north, Iceland, the Faroe Islands (north of Scotland), the Azores (west of Portugal), the Madeira Islands (west of Morocco), and Spain's Canary Islands (off the coast of Western Sahara). The Canary Islands are now considered to be an integral part of Spain.

1.1. BATHYMETRY

The bathymetry of the North Atlantic is dominated, first, by the continental margin, consisting of the continental shelf, slope and rise. [3] Second, there is the Mid-Atlantic Ridge, extending south of Iceland to and beyond the North Atlantic's limits, well into the South Atlantic. Also, on the western side of the ocean, the Grand Banks off Newfoundland extend more than 200 miles (370 km) offshore. In addition, there are Georges Bank off Massachusetts and the Scotian Shelf off Nova Scotia, both of which extend to about the 200-mile limit offshore.

On the eastern side of the North Atlantic, both the Baltic and North Sea are underlain by shelves, which extend north of the Shetland Islands, as well as shelves, less than 200 miles wide, to the west of the United Kingdom and Ireland. All of these shelf areas are valuable fishing grounds and, in two cases, for offshore oil and gas (hydrocarbons) as well.

The Mid-Atlantic Ridge is a submarine mountain system, with an average height of about 10,000 feet (3,300 meters). A few peaks emerge as islands, such as in the Azores. There are earthquakes and active volcanoes along the Ridge, as well as a great rift, that is continually widening and filling with molten rock from the earth's interior. As a result, the Western and Eastern Hemispheres are slowly moving away from each other.

Among the other features of the North Atlantic sea floor are sea mounts, small ridges, banks, troughs, and abyssal plains.

1.2. OCEAN CIRCULATION

The major ocean currents of the North Atlantic tend to move in a clockwise fashion, driven primarily by the prevailing winds. From the Straits of Florida, a warm current, the Gulf Stream, flows north and north-east along the coast of the U.S. until, south of Cape Cod, it turns more easterly to cross the ocean and is renamed the North Atlantic Drift. One branch of the Drift passes off Ireland, the U.K. and Norway, warming the climate of these areas, since the prevailing winds at these latitudes is largely westerly, moving from the ocean to the land. Another branch of the Drift flows southward past Spain, Portugal, and North Africa as the relatively cool Canary Current. To the south of Western Sahara, the current turns westward as the North Equatorial Current, to again cross the ocean, with one branch eventually turning northward past the Straits of Florida to form the Gulf Stream.

One additional current to note is the cold Labrador Current, flowing south and south-west along the coast of Canada, between the mainland and the Gulf Stream.

1.3. LIVING MARINE RESOURCES

The North Atlantic has long been an important commercial fishing area, but the heavy exploitation, particularly in recent decades, has resulted in certain stocks being seriously overfished. Among these are the cod and the North Sea herring.

The two principal fishing areas are Canada's Grand Banks and the North Sea; the latter is bordered by seven countries, leading in the past to serious fishing disputes. Other important areas are the waters around Iceland, the Baltic Sea, the west coast of Norway, north-west Spain, and the Georges Bank between New England and Nova Scotia.

Cod is the most prevalent of the species, found in all areas of the North Atlantic and is, in most grounds, considerably overfished. Associated with the cod are haddock, another valuable food fish, although the supply of haddock is a good deal less than that of cod. Another species present throughout the North Atlantic is herring, particularly in the European waters; here again, the resource has been much overfished.

Salmon, an anadromous species inhabiting both salt and fresh waters, appear on both sides of the North Atlantic, as do flounders, mackerel and redfish. Menhaden, an industrial fish used for fishmeal, is taken off the east coast of the U.S. Other important species include whiting (particularly in the North Sea), hake, and sardines. Tunas are also caught on both sides of the North Atlantic, particularly in the south and central portions, and whaling is still carried on by Norway and Iceland. Seals support significant fisheries in the northern parts of the ocean.

Shellfish are important more for their value than their volume and they, too, are found throughout the shallower areas of the North Atlantic. The term "shellfish" is taken here to include both molluscs (scallops, clams, oysters, snails, squid, etc.) and crustacea (shrimp, crayfish, lobsters, crabs). One problem with shellfish is that they are particularly sensitive to pollutants, resulting in serious health problems due to high coastal population levels and poor or non-existent environmental controls.

1.4. OFFSHORE OIL AND GAS

Canada's Grand Banks and the North Sea -- the two most productive fishing grounds in the North Atlantic -- are also the areas of the principal oil and gas exploitation.

Production in the North Sea began in 1959 with the discovery of a large gas field off The Netherlands; this was followed by the opening of other gas fields off the Dutch and British coasts. Eventually, exploration and exploitation moved north to oil fields between Denmark and Scotland, and then farther north to more oil and gas developments in the waters between Scotland and Norway. Tremendous advances in offshore technology have occurred over the past four decades in the handling of sensitive environmental and other conditions. Exploration and development in even deeper waters can be expected.

In the case of Canada, production takes place in the Hibemian field off the south-east coast of Newfoundland. It should be noted that with the decline in the fisheries resources of the Grand Banks (including a moratorium on the harvesting of the northern cod), Newfoundland has been experiencing a severe economic decline, and the development of the oil field has had a meaningful impact on the area. Also, the Hibemian field operations have been moving farther offshore and may soon advance seaward of Canada's 200-mile limit -- the first offshore hydrocarbon operations in the world to do so.

2. Historical Background

The early history of the North Atlantic is characterised by the discovery and settlement of lands west of Europe by Norse and other groups. Iceland was visited by Irish monks before the 9th century, but they abandoned the territory with the arrival of Norse settlers (850-857). Greenland was discovered by the Vikings about 982.

At approximately the same time the Vikings apparently came to Newfoundland; over four centuries later in 1472, the Portuguese came to the area followed by John Cabot of England, and by the fishermen of several European countries who began exploiting the rich fisheries of the Grand Banks. By 1610, Newfoundland had its first permanent settlement.

Christopher Columbus, an Italian, sailed from Spain in August, 1492 and two months later landed on Watling Island in what is now The Bahamas, thus opening the way for colonisation by the Europeans. In 1510 Pizzaro, one of the Spanish conquistadors, landed in Colombia and began an intensive search for gold and silver in Colombia, Ecuador, and Peru. Three years later, Ponce de Leon, another Spanish explorer, landed in St. Augustine, Florida; the town, however, was not permanently settled until 1565, but it is still the oldest settlement in the U.S. Other settlements followed - Jamestown (1607), Plymouth (1620) and gradually the New World came to be populated and exploited by European émigrés.

The European maritime countries also looked south toward Africa and gradually established coastal trading stations where they traded European goods for slaves and gold. In 1488, Bartholomew Dias rounded the Cape of Good Hope at the southernmost tip of the continent and, three years later, Vasco da Gama reached the East African coast; a year later, he landed in India. Thus began a thriving trade involving Portugal, Holland, England and France, and other maritime powers.

The North Atlantic served not only as a start of voyages to distant continents; it was also the scene of a great deal of intra-European merchant shipping, as well as fishing. There was a long contest here between the concepts of "free" and "closed" seas. By the end of the 13th century, it had been established in Norwegian law that foreigners were not entitled to sail north of Bergen without a royal licence. Eventually, Norway passed under Danish control; until the end of the 16th century, Denmark maintained a close trade and fishing monopoly with its northern territories, and prevented other countries from utilising the waters of the Norwegian Sea.

The Dutch, however, with their expanding East Indies holdings, favoured a free seas policy. In 1605 Grotius, a young Dutch lawyer, published *Mare Liberum,* in which he held that the sea could not be subjected to private ownership (involving fishing and navigation rights). Over time, the free seas concept won out, although modified by various bands of national jurisdiction along a country's seacoast.

Commercially, as far back as the 13th century, certain North German towns formed leagues in order to obtain mutual security, exclusive trading rights and, wherever possible, trade monopolies. The most famous of these was the Hanseatic League, founded in 1241 by the German ports. Gradually, the League expanded to include Bremen, Bruges, Bergen and other ports; it remained active for nearly four centuries.

3. The 18th and 19th Centuries

By 1700, the North Atlantic was experiencing both increased maritime commerce as well as immigration to the New World. Britain was the dominant maritime power, and British exports, following the growth of the industrial revolution there, led to greater intra-European and transatlantic shipping. Immigration, particularly to eastern North America, began to grow (and continues to grow into the 20th century) as Western Europeans -- many of them landless or the urban poor -- sought to take advantage of the rich resources they had heard about in the New World.

During the American Revolution (1775-1783) the British were highly dependent on the North Atlantic as a supply route to their troops fighting the colonists. They also utilised the waterway 30 years later in the War of 1812 with the United States. The war broke out in part because of British restrictions on American trade with neutral countries (much of Europe at that time was engulfed in the Napoleonic Wars), and also because of Britain's impressment of American seamen engaged in this neutral trade. After three years the war ended, but the final treaties resulted in neither side gaining significant concessions, territorial or otherwise.

During the 19th century, European powers continued to expand their maritime holdings, particularly in Africa, but also Southeast Asia and the Pacific islands. The traditional maritime powers were joined in the latter part of the century by Germany, Italy, and Belgium, and all utilised the North Atlantic as part of their overseas routes.

The North Atlantic in the 19th century was also the scene, along with other ocean bodies, of the beginnings of the science of oceanography. Co-ordinated measurements began of winds, currents, water temperatures and salinity, and there were studies of fish, marine mammals, marine plants, and other living phenomena of the marine environment. From 1831 to 1836, Charles Darwin served as a naturalist on the *H.M.S. Beagle*, collecting specimens and observing changes in both living and nonliving marine objects. Forty years later, Britain's Charles Thompson served as director of a scientific staff aboard the *H.M.S. Challenger*, which conducted a three-year cruise starting in December, 1872. The ship travelled about 69,000 miles (128,000 km) and established over 350 observation stations around the world.

4. The 20th Century

The crowning event of the early 20th century in the North Atlantic was World War I (1914 - 1918) where, during the early years, the British and German navies clashed. At the Battle of Jutland (1916) off the west coast of Denmark, the high seas fleet of Britain met the German Grand Fleet. Although the British suffered heavy losses, the German fleet retreated to its home base, and German capital ships never again ventured out into conflict. Rather, Germany intensified the attacks of its submarine U-boats against vessels bringing supplies, including armaments, to the British. In January, 1917 the German submarines began attacking the merchant ships of neutral countries, including the United States. In April, 1917 the U.S. entered the war against Germany and its warships began convoying groups of merchant ships loaded with supplies for Britain across the North Atlantic.

Just over two decades later, the North Atlantic was again a scene of war. World War II (1939-1945) initially pitted Britain and France against Germany; the Soviet Union, invaded by Germany in 1941, became an ally of the British and French. By the summer of 1940, France had surrendered to the Germans, leaving the British beleaguered.

German submarines - now larger and more technologically advanced than twenty years earlier - again sought to blockade Britain, so that food, military supplies, and other necessities, particularly from the United States and Canada, could not reach Britain. In 1941, the U. S. leased to Britain 50 overage destroyers to help combat the threat of German U-boats.

The situation in the North Atlantic changed dramatically in December, 1941 when Japan attacked the U.S. fleet at Pearl Harbor, and the United States declaration of war against Japan was followed by Germany's declaration of war against the U.S. An escorted convoy system was again employed across the North Atlantic to Britain; the embattled Soviets also required supplies. Consequently, one part of the overseas convoy system continued on to the Soviet port of Murmansk on the Arctic coast.

In the years since the end of World War II, the North Atlantic basin has been free of armed conflict between countries, but the area, for nearly half a century, was the scene of intense naval rivalry in the cold war between the United States and the Soviet Union. The naval build-up began around 1946 and (for the Soviets) started to de-escalate after the break-up of the Soviet Union in 1989. Within a few years after that event, Russia - the primary successor to the old Soviet Union - began to experience severe economic problems and could no longer support a maximum naval effort.

One measure of naval rivalry may be seen in the number of aircraft carriers and submarines the two powers deployed in 1991, at the height of the naval build-up. At that time, the U.S. had 12 aircraft carriers, while the Soviets had five. The U.S. maintained 110 submarines to the Soviets' 239. Some of the undersea craft of both fleets were ballistic-missile-carrying submarines, designed to deploy missiles with a maximum range of 4,000 miles, and targeted on strategic points in the adversary's territory. Although these and other warships could operate in all oceans, a large number were concentrated in the North Atlantic.

The North Atlantic has been important for both war-related and peaceful advances in history. The countries bordering the North Atlantic became leaders in science, technology, and the humanities. They also became explorers and conquerors of much of the Third World, although without, for a considerable period of time, transferring their leadership knowledge to the people and institutions of the Third World.

The countries adjacent to the North Atlantic are still members of a "rich man's club" in terms of trade, technology, and commodities. But the intricate systems many of them employ to remain competitive may, in time, be threatened in a globalised world by depressions, inflations, and other negatives that they, in spite of their technological prowess, may be unable to stem.

5. Managing the North Atlantic Ecosystem

The North Atlantic, as noted earlier, is a semi-enclosed water body, which opens to the Central and South Atlantic on its southern border. Most of its marine plants and animals are in an ecosystem -- a term signifying a self-sustaining community. In this chapter, the term "ecosystem" is expanded beyond its ecological definition to include also the coastal areas of the ocean basin and the human activities, such as shipping, which are part of the geographical community.

"Management" is the orderly processing of people and events, including planning, the setting of priorities, the rejection of certain proposals, and the implementation of decisions.

Under international law today, the coastal state has exclusive jurisdiction over its internal waters (bays, estuaries, etc.), and virtual jurisdiction over its territorial sea, extending for most states out to 12 miles from its baseline along the coast. In these zones, no foreign fishing is permitted without prior international agreement.

Seaward of the territorial sea is an exclusive economic zone (EEZ) extending up to 200 miles from the baseline from which the breadth of the territorial sea is measured. Within this EEZ, the coastal state also has jurisdiction, subject to certain conditions, such as, should the coastal state be unable to harvest the total allowable catch of its fisheries resources, it should permit foreign fleets to have access to the surplus, making possible the exploitation of the maximum sustainable yield as modified by environmental and other factors.

In this section, we shall consider various activities associated with the North Atlantic ecosystem.

6. Living Marine Resources

The exploitation and management of living marine resources has several basic components. One is conservation -- how to maintain (and perhaps improve) the biological condition of the stock, or association with stocks. A second is allocation - who will have access to what portions of the stocks, or even access to any harvesting of the stocks in a given area? A third component is:

Jurisdiction - what country or international body has the authority to promulgate the rules and regulations involving fisheries exploitation in a particular area? Since all coastal states are now entitled to an exclusive economic or fisheries zone up to 200 miles off their coasts, most of the coastal waters of the world are now within some country's limits. And what of the high seas beyond the 200-mile limits? Who has authority over the exploitation of living resources there?

In past years, conservation methods have been based on the concept of maximum sustainable yield (MSY), that is, the maximum volume (weight) of a stock or group of stocks that can be harvested over a year, or other period of time, without endangering the ability of stock to replenish itself through the production of new recruits. Scientific surveys, complex theoretical models, the historical records both of the harvest of the stock in question and of its principal predators and prey -- these and other forms of evidence have been used by fisheries scientists to determine MSYs But it eventually became evident that the MSY concept was a highly complex phenomenon, involving such factors as fluctuations in water temperatures, water circulation, the year-classes of fish, degradation or destruction of nursery grounds, and improper harvesting techniques. One way to identify overfishing, that is, exceeding the MSY over time, is to watch the statistics of catch per unit effort for a particular stock. If this catch declines over a period of time, the stock is being harvested beyond its MSY.

A variation of MSY is total allowable catch (TAC), which means the maximum sustainable yield as modified by relevant environmental and economic factors. Thus, within its EEZ, a coastal state may first determine the maximum sustainable yield of its fisheries resources by objective scientific criteria, and then adjust this, if necessary, because of environment and/or economic criteria. Such a definition leaves a great deal of leeway to the coastal state.

Allocation is basically of two varieties. One involves the question of which vessels have access to a particular fishery. Assuming that foreign fishermen are excluded from fishing within the state's EEZ, there is still the issue of which of the vessels, registered within the coastal state, may partake of the fishery. At one end of the spectrum is a "free" or "open" fishery, under which all registered vessels may engage in the harvest. Although subject to certain conservation regulations, such as a minimum mesh size for their nets, trawls, etc., there may be closed areas or seasons for the fishery, or a quota on the amount of catch of a certain species which may be landed from a single trip.

At the other end of the spectrum are "limited entry" systems in which a predetermined number of licences will be issued for a given fishery. This should lead to greater fishing efficiency and to greater income for the individual fishermen. But a fundamental issue here is how to allocate the licences among a potentially large group of competing fishermen -- by selling to the highest bidder, by auction, by distribution according to historical usage of the fishery?

There are variations in the limited-entry scheme, one of them being the use of individual transferable quotas (ITQs) which can, at times, be moved from one vessel to another. But some of the basic problems here correspond to those of other limited-entry systems. How should ITQs be implemented in a multi-species or multiple stock fisheries? How should a mix of species be managed where incidental catches of non-target species are common? In allocation issues, as in those of conservation, there are still a considerable number of uncertainties to be worked out.

Regarding the North Atlantic area basin, there are principally three geographic areas with specific problems. One is the Northeast Atlantic, particularly the North and Baltic Seas, the Norwegian Sea, and the waters to the west of the British Isles. Second is the Northwest Atlantic, whose fishing grounds are shared by Canada and the United States. Third are the high seas beyond the 200 mile limits.

7. Northeast Atlantic

The principal fishing area here is the North Sea which, as noted earlier, is bordered by seven coastal states. As each of the littoral countries in the years after World War II expanded its fishing capabilities, considerable congestion and overfishing occurred. Eventually, there was a marked decline in stocks of herring, cod, mackerel and sole, with the result that in 1983, all but two of the bordering states initiated the Common Fisheries Policy (CFP). The two holdouts were Norway and Sweden. Norway abstained from the CFP because of its extensive fisheries in the North and Norwegian Seas, over which it wished to retain its control. Sweden does relatively little fishing in the CFP area.

Under the CFP, a total allowable catch is determined each year for certain commercially valuable stocks in the North Sea. Once this is done, quotas of the TAC are assigned to each of the CFP member states. Over the years, the results of the CFP efforts have not been too heartening since, for a number of the "controlled" stocks, there have been declines in catch-per-unit effort statistics. One complicating factor is the Spanish, who have by far the largest fishing fleet in Western Europe (nearly 19,000 vessels with 85,000 crew members). To date, this fleet has not been part of the CFP, but pressures for admittance are mounting, and within the next few years the moratorium on Spanish harvesting will gradually be modified.

East of the North Sea the fisheries of the Baltic Sea are under the control of the International Baltic Sea Fishery Commission, to which all Baltic Sea littoral states belong, with the exception of Denmark and Germany, whose coastal Baltic waters form a part of the CFP. The Commission determines the TACs of four species -- cod, haddock, herring, and sprat -- and allocates shares of the TAC to the member states. In recent years, supplies of cod and haddock have declined considerably.

8. Northwest Atlantic

In this part of the ocean two countries, the United States and Canada, have jurisdiction over the fisheries out to the 200 mile limit. There has been relatively little co-operation between the two states regarding the management of their respective fisheries. In 1979, a new international organisation, Northwest Atlantic Fisheries Organisation (NAFO), was formed designed primarily to regulate fishing in the international waters seaward of the 200-mile zone; however, the organisation has not been effective. Foreign fishing for "straddling stocks" has been a particularly vexing problem for Canada, where the "nose" and "tail" of the Grand Banks off Newfoundland extend beyond the 200-mile limit; over these two continental shelf areas, there are rich fishing grounds.

Canada, during the 1990s, has had considerable difficulty maintaining its stocks of Northern cod and turbot (Greenland halibut). The continued decline in cod stocks prompted the Canadian government in 1992 to impose a moratorium on cod fishing within its exclusive fishing zone, leading to the unemployment of some 50,000 Canadian fishermen and plant workers. Yet Spanish and Portuguese fleets continued to fish for cod and turbot in the nose and tail of the Grand Banks, leading to some ugly encounters, particularly between the Canadians and the Spanish, beginning in 1995. The enactment by the United Nations in that same year of the United Nations Agreement on Highly Migratory and Straddling Fish Stocks should eventually provide some relief to this highly charged situation. The Agreement provides that both coastal and distant-water fishing states shall co-operate in the conservation and management of straddling stocks through the use of available scientific data and with provisions of dispute settlement mechanisms.

9. The High Seas

Turning to North Atlantic high seas fisheries (with the exception of the nose and tail of the Grand Banks), it should be noted, first, that there are three main species of commercial importance -whales, tunas, and swordfish. All three are found in considerable quantities within many 200-mile zones as well.

Whaling is principally governed by the International Whaling Commission (IWC), which began operations shortly after World War II. The Commission, which now has about 40 members, was designed to effect a system of international regulations for the whale fisheries, and to ensure effective conservation of the whale stocks. These objectives were to be met through data acquisition and assessment and, based on scientific findings, the adoption of regulations for the conservation and utilisation of whale resources.

Unfortunately, these goals have never been realised. Since the IWC membership included pro-whaling states (e.g., Norway and Japan), anti-whaling states (United States), and countries with no whaling experience whatsoever (Finland and Switzerland), it is difficult to win adoption by the IWC of the scientists' findings. After years of steadily declining catches (particularly in the Antarctic waters), the IWC, in 1986, instituted a complete moratorium on all whaling, except for subsistence fishing by indigenous peoples (such as the Eskimos), or whales taken for scientific purposes. Canada withdrew from the IWC in 1982, and Iceland did ten years later. In 1993, Norway, an IWC member, resumed the commercial whaling for minke whales in the Northeast Atlantic and the Arctic Ocean. The minke whale is relatively small and, according to some experts, the Northeast Atlantic stock should be large enough (perhaps 60,000 animals) so that a modest harvest could be justified.

A second, highly migratory species are the tunas, particularly albacore and yellowfin in the North Atlantic. These are regulated by the International Commission for the Conservation of Atlantic Tunas (ICCAT), which collects and disseminates scientific information. On the basis of the data received and analysed, the Commission recommends conservation measures for all tuna stocks of the area.

The Commission also has the responsibility of regulating Atlantic swordfish, a species which in the 1960s averaged more than 250 pounds per fish and, which today, averages about 90 pounds. Overfishing has caused the amount and value of the commercial swordfish catch in the past decade to decline more than half, and the recreational fishery is virtually extinct. If the current trend continues, there will come a day when swordfishing in the North Atlantic is no longer a viable commercial proposition.

There are other problems. Overfishing of cod, haddock, and flounder have virtually decimated these species on Georges Bank off southern new England. There is little, if any, co-operative management of fisheries on the Bank. Red tides *(Pfiesteriapiscidia)* are threatening fish and shellfish in coastal waters. These, and other phenomena, jeopardise the abundance of fish and shellfish in the coastal waters of the region.

10. Mariculture

With the gradual decline in capture fisheries in the North Atlantic, increasing attention is being paid to fish farming of mariculture, one of the world's fastest growing food industries. Fish are grown in pens or other devices, where controls are kept on disease, weight, reproduction, and other factors. Among the most favourite species are salmon, shrimp, catfish, and various forms of shellfish. For example, up to half the salmon consumed in the United States is now of the farmed variety.

One problem of fish farming is the environmental damage it may cause. Mangrove swamps and other intertidal areas must be destroyed for some of the mariculture projects. Effluents and drugs used to control disease may spread into nearby waters. But whatever the ill effects, the mariculture industry in the North Atlantic is continually expanding, and scientific efforts are improving to fashion a product which, in terms of texture, taste, and size, is superior to the wild variety of certain species.

11. Marine Pollution

The waters of the North Atlantic -- particularly coastal belts, bays, and semi-enclosed seas, are liable to pollution from two categories of agents -- land-based sources and vessel-source pollutants. The land-based sources involve primarily direct outfalls and polluted rivers entering the sea, as well as runoff from agricultural lands (often covered by insecticides), from streets and parking lots, discharges from sewage treatment plants, disposal of plastics, etc. In most cases of land-based pollutants, there is little the international community can do to remedy the situation, except in the case of multi-national agreements.

The International Maritime Organisation (IMO) has taken a number of steps to combat vessel-source pollutants. Over the past 40 years, it has adopted conventions relating to pollution prevention, as well as navigation efficiency and marine safety. Among the agents of vessel-source pollution are chemicals, sewage, radioactive wastes and, most important, oil. Some oil discharges are also generated by offshore oil operations. There are various standards the IMO has set within the offshore jurisdictional zones off a state's coast, including regulations on compensation a coastal state can claim for damage resulting from intentional or operational discharge of pollutants within these zones.

There have been a number of international agreements on pollution in the North Sea (such as the 1969 Agreement for Co-operation in Dealing with Pollution of the North Sea by Oil), and in the Baltic (the 1974 Convention on the Protection of the Marine Environment of the Baltic Sea Area), but such agreements are largely for scientific studies and lack specific anti-pollution measures. Nor is the rest of the North Atlantic effectively covered in terms of environmental protection and preservation, except for the 1975 London Dumping Convention which contains a "black" list of substances, completely banned from dumping in the oceans, and a "gray" list of substances that may be dumped only if a prior special permit has been obtained.

12. Commercial Shipping

The North Atlantic is the site of a great many shipping routes, not only between North America and Northwest Europe, but also between other ocean basins, such as the Mediterranean, the South Atlantic and the Pacific, and the eastern and western coasts of the North Atlantic. There are several very congested points -- the English Channel area, the approaches to New York City, the Strait of Gibraltar, the Florida Straits, and the waterways connecting the Baltic and North Seas. Many types of cargoes are handled, such as dry bulk and liquid, general cargoes, containers, refrigerated ships, etc.

Although each North Atlantic port, or group of ports, has its own specific rules and regulations for handling vessels and cargoes, there are certain general conventions, promulgated by IMO, which apply world-wide. These conventions cover three particular issues:
- protection against pollution, particularly by oil;
- vessel safety; and
- liability for damages caused by vessels.

Two conventions relating to vessel safety are the Convention on International Regulations for Preventing Collisions at Sea, and the 1974 Safety of Life at Sea (SOLAS). One feature of the 1972 convention is the establishment of Traffic Separation Schemes -- a provision which might have prevented the 1956 collision south of Nantucket of the Italian flagship, '*Andrea Doria*', and the Swedish Stockholm ', where 44 lives were lost. And despite IMO's actions, there have been, of course, a series of oil tanker accidents in the North Atlantic, including the *'Torrey Canyon'* (1967) off Brittany, the *'Argo Merchant'* (1976) off Nantucket, and the *'Amoco Cadiz'* (1978) off the coast of France.

13. Other Activities

There is considerable scientific research in the North Atlantic, both in coastal waters and the deep ocean. A number of important oceanographic institutions, laboratories, and educational facilities are located in the area, and the level of scientific and technological output from these facilities is probably unmatched anywhere in the world. The topics handled include physical, chemical, biological, and geologic oceanography, and there is close interaction among the various marine science centres.

Marine recreation is an important activity in this generally affluent basin, whether it be sports fishing, sail or power boating, or other forms of pleasure seeking. Often, recreational fishermen compete with commercial fishermen for the same species, such as tunas, dolphins, bluefish, flounders, and salmon. Likewise, recreational boaters may, at times, interfere with the movements of commercial fishing and of fishing boats.

Coastal zone management is well advanced, particularly in the United States and Canada, and involves designing rules and promoting plans for both the short-term and long-term development of the narrow belt of waters along the coast, together with the strip of land immediately landward of the tide line. This has become a highly complicated activity, involving various jurisdictions and competing interests. Some North Atlantic states feel that simple land-use and sea-use planning are sufficient for their needs, and that the superposition of a complex coastal zone management layer in this transition area is unnecessary.

14. Potential Future Trends in the North Atlantic

Any attempt to forecast future trends in the North Atlantic basin must be based on two assumptions:
1. that no war or other major catastrophe occurs within the foreseeable future; and
2. that the general level of prosperity continues to prevail. Technology will continue to play an important role in the use and management of the marine environment, and the countries and industries of the area will continue to seek maximum benefits from utilising the resources of that environment.

To start with, the fishing industry will undergo significant changes. The catches of traditional, high-value food fish, such as cod, haddock, flounder, and mackerel, will continue to decline, in some cases to the point where they are commercially extinct. The stocks of lesser value food fish (e.g., butterfish, squid, bluefish) are not endangered and are largely exported to Asia. Meanwhile, mariculture will expand considerably, particularly for salmon, shrimp, and molluscs. New technological developments will facilitate the mariculture growth and reduce its pollution risks.

A second technological gain will be the general introduction of the Satellite Positioning System (SPS). With a hand-held computer, the crews of ships can determine their position on the ocean surface to within a few feet. Consequently, there is no technical reason why narrow sea lanes cannot be extended both within the territorial seas and the EEZs of coastal states, as well as through high seas.

Continuing on with shipping, there may well be land bridges developed in Europe for container traffic using rail or road connectors from such ports of Bremerhaven, Rotterdam, Antwerp, and Le Havre into Central and Eastern Europe and Russia. Such containerised land bridges already exist in the United States and Canada, and their development in Europe could add to the benefits expected from the gradual economic integration of that region.

There will be increased pressures in the coming years on the coastal zones on both sides of the North Atlantic, and growing problems of private vs. public access to the water. Demands will grow for public parks, beaches, and marinas. This leads to yet another trend, a growing need for zoning, both in the land and water areas of coastal zones. With the use of SPS, water areas can be clearly identified and charted.

Certain it is that technology will soon permit ever-deeper drilling for oil and gas on the continental shelves on both sides of the North Atlantic. But this also brings continued issues of destruction of natural undersea habitats, the routes of pipelines on the sea bottom, and reception and storage facilities on the shoreline. Surrounded by a rim of crowded, highly developed countries, the North Atlantic is a scene of a plethora of intense and vocal debates -environmentalists vs. citizen groups vs. business interests, nationalists vs. internationalists, traditionalists vs. technical innovators. Of all the ocean basins in the world, the North Atlantic has probably the highest incidence of marine-related confrontations anywhere.

References and Notes

Abrahamsson, B.J. (1989) International Shipping Developments, Prospects and Policy Issues, *Ocean Yearbook 8,* 15 8-176.

Bimie, P. (1 99 1) Problems Concerning Conservation of Wildlife in the North Sea, *Ocean Yearbook* 9, 339-369.

Bjorndal, T. and Conrad, J.M. (1998) A Report on the Norwegian Minke Whale Hunt, *Marine Policy* **22**, 161-175.

Couper, A., ed. (1 983) *The Times Atlas of the Oceans,* Times Books, Publishers, London.

Day, D. (1995) Tending the Achilles 'Heel of NAFO: Canada Acts to Protect the Nose and Tail of the Grand Banks, *Marine Policy* 19, 257-271.

Dzidzomu, D. (1998) Four Principles in Marine Environment Protection: A Comparative Analysis, *Ocean Development and International Law Journal, 29 pp.91-125.*

Fernandes, L., et al. (1 996) A Conceptual Framework for Measuring the Sustainability of the Use of the North Sea, *Ocean Yearbook* 12, 358-378.

Karagiannakos, A. (1 996) Total Allowable Catch (TAC) and Quota Management System in the European Union, *Marine Policy* **20,** 234-249.

Knauss, J. (1 997) The International Whaling Commission, *Ocean Development and International Law Journal* **28,** 78-89.

McDorman, T.L. (1998) Canada and Whaling: An Analysis of Article 65 of the Law of the Sea Convention, *Ocean Development and International Law Journal* **29,** 179-18 5.

Meltzer, E. (1994) Global Overview of Straddling and Highly Migratory Fish Stocks: The non sustainable Nature of High Seas Fisheries, *Ocean Development and International Law Journal* **25,** 255-345.

1. A territorial collectivity is administered by a directly elected General Council, with the French Government represented by a Prefect. It is represented in the National Assembly in Paris by one Deputy and in the Senate by one Senator.

2. Unless otherwise noted, all mileage figures are in nautical, not statute, miles. One nautical mile (6,076 feet) equals 1. 15 statute miles, or 1,852 kilometers.

3. Within the EEZ, a coastal state not only has jurisdiction over fisheries, but also over marine scientific research, protection and preservation of the marine environment, the exploration and exploitation of seabed minerals (including hydrocarbons), and the establishment and use of artificial islands. Some states, such as Canada, claim only a 200-mile exclusive fishing zone.

CHAPTER 12. THE MEDITERRANEAN AND BLACK SEAS

A. VALLEGA

1. The Large Mediterranean Marine Region

The Mediterranean Sea and the Black Sea occupy some three million square kilometres of marine space. They are supposed to be relics of the ancient Tethys ocean, whose creation dates back to 225 million years ago (Triassic period, Mesozoic era), and whose compression is thought to have initiated 38 million years ago (Oligocene period, Cenozoic era) because of the colliding shift of African and Euro-Asian plates. In light of the common origin of these seas, it would be appropriate to consider them together and to propose, at least as for study purposes, the concept of the Large Mediterranean Marine Region (LMMR), meant as a unique enclosed marine space, linked to the Atlantic Ocean by the Strait of Gibraltar (5 km long, 13 m minimum width, 500 m average depth), and to the Red Sea and Indian Ocean by the Suez Canal (161 km long, 90-100 m surface width, 16.1 m maximum depth). This region would include three distinct sub-regions:

- the Mediterranean Sea embracing the Tyrrhenian, Adriatic, Ionian and Aegean seas, plus the Sicilian Channel and the water interposed between the Sardinian and Spanish coasts;
- the marine interface between the Mediterranean and the Black Sea embracing the Sea of Marmara, the Dardanelles and Bosphorus straits;
- the Black Sea area together with the Sea of Azov, and the connecting corridor consisting of the Kerchensky Strait. The Mediterranean Sea comprises some 84.4 per cent of the surface and the 87 per cent of the water volume of the LMMR.

Table 1. The Mediterranean and Black Sea. General Data.

Seas	Sea area	Greatest distance	Greatest distance	Maximum depth	Water volume
	000 sq km	N/S km	E/W km	m	000,000 cu km
Mediterranean Sea (A)	2,505	960	3,700	4,846	3,653
Black Sea (B)	461	600	1,100	2,245	547
Total (C)	2,966	-	-	-	4,200
B:A·100	18.4	66.6	29.7	46.3	15.0
A:C·100	84.4	-	-	-	87.0
B:C·100	15.6	-	-	-	13.0

Adapted from [1,2]

The Mediterranean Sea is bordered by 21 countries including 5 African and 3 Asian states, whilst the Black Sea has only 6 countries bordering it.

2. Articulated Physical Contexts

The plate tectonic dynamics, which has involved the LMMR, has not only been characterised by the convergence of the African and Euro-Asian plates, but also by the rotation of the Arabian plate and that of the Turkish, and Sardinian-Corsican micro-plates. This plate shift has caused geological stress in many areas, particularly in those located along the Anatolian fault extending from the Black Sea to the Central Mediterranean. Stress is shown by a linear concentration of volcanoes and widespread seismic activity. [3]

Despite this common origin and subsequent tectonic involvement, the Mediterranean and Black Sea have different ocean floor features. A submarine ridge extending from Sicily to the African side divides the Mediterranean Sea into two distinct eastern and western parts. The western Mediterranean includes three submarine basins - Alborán, Algerian, and Tyrrhenian - separated from each other by submerged ridges, and endowed with large abyssal areas. The eastern Mediterranean mainly includes the Ionian Basin, north-west of which is the Adriatic Sea, and the Levantine Basin, north-west of which is the Aegean Sea with wide abyssal areas. Essentially, the Mediterranean environment is based on extensive, morphologically articulated abyssal areas, and on narrow continental shelves, except for the northern and central Adriatic Sea, and the Gulf of Gabes on the eastern coast of Tunisia.

Mediterranean hydrodynamics is driven by three layers of water masses: surface, intermediate, and deep. In its western basins, the surface layer has a remarkable varying thickness, from 250 to 1,000 ft, fundamentally determined by a minimum temperature at its lower limit. Shifting seawards, the minimum temperature is generally absent, and a layer of low-temperature decrease is found instead. The intermediate layer, situated at depths between 1,000 and 2,000 ft, is infused with warm, saline water from the eastern Mediterranean, and reaches its highest temperature and salinity at 1,300 ft. The deep layer, where the great bulk of Mediterranean water is contained, occupies the remaining zone between the intermediate layer and the bottom, and generally consists of very homogeneous water.

The Mediterranean Sea receives only about one-third of the amount of evaporated water lost from rivers. In summer, the Mediterranean surface water becomes increasingly saline through the intense evaporation, and its density increases correspondingly. It therefore sinks, and the excess of this denser bottom water emerges into the Atlantic Ocean, over the sill forming the shallow Straits of Gibraltar, as a westward sub-surface current below an inward current. The inflow water descends to 230 or 260 ft below the surface. The Mediterranean has been metaphorically described as breathing, for it inhales surface water from the Atlantic and exhales deep water in a counter-undercurrent.

As a consequence of salinity and density, the Mediterranean has a continuous inflow of surface water from the Atlantic Ocean. After passing through the Straits of Gibraltar, the main body of incoming surface water flows eastward along the North African coast. This current, which is very powerful in summer when evaporation in the Mediterranean is at a maximum, is the most constant component of the local circulation. This inflow of Atlantic water loses its strength as it proceeds eastward, but it is still recognisable as a surface movement in the Sicilian channel and even off the Levant coast. This anti-clockwise circulation pattern generates separate water movement in both eastern and western basins. Due to the complexity of the northern coastline and the numerous islands, many small eddies and other local currents form essential parts of the circulation. Although tides may only be significant in the Gulf of Gabes and in the northern Adriatic, they do further complicate the currents in narrow channels like the Straits of Messina.

Turning to the eastern part of the LMMR, it should be noted that the Black Sea bottom is mostly composed of the Euxine abyssal plain, and therefore its continental shelves are narrow, except for the northern and western sides, from which shelves extend both westward (Sea of Marmara) and eastward (Sea of Azov). A cardinal feature of the Black Sea is that oxygen is dissolved only in the upper water levels. Below a depth of about 230 to 330 ft at the centre, and 330 to 500 ft near the edge, there is no oxygen and the sea is contaminated by hydrogen sulphide, bringing about a saturated, gloomy, "dead" zone frequented only by adapted bacteria.

The main current flows anti-clockwise, with its branches forming gyres (eddies) and sometimes large closed rotations. It flows relatively slowly on the surface in the open sea, but reaches 16 to 20 in (40 to 60 cm) per second near the shoreline, while its speed is a mere inch or so per second in the depths. Flows in the Bosphorus are complex, with Black Sea surface water going out and deep, saltier water coming in from the Sea of Marmara. Surface winds are strong influential factors, especially in the shallow sill, or threshold, between the two basins. Besides, they also affect flows to and from the Sea of Azov through the Kerch Strait. Water exchange through the Bosphorus is slow, and a complete recycling of Black Sea waters takes about 2,500 years.

The overall water balance of the sea results from a combination of the factors of precipitation, inflow from the continental mass and the Sea of Azov, surface evaporation, and outflow through the Bosphorus. The annual water level, therefore, varies according to factors influencing these components. Vertical intermixing of water, except at or near the wind-whipped surface, is limited because of the compact, and hence stratified, nature of the sea. Hundreds of years are required to bring water in a cycle from depth to surface, although there is some limited bottom turbulence caused by the warmth of the Earth's crust and by chemical reactions in the seabed.

According to literature, essentially research and consequent scenarios designed in the framework of the UNEP Mediterranean Action Plan [12,13], the Mediterranean Sea is not expected to undergo great sea-level variations because of climatic change. By mid-21st century, the sea-level is estimated to rise between 20 and 40 cm, with variations reaching their thresholds particularly in lowlands, including deltas, such as the Nile, Po and Ebro. [4,5,6,7,8] It goes without saying that, despite forecasts of minimum variations, coastal management programmes should consider carefully the impacts from land uses on the bio-geochemical cycles, such as coastal erosion and extreme events, because most coastal belts are subject to intense population and strong human pressure. The Black Sea has been investigated on a smaller scale, and some scenarios expect the sea-level to rise from 1 to 4 m by the end of the 21st century. [9,10] A large discrepancy among predictions does reflect the need to adopt a more comprehensive approach which considers the Mediterranean and Black seas jointly.

3. Variability in Ecological Conditions

The environmental conditions of the Mediterranean Sea and Black Sea essentially differ because of the different role of factors influencing the ecological conditions, especially in the coastal environment. [11] Fundamentally, the Mediterranean is affected by the coastal resource use and navigation. The river runoff, such as in the Ebro, Rhone, Po and Nile mouths, is ecologically influential but not to such an extent to being regarded as a key problem. On the other hand, the Black Sea is highly influenced by river runoff, mostly on account of two circumstances. First, many cities and industrial areas are located along the rivers flowing into the marine waters e.g. Danube, Dnestr, South Bug, Dnepr, Don, and they discharge large amounts of solid and liquid waste. Secondly, the hydrological properties of these rivers are extensively exploited to produce energy, and therefore many dams have been built along their course influencing the local ecosystems, as well as the conditions of brackish and marine coastal water. These factors may be regarded as the key reasons why the essential problem of the Black Sea is eutrophication caused by excessive nutrients flowing into its water. [14]

The ecological risk jeopardising the Mediterranean was socially and politically perceived as critical in the first half of the 1970s, and led the Mediterranean states to adopting the Mediterranean Action Plan (MAP, 1975), under the auspices of the United Nations Environment Programme (UNEP), and the Barcelona Convention on the Protection of the Mediterranean Sea against Pollution (1976). Investigations carried out during the 1970s and early 1980s within this co-operation framework concluded that 'for the load of organic matter, expressed in terms of biochemical and chemical oxygen demand, about 60 of 65% of the total load is generated from coastal sources while the rest is due to river discharges. It is estimated that 635,000 tonnes of petroleum hydrocarbons reach the Mediterranean Sea per year out of which 330,000 tonnes are spilled oil from tankers, ballasting and loading operations, bilge and tank washing, 170,000 tonnes as land-based discharges (160,000 from municipal and 110,000 from industrial sources) and 35,000 tonnes as atmospheric deposition. Eutrophication, as reported for embayments and estuaries around the Mediterranean, is mostly associated with the release of untreated domestic/industrial wastewater. Eutrophication is a local rather than regional problem in the Mediterranean Sea. It is frequent in the north Adriatic, Izmir Bay, Elefsis Bay, the lagoon of Tunis and in all such areas where the rate of input of domestic-industrial waste water exceeds that of the exchange with the open sea. [15]

The ecological stress affecting the Black Sea became the focus of intense discussions in the late 1980s and early 1990s.[16] According to Vinogradov and Simonov (quoted by Mandych *et al.* 1992: 196), the ecological conditions of the northern and eastern sides of this sea, approximately coinciding with the northern part of the region, are essentially due to the following six causes considered in order of decreasing importance:

(i) natural and anthropogenic changes of river runoff,
(ii) increase of biogenic and organic matter income to the sea with river runoff,
(iii) income of pollutants of different origin with river runoff,
(iv) direct pollution of the sea surface with petroleum and wastes of economic activities,
(v) strong pollution of straits connecting the Black Sea with the Mediterranean, and
(vi) expansion of trawling. Unfortunately, such a comprehensive view [37] cannot be found in literature as regards the southern and south-western sides, approximately coinciding with the subtropical part of the region. Hence, an unbalanced assessment has characterised the two climatic areas of the temperate north and the sub-tropical south which comprise the Black Sea and Sea of Azov [38]

The Odessa area has been affected by a strong impact caused by municipal waste, the enormous runoff of Dnepr and South Bug rivers, and maritime navigation. Here 'the shallowness and negligible exchange of water with the central part of the sea stimulates the development of hypoxia in their bottom waters'.[17] This evaluation is confirmed by Keondijan *et al.*,[18] who stress that 'the chronic pollution of the shelf zone by industrial and domestic wastes, toxic chemicals, oil and petroleum products as well as increased content of easily soluble organic and biogenic substances coming by the rivers inflow have caused great reduction of the number of biological species in the bay waters'. Waters facing the Danube delta and Dnepr deltas, essentially influenced by river runoff, were classed as 'polluted' and 'dirty', whilst the waters of the Karkinitski and Kalamitski gulfs, widely influenced by industrial waste and drainage water, were classed as 'very dirty'. The Sevastopol Bay, affected by dumping and merchant and naval carriers, was claimed to be 'polluted' and 'dirty'. The South coast of the Crimea, Krasnodarsky Krai and Pitsunda Point, essentially affected by municipal waste, were considered as 'polluted'. Very demanding ecological problems, caused by industrial installations, mostly consisting of petroleum products and municipal waste, characterise the seaport water of Sukhumi, Poti and Batumi areas. These circumstances lead to the conclusion that 'the contemporary state of the Black Sea, the natural environment of its coasts has such stage of unfavourable changes for Man's activities that the northern areas can be considered to be the region of an ecological calamity'.[17]

4. The Changing Geopolitical Context

Sea use implementation in various parts of the LMMR has been influenced more by geopolitical than ecological conditions. Indeed, in this respect, the early 1990s may be regarded as a watershed in both the Mediterranean and Black Sea. The changes should not be underestimated; the Mediterranean Sea was no longer affected by the Cold War, while the Black Sea no longer served as a base for the Soviet Union's naval strategy involving the Mediterranean, Red Sea and Indian Ocean.

During the Cold War, the Soviet Union used the Black Sea to pursue two strategies: (i) a *naval strategy,* according to which their bases along the Black Sea rivers, mostly on the western side of the basin, enabled their warships to reach the Mediterranean, Atlantic and Indian oceans; and (ii) an *economic strategy*, according to which the Black Sea seaports near the big river mouths, such as Constance (Danube), Odessa (Dnestr and Dnepr), and Rostow (the Don-Donez system), could be utilised as interfaces between the river and maritime transportation, served by both bulk and container ships, and could therefore act as nodes for the southward and eastward Soviet merchant routes. The naval strategy has declined since the collapse of the Soviet Union, whereas the economic strategy has not only persisted but has become increasingly important. Since the liberalisation of international trade, the Imperial route of Russian power, connecting the Suez Canal and Red Sea to the Indian and Pacific Oceans, has acquired an attractive new role. The interaction between the naval and economic strategies and, in some historical periods, the leading role of the latter were emphasised by Vigarié (1976). [21] In this respect, Tietze, 1992 [22] highlighted the so-called Pontus-based Russian strategy where the combined use of river and maritime ships has induced Russia to link the Baltic Sea in the north, with the Caspian and Black Seas in the south. On the whole, there is an extensive strategic stage whose role in international trade will be implemented according to the following factors: technological (the improvement of sea-river vessels), economic (the development of the former Soviet area), and political (the mitigation of stress and disorder in the regions surrounding the Black Sea.

At present, this view is somewhat utopian, because the Black Sea and Caspian regions have been affected by intense geopolitical stress. The Islamic world has exerted great pressure on the Caucasian area where the Islamic republics - Kazakstan, Turkmenistan and Azerbaijan - form a geopolitical wedge turned to the Russian space. At the end of the 20th century, this spatial process has been highlighted by the uprising of the Dagestan region against Russia to establish an independent Islamic state. Geopolitical stress has also been caused by conflicting relations between Russia and some important components of the past Soviet Union, such as Georgia, Ukraine and Moldavia. Since the early 1990s, Georgia has been dealing with Russia-supported movements fighting the local government in the hope of uniting Southern Obsezia to Northern Obsezia. Ukraine has coped with the Crimean separatist movements, and Moldavia has faced the Russia-supported separatist movements in the Transdnestr area. As a result, the regions facing the ancient Pontus Eusinus may be divided into two areas: the south-east subject to Islamic pressure, and the south and north where conflicts between Russia and other states have taken place and spread. Despite its cultural and historical causes, this geopolitical stress is primarily concerned with the role of the maritime space, which is an essential gateway of Russia towards the westward (Mediterranean and Atlantic) and eastward (Red Sea and Indian Ocean) trading areas. Competition arising from the prospect of exploiting the offshore oil and gas fields and marine biomass resources has also had a strong potential role.

During the Cold War, the Black Sea served as a base for the Soviet Union to confront the Western World, whilst the Mediterranean Sea was extensively, and at times dramatically, subject to the consequences of that confrontation. Their waters had systematically been garrisoned by NATO and Warsaw Pact fleets till the collapse of the Communist Empire. Under these traumatic conditions, the Mediterranean was centre-stage in the attempt to foster collaboration between the developed and developing worlds. In this respect, the United Nations and the European Economic Community (EEC) played a leading role, but results were achieved above all because most northern Mediterranean countries, such as Spain, France and Italy, benefiting from a long tradition of cultural interaction, managed to form constructive relations with countries on the Asian and African sides (apart from Libya), especially with the Maghreb area (from Morocco to Egypt). The Mediterranean was also at the centre of the confrontation between the Arab and Western World and had to deal with a long period of terrorism and guerrilla-war. In this climate, despite the frustrating Cold War atmosphere, two events had a crucial influence on establishing collaboration. First, the Suez Canal, which had been closed in 1967 as a consequence of the third Arab-Israeli war, was reopened in 1975, thereby restoring the Mediterranean as the gateway for east-west maritime routes, and enabling all the Mediterranean countries to trade with the outside world. Consequently, world containership fleets, including carriers sailing round-the-world routes, and which have been growing ever since the late 1960s, began to regard this sea not only as a key space to unload containers for inland Europe but also as a passage space to transit to the Atlantic Ocean (westbound routes) and Indian Ocean (eastbound routes). Second, mainly thanks to EEC efforts, the north and south began co-operating and relations between Europe and the Arab World improved. This process was reinforced in the 1980s when Greece and Spain joined France and Italy to become EEC members (1981), therefore re-shaping the European political basis on which inter-regional trade and investments could be encouraged.

Although the east-west confrontation in the Mediterranean suddenly declined after the collapse of the Soviet Union (1991), there were fresh outbreaks of regional conflicts, especially in Yugoslavia (1991), which were the cause of renewed stress. As a result, the Mediterranean waters have continued to be sailed by naval fleets, which have somewhat disturbed the prospect of optimising merchant and passenger traffic, and have continued to be characterised by recreational sea uses in some parts of the basin, particularly in the Ionian and Adriatic seas. The post-Cold War era has also been marked by the rise in the illegal maritime trafficking of drugs, arms, nuclear materials, and immigrants. Consequently, maritime stress throughout the LMMR caused by global factors has been replaced by regional stress calling for international intervention. It was hardly surprising when NATO military forces went into action for the first time, not during the Cold War, but during the Kosovo conflict (1999).

Undoubtedly, the evolving geopolitical conditions of the LMMR should require a more detailed investigation than can be dealt with here. In particular, the role of the maritime jurisdictional zones should be regarded as a key factor influencing the prospect of optimising the use of marine resources. In this respect, the Mediterranean and Black Seas are marked by very different conditions. As can be seen from Table 2, all the Mediterranean coastal and island countries claimed their 12 nm territorial seas, except for Greece and Turkey which claimed 6 nm zones in order to leave a corridor for ships sailing between the Aegean and Marmara Seas, and Syria which claimed an excessive 35 nm zone. [23,24] Only a few Mediterranean countries claimed their contiguous zone and fishery zone.

As regards coastal waters, the most important political process has consisted of the establishment of continental shelves by bilateral agreements, which are required by international law of the sea where local states are endowed with opposite and contiguous baselines distant less than 400 nm. This initiated in the late 1960s when Italy and the former Yugoslavia reached an agreement on the subdivision of the northern and central Adriatic Sea, therefore giving birth to the offshore oil and gas industry in the Mediterranean. Bilateral conventions followed, being essentially concerned with the maritime areas surrounding Italy and facing Libya. As a result, the present framework is marked by an imbalance between the central Mediterranean, on the one hand, which is almost completely covered with agreed continental shelves, and the rest of the Mediterranean, where these zones are few (western Mediterranean) or absent (eastern Mediterranean). Morocco and Egypt proclaimed their intention to establish their exclusive economic zones, but this was not feasible since there had been no bilateral agreement, which is legally necessary because the baselines of the opposite and contiguous states are less than 400 nm. During the 1970s, the EEC claimed on their part to be against the establishment of these kinds of zone in the Mediterranean therefore determining the adverse approach from its member states. Nevertheless, recently Spain has established an Exclusive Fishery Zone.

Substantially, the jurisdictional conditions of the Black Sea do not differ from those of the Mediterranean because, in the former space too, the coastal states have baselines of less than 400 nm, and so the continental shelves and exclusive economic zones may be established only by adopting agreements. Continental shelves were agreed between Turkey, Russia and Georgia. Consequently, the central and southern parts of the Black Sea were divided into two jurisdictional domains, the western one influenced by collaboration triggered by the European Union (EU, established in 1993), and the eastern one troubled by geopolitical stress involving the former Soviet Union space.

Table 2. The Mediterranean Sea. National maritime jurisdictional zones

State	CR	TS	CZ	EEZ	EFZ	CS
Albania	(*)	12	-	-	-	200m/EXP
Algeria	1996	12	-	-	50	-
Bosnia and Herzegovina	1994	12	-	-	-	200m/EXP
Croatia	1995	12	-	-	-	200m/EXP
Cyprus	1988	12				EXP
Egypt	1983	12	24	(1)	-	200m/EXP
France [3]	1996	12	24	-	-	-
Greece [3]	1995	6/11 [5]	-	-	15	200m/EXP
Israel	(*)	12	-	-	-	EXP
Italy [3]	1995	12	-	-	-	200m/EXP
Lebanon	1995	12	-	-	-	-
Libyan Arab Jamahiriya	-	12	-	-	-	-
Malta	1993	12	24	-	25	200m/EXP
Monaco	1996	12	-	-	-	-
Morocco	-	12	24	(2)	-	-
Slovenia	1995	12	-	-	-	200m/EXP
Spain [3]	1997	12	200	-	(5)	200m/EXP
Syrian Arab Republic	(*)	35	41	-	-	200m/EXP
Tunisia	1985	12	-	-	--	-
Turkey	(*)	6	-	-	-	-
Yugoslavia	1986	12	-	-	-	200m/EXP

Notes
[1] Egypt stated its intention to claim EEZ but, as a matter of fact, this could only be done as a result of bilateral agreements.
[2] Morocco stated its intention to claim EEZ. It was done on the Atlantic but not Mediterranean side where bilateral agreements are required.
[3] In 1977 the EEC decided that their Mediterranean member States should not claim, or agree on, EEZ.
[4] The 10-mile limit applies for the purpose of regulating civil aviation.
[5] In 1997 Spain claimed a exclusive fishery zone in the Mediterranean;
(*) The state did not sign the Convention.

Abbreviations
CR: Ratification of 1982 UNCLOS; TS, territorial sea; CZ, contiguous zone; EEZ, exclusive economic zone; EFZ, exclusive fishery zone; CS, continental shelf; EXP, exploitability; 200nm/EXP, depth (200 m) plus exploitability; nm, nautical mile.

Source: [25,26].

Table 3. Continental shelves of the Mediterranean Sea. Bilateral agreements.

Year	States	Marine area of interest
1968	Italy—Yugoslavia	Northern and Central Adriatic Sea
1971	Italy—Tunisia	Channel of Sicily
1974	Italy—Spain	Western Mediterranean, Sea of Sardinia
1977	Greece—Italy	Ionian Sea
1984	France—Monaco	Western Mediterranean
1970s-1980s	Libyan Arab Jamahiriya—Tunisia [1]	Central Mediterranean
1986	Libyan Arab Jamahiriya—Malta [2]	Central Mediterranean
1992	Albania—Italy	Southern Adriatic

[1] The establishment of the continental shelf was marked by this sequence of events. *1977*: special arrangement for the submission to the International Court of Justice on the difference in the design of the continental shelf boundary; *1982*: International Court of Justice, judgement; *1985*: International Court of Justice, application to revise and interpret the 1982 judgement in the case concerning the continental shelf.

[2] In 1976 a special arrangement for the submission to the International Court of Justice on the difference in the design of the continental shelf boundary was reached. In 1985 the judgement of the Court of Justice was enunciated. In 1986 the subsequent agreement between Libya and Malta was achieved.

5. Sea Use Development

As already mentioned, 21 states, including two island countries (Cyprus and Malta), surround the Mediterranean Sea, and 6 states border the Black Sea. Turkey is the only state facing both these seas, as well as the Marmara Sea. As an estimate, the coastline length of the whole region is approximately 57,000 km, 88 per cent of which belongs to the Mediterranean. The Land Coastal Pressure (LCP), meant as the ratio between the land area and coastline length, may be estimated at about 170 sq. km to each coastline kilometre. However, as Table 4 shows, the LCP of the Black Sea is almost double that of the Mediterranean Sea. Instead, Coastal Human Pressure (CHP), indicating the number of people living in the surrounding sea areas per coastal kilometre, may be estimated to be about 7,500 in the whole LMMR. Yet, this pressure in the Black Sea (about 18,000 inhabitants) is three times more than in the Mediterranean (about 6,000 inhabitants). Despite being merely approximate, these indicators emphasise that the Black Sea has been subject to much higher pressure than the Mediterranean. [39] Furthermore, the population on the Black Sea is distributed much more diversely than in the Mediterranean. The major concentrations of residential, agricultural and industrial settlements may be found on the northern side, from the Danube delta to Georgia. [27] In particular, 'the shores of the Black Sea are among the most densely populated in Russia. Throughout the area there is a rather dense network of small and medium rural and resort settlements, together with many towns of various sizes. The population on the Black Sea coast on the whole grows, mainly due to health-resort and industrial towns; the rural population on average is stable or declining'.[28]

Table 4. Coastal pressure indicators

Seas	Land Surface (a)	Inhabitants (b)	Coastlines (c)	LCP (a:c)	CHP (b:c)
	'000 sq. km	thousands	km		
Mediterranean Sea	7709.7	314, 976	50,743	151.9	6,207.2
Black Sea	1899.8	120,760	6,670	284.8	18,104.5
Total	9609.5	435,736	57,413	167,3	7,589.5

The countries facing these seas are marked by very different economic conditions, which have exerted a strong influence on coastal resources exploitation and use development. The gross domestic product (GDP) per capita, and the human development indicator (HDI) as designed by the UN Development Programme (UNDP), may provide a preliminary view of the wide economic diversity. The poorest Mediterranean country (Albania, less than 3,000 USA dollars) has a per capita product seven times less than that of the richest country (France, more than 21,000 USA dollars) The richest Black Sea country (Turkey, more than 5,000 USA dollars) has a per capita product equal to 25 per cent of that of the richest Mediterranean country. Finally, the poorest Black Sea country (Georgia, 440 USA dollars) has a per capita product fifty times less than that of the richest Mediterranean country. When attention turns to HDI, an indicator conceived to estimate the quality of life, the framework changes significantly. The lowest ranking Mediterranean countries are not Balkan, but are two Arabian states, Morocco and Egypt, whose human development level is estimated as about 60 per cent of that of France, which, as far as quality of life is concerned, is statistically the leader in the global Mediterranean region.

The political efficiency of the Mediterranean and Black Sea may be assessed tentatively by comparing, as UNDP proposes, the GDP measuring wealth production with the HDI evaluating social conditions. When the HDI of an individual state ranks higher than its GDP, then its economic and social system may be regarded as efficient; inversely, the opposite deduction may be drawn when the HDI ranks lower than its GDP. The Mediterranean and Black Sea frameworks are marked by very different conditions. As regards the Mediterranean, the whole Maghreb, Syria, and to some extent Italy, are characterised by rather inefficient systems, while conditions seem to be poles apart on the opposite side - especially France and Spain - which are marked by much more efficient systems. Except for Turkey, all the Black Sea countries would have been successful in ensuring better conditions than those allowed by their GDP.

Bearing this framework in mind, the sea uses may be considered by distinguishing the coastal and deep-sea contexts.

Currently, there is no sufficient basis for implementing a homogenous methodological approach towards use development in the Mediterranean and Black Sea. To compensate this gap, it is worth noting that the Mediterranean Sea entered a post-modern stage in the 1970s, heralded by the decline of conventional uses, and rise of a small group of uses which have become increasingly important. In the 1990s, a similar process began in the Black Sea, essentially triggered by the transition from a centralised to a capitalist economic organisation. As a preliminary approach, the framework shown in Table 5 may be considered, which compares the implementation of uses in the Mediterranean with that in the Black Sea. Merchant transportation, tourist facilities, and oil and gas exploitation are significant uses highlighting the gap between these two seas. This alone may not necessarily lead to envisaging pessimistic scenarios. On the contrary, the gap between the two sub-regions of the LMMR may be grounds for implementing economic interaction - which had in fact persisted in the past despite the negative conditions caused by Cold War - and for designing and operating new prospects of co-operation. Nevertheless, in order to realise these conditions, the geopolitical framework has to improve, and the geopolitical stress in the Black Sea, in particular, has to diminish.

Bearing these general features in mind, some uses may be considered more concisely.

Maritime transportation—Shifting from the Mediterranean to the Black Sea, bulk transportation acquires different patterns. The Mediterranean marine space is sailed by bulk carriers (largely oil, coal, iron and bauxite ships) coming from the Indian Ocean through the Suez Canal and, less extensively, from the Atlantic through the Straits of Gibraltar. They all head towards the main EU seaports, which are mostly Spanish, French and Italian, therefore involving the north-western Mediterranean waters. By contrast, the Black Sea has always played the role of an exporting seaport area and, consequently, it has generated traffic flows along three routes directed towards the Mediterranean seaports on the EU side, the Atlantic Ocean and the Indian Ocean. The environmental impacts from bulk transportation were so strong that the International Convention for Prevention of Pollution from Ships (MARPOL, 1973-1978) declared both seas were special areas where strict measures had to be adopted to protect coastal and deep-sea ecosystems.

Container traffic is currently the most important component of merchant navigation and different features have taken shape in the Mediterranean and Black seas. As regards the routes, the Mediterranean has been characterised by three patterns:
- *transit routes*, serving traffic flows from the Red Sea and Indian Ocean towards the Atlantic Ocean, and traffic flows crossing the Mediterranean in the opposite direction, from west to east;
- *deep-ocean and mid-ocean routes*, linking the Mediterranean seaports, essentially those located on the EU side, to extra-Mediterranean seaports;
- *transhipment routes*, consisting of short-sea intra-Mediterranean itineraries around a limited number of seaports (mainly Algeciras, Gioia Tauro and La Valletta), where full deep-ocean containerships unload containers, which are then reloaded on mid-ocean carriers and transferred to their final Mediterranean destinations.

Table 5. The framework of key coastal and deep-ocean uses. A basis for comparative analyses.

Key coastal and deep-ocean uses	Mediterranean Sea	Black Sea
Ecosystem conservation [*]	●	◗
Coastal agriculture and livestock	✪	✪
Fisheries	●	●
Aquaculture	●	○
Urban settlements	●	◗
Land transportation facilities	●	◗
Air navigation facilities	●	◗
Land tourist facilities	●	◗
Merchant seaports	●	◗
Naval seaports	●	✪
Coastal industries	●	◗
Coastal thermoelectric plants	●	◗
Merchant navigation	●	◗
Naval navigation	✪	✪
Recreational navigation	●	◗
Oil and gas exploitation	✪	○
Seabed telephone cables	✪	○
Research	✪	◗
Cultural heritage protection	○	○

[*] Parks, reserves, habitats and species protection.
●, highly developed;
✪, moderately developed;
◗, less developed;
○, expected to be operated.

In comparison with the Mediterranean, the Black Sea routes are much simpler since they cross the Sea of Marmara to reach the Mediterranean, where they diverge westwards (Atlantic Ocean) and eastwards (Indian Ocean). These routes are sailed by carriers in the opposite direction supplying the finished goods from countries on the Black Sea. Some maritime transportation is linked by river transportation, essentially along the Danube. The seaport container traffic is much more extensive in the Mediterranean than in the Black Sea. During the late 1990s, most traffic has been handled in Algeciras, followed by Ligurian seaports (Genoa and La Spezia), and Barcelona. In relation to the Mediterranean, container traffic in the Black Sea has been insignificant. Such a gap might be mitigated if the Black Sea enters an economic development stage.

Tourism and recreational uses - During the last two decades, tourist and recreational uses have evolved in the Mediterranean bringing about three concurrent processes:
- the implementation of mass tourism;
- the diffusion of élite tourist resorts, such as high class coastal and island villages chiefly linked to the implementation of yachting, sailing vessels and recreational harbours; and
- cruise tourism. Quantitative and qualitative improvement of this pivotal sea use has largely benefited from increasing internationalisation of Mediterranean tourism. For a long time, similar development had been impeded in the Black Sea because of geopolitical constraints (Soviet domain in most coastal belts), and economic factors (lack of high standard recreational facilities, except for a few coastal areas). Despite the persisting geopolitical stress, significant progress has been made in this area during the 1990s, exemplified by the implementation of cruise itineraries as a prolongation of those in the Mediterranean.

Offshore oil and gas - Compared to the North Sea and the Gulf of Mexico, the LMMR may be considered as a space moderately endowed with offshore oil and gas fields. Resources have been found in a small range of areas: the western Mediterranean, especially in the marine spaces off Castellon; the Sicilian Channel, namely the subsoil between Sicily, Tunisia and Libya; most Adriatic and Ionian seas; the Aegean Sea, particularly in the continental shelf between Greece and Turkey; and the Black Sea. The exploitation of gas fields began in the Adriatic Sea during the 1970s, and has recently been followed by that of oil fields in the Ionian and Sicilian continental shelves. The exploitation of fields in the Black Sea is also expected. Increased exploitation in the Black Sea and the Mediterranean seabeds would subsequently strengthen the role of the LMMR as a second offshore oil and gas area in Europe, after the North Sea.

6. Regional Co-operation

During the first half of the 1970s the Mediterranean Sea was being piloted as a marine space. It was used as a reference with the aim of designing and putting into effect the environmental protection of marginal, semi-enclosed and enclosed seas, i.e. which form part of the ocean mostly prone to human pressure and stresses brought about by resource use. At that time, the Mediterranean was regarded as one of the most endangered seas to the point that UNEP decided to test the practicability of its Regional Seas Programme (RSP, convened in 1974) in this basin. The Mediterranean Action Plan (MAP), launched in 1975, and the subsequent Convention on the Protection of the Mediterranean Sea against Pollution (1976), were the initial steps of the far-sighted and ambitious RSP. It was also the first time in history that a regional sea was dealt with by a comprehensive approach shared by its coastal and inland states, and the first occasion where environmental protection-aimed principles, enunciated by the 1972 UN Conference on the Human Environment, could be carried out on a regional scale. [29]

Table 6. Evolution of MAP organisation

Year	MAP body	Place
1975	Adoption of MAP by 16 states	Barcelona, Spain
1976	Adoption of the Convention on the Protection of the Mediterranean Sea against Pollution by 16 states and European Economic Community (EEC).	Barcelona, Spain
1976	Launching of MEDPOL, Co-ordinated Mediterranean Pollution Monitoring and Research Programme	Barcelona, Spain
1976	Establishment of the Regional Oil Combating Centre (ROCC), today Regional Marine Pollution Emergency Response Centre (REMPEC)	La Valletta, Malta
1977	Establishment of the Blue Plan/Regional Activity Centre (BP/RAC)	Sophia Antipolis, France
1977	Establishment of the Priority Actions Programme/Regional Activity Centre (PAP/RAC)	Split, Croatia
1979	Establishment of the Mediterranean Trust Fund (MTF)	==
1980	Establishment of the MAP Co-ordinating Unit	Geneva, Switzerland
1982	Removal of Co-ordinating Unit	from Geneva, to Athens, Greece
1987	Establishment of the Specially Protected Areas/Regional Activity Centre (SPA/RAC)	Tunis, Tunisia
1989	Establishment of the Secretariat of the Protection of Coastal Historic Settlements Centre (HSC)	Marseilles, France
1993	Establishment of the Environment Remote Sensing/Regional Activity Centre (ERS/RAC)	Scanzano, Italy
1996	Establishment of the Mediterranean Commission on Sustainable Development (MCSD).	Athens, Greece

Both MAP and the 1976 Barcelona Convention were the initial steps of a long and quite intense process, which has evolved through two phases. The 1975-1995 phase was inspired by the principles of the 1972 UN Conference, and consisted of the implementation of the legal endowment by adopting protocols to the 1976 Convention, together with the creation of technical means by establishing the Regional Activity Centres (RACs) and programmes guided by the Co-ordinating Unit of MAP. [30] The second phase initiated in 1995 with the amendment of the 1976 Convention, which led to adopting the Convention on the Protection of the Mediterranean Sea and the Coastal Region, and the launching of MAP Phase II. These two concurrent legal and organisational steps were undertaken to respond to the inputs from the 1992 United Nations Conference on Environment and Development (UNCED) triggering the passage from an environment-concerned to an ecosystem-referred and sustainable development-aimed strategy. The sequence of steps of this long process are shown in Table 7.

Table 7. The Barcelona Convention system

Convention and protocols	Adoption and place	Status development [*]
Convention for the Protection of the Mediterranean Sea against Pollution Responsible: The MAP Co-ordinating Unit	Barcelona, Spain	a) 1976 b) 1978 c) 22
Dumping Protocol Protocol for the Prevention of Pollution of the Mediterranean Sea by Dumping from Ships and Aircraft Responsible: The MAP Co-ordinating Unit, Athens	Barcelona, Spain	a) 1976 b) 1978 c) 22
Emergency Protocol Protocol concerning Co-operation in Combating Pollution of the Mediterranean Sea by Oil and Other Harmful Substances in cases of Emergency Responsible: REMPEC, Malta	Barcelona, Spain	a) 1976 b) 1978 c) 22
Land-based Protocol Protocol for the Protection of the Mediterranean Sea against Pollution from Land-Based Sources Responsible: MEDPOL, Athens	Athens, Greece	a) 1980 b) 1983 c) 22
SPA Protocol Protocol concerning Specially Protected Areas and Biodiversity in the Mediterranean [**] Responsible: Specially Protected Areas, Regional Activity Centre, Tunis	Geneva, Switzerland	a) 1982 b) 1986 c) 22
Offshore Protocol Protocol for the Protection of the Mediterranean Sea against Pollution resulting from Exploration and Exploitation of the Sea-Bed and its Subsoil Responsible: MEDPOL, Athens	Madrid, Spain	a) 1994 b) not yet c) 11
Convention on the Protection of the Marine Environment and the Coastal Region of the Mediterranean Sea Resulting from the amendment of the 1976 Convention	Barcelona, Spain	a) 1995 b) not yet c) 22
Hazardous Waste Protocol Protocol on the Prevention of Pollution of the Mediterranean Sea resulting from the Transboundary movements of Hazardous Wastes and their Disposal. Responsible: not yet decided.	Izmir, Turkey	a) 1996 b) not yet c) 11

[*] *Codes*: a) approval, b) entering into force, (c) number of ratification (re: 1996).
[**] This protocol results from the amendment of the Protocol on Specially Protected Areas by the Ninth Ordinary Meeting of the Barcelona Convention (1995).
From [31,32].

In this context, a main concern of MAP has consisted of experimenting with coastal management in the developing Mediterranean coastal areas. [33] These actions have been so intense as to induce the Mediterranean states to include coastal management in the 1995 Convention, and to encourage the MAP Co-ordinating Unit to strengthen efforts aimed at operating integrated programmes. The steps taken in this area are outlined in Table 8.

During the 1990s, some twenty years after the birth of the Mediterranean co-operation mechanism, the Black Sea became a focus of international discussion, because of its environmental degradation. Its ecological constraints were so awful and worrying, and the media pressure and public opinion was so intense that, on June 21 1997, the *International Herald Tribune* published a report: 'Flooded with pollutants from land, sea and air, racked by civil war on one coast, packed with tourists on the others; trapped in the middle of one of the world's great political and economic transformations: The Black Sea, the dirtiest in the world, is dying an agonising death. In merely 30 years, the sea has degenerated from one of the world's most productive bodies of water to a toilet bowl for half of Europe. It has become a dumping ground for vast quantities of phosphorous, inorganic nitrogen, oil, mercury, and pesticides generated by the 160 million people in its basin. The result is a body of water that is fast becoming devoid of oxygen, and the fish and plant life that need oxygen to flourish'. [40] At that time, the consequences of mismanagement of the Black Sea during the Communist empire became self-evident. In 1996, the actions convened in the international arena led the coastal states to adopt the Strategic Action Plan for the Rehabilitation and Protection of the Black Sea, which should be intended as the preliminary step to convening an Action Plan in the framework of the UNEP Regional Seas Programme.

The consideration of LMMR leads to wondering, as a key question, whether effective co-operation could be established between the Mediterranean Sea and Black Sea, essentially with the objective of optimising ecological management and economic development in both areas. Some local conditions and influence from the external environment require increasing attention. As regards local conditions, there exists a basic difference between the two seas: as its main need, the Black Sea has to deal with the ecological consequences brought about by long mismanagement, while the impact caused by increasing human pressure, particularly from recreational uses, has been the key issue of the Mediterranean. As regards the external factors, strong influence has been caused by the UNEP action plans, concerned with the Mediterranean and Black Sea, have been convened in different times—the latter twenty years after the former—and that the Mediterranean coastal and island states have made efforts to strengthen co-operation while the Black Sea states have been far from such a co-operation-oriented attitude.

Besides that, in recent times the EU strategy has been directed to establish and implement increasingly wider collaboration eastwards—namely towards Central and Eastern Europe. By contrast, the EU has developed weaker co-operation southwards, namely towards Maghreb and Mashrek areas, and has been quite reluctant to strengthen co-operative linkage south-eastwards, namely towards the Black Sea and the Caspian Sea, where the geopolitical framework has not been so catalyst. To evaluate these factors in depth, it should also be taken into account that the United Nations has only the role of stimulating collaboration and address funds to developing countries, basically through the World Bank.

Table 8. Mediterranean Action Plan. Coastal area management programmes

Experimental, preparatory activity and pilot projects

1987--1988, pilot projects. Main tasks: assessment and management of individual resources, integrated planning studies, capacity building.

Places: Kastela Bay (Croatia), Izmir Bay (Turkey), Island of Rhodes, Syrian Coast

Pre-take off phase

1989--1993, Coastal Area Management Programmes (CAMPs) - 1st generation:

The Syrian Coast, Syria	1989—1992
The Izmir Bay, Turkey	1989—1993
The Kastela Bay, Croatia	1989—1993
The Island of Rhodes, Greece, 1st phase	1990—1993

Take off phase

1993--1995, Coastal Area Management Programmes (CAMPs) - 2nd generation:

The Albanian Coast, Albania	1993—1995
The Island of Rhodes, Greece, 2nd phase	1994—1996
The Region of Fuka, Egypt	1994 onwards
Sfax, Tunisia	1995 onwards

Drive to maturity

1995 onwards, Coastal Area Management Programmes (CAMPs) - 3rd generation:
Preparation of programmes for Algeria, Israel, Lebanon, Malta, Morocco. Expected to start in 1997.

Adapted from [31].

To evaluate the consequences of these factors, it should be borne in mind that the UN system, despite the resources addressed by The World Bank to the Mediterranean and Black Sea developing countries, is not in a position to allocate funds in the Black Sea comparable to that that may be allocated by the EU. Consequently, if the EU persists in its weak interest, it is unlikely that the Black Sea—despite efforts from the UN system—will develop to the point to reduce its gap to the Mediterranean substantially. Fortunately, the UE has given some signals to change its approach, which may prefigure actions oriented to fund actions aimed at protecting the environment and stimulating multi-national co-operation in the area.

These considerations lead to believing that, when centred on the prospect of co-operation between the Mediterranean and Black Sea, scenarios on the LMMR should postulate that the local conditions ought to be the most influential factors. Essentially, only where negative geopolitical conditions are removed could actions be expected to be carried out in order to set up effective changes. Nevertheless, the Black Sea is such an important area upon which stability in wide economic and political European and extra-European arenas depends, that the Western World will be encouraged anyhow to adopt special actions to give impetus to the ecological improvement and economic development. All in all, there is an uncertain future ahead and any possible scenarios are hard to predict.

Acronyms

BP	Blue Plan
CAMP	Coastal Area Management Programmes
CHP	Coastal Human Pressure
cm	Centimetre
CS	Continental Shelf
cu	Cubic
CZ	Contiguous Zone
EC	European Community
EEZ	Exclusive Economic Zone
EFZ	Exclusive Fishery Zone
ERS	Environment Remote Sensing
EU	European Union
GDP	Gross Domestic Product
HDI	Human Development Indicator
HSC	Historic Settlements Centre
in	inch(es)
IPCC	Intergovernmental Panel on Climate Change
km	kilometre
LMMR	Large Marine Mediterranean Region
LPC	Land Coastal Pressure
MAP	Mediterranean Action Plan
MARPOL	International Convention for Prevention of Pollution from Ships
MCSD	Mediterranean Commission on Sustainable Development
MEDPOL	Mediterranean Pollution Monitoring and Research Programme
MTF	Mediterranean Trust Fund
NATO	North Atlantic Treaty Organization
nm	nautical mile
RAC	Regional Activity Centre
REMPEC	Regional Marine Pollution Emergency Response Centre
ROCC	Regional Oil Combating Centre
RSP	Regional Seas Programme
SPA	Specially Protected Areas
sq km	square kilometre
TS	Territorial Sea
UNCED	United Nations Conference on Environment and Development
UNDP	United Nations Development Programme
UNEP	United Nations Environment Programme

References

1. Couper, A.D. (ed.),*The Times Atlas of the Oceans*. The Times Books, London, 1983.

2. The New Encyclopaedia Britannica, Chicago, 1991

3. Vallega A., *Fundamentals of Integrated Coastal Management*. Kluwer, Dordrecht, 1999.

4. Flemming N.C., Predictions of Relative Coastal Sea-Level Change in the Mediterranean Based on Archaeological, Historical and Tide-Gauge Data. In *Climatic Change and the Mediterranean*, vol. 1, eds L. Jeftic, L.D. Milliman and G. Sestini, Arnold, London, 1992, pp. 247-81.

5. Jelgersma S. and Sestini G., Implications of a Future Rise in Sea-Level on the Coastal Lowlands of the Mediterranean. In *Climatic Change and the Mediterranean*, vol. 1, eds L. Jeftic, L.D. Milliman and G. Sestini, Arnold, London, 1992, pp. 282-303.

6. Milliman J.D., Sea-level Response to Climatic Change and Tectonics in the Mediterranean Sea. In *Climatic Change and the Mediterranean*, vol. 1, eds L. Jeftic, L.D. Milliman and G. Sestini, Arnold, London, 1992, pp. 45-57.

7. Pirazzoli P.A., Sea-level changes in the Mediterranean. In *Sea-level changes*, ed. M.J. Tooley & I. Shennan, Blackwell, Oxford, UK, 1987, pp. 152-81.

8. Nicholls R.J. and Hoozemans F.M.J., Vulnerability to sea-level rise with reference to the Mediterranean region. In *Proceedings of the Second International Conference on the Mediterranean Coastal Environment MEDCOAST'95*, ed. E. Özhan, Middle East Technical University, Ankara, vol. 2, 1995, pp. 1199-213.

9. Kaplin P.A., Porotov A.V., Selivanov A.O., and Yesin N.V., The North Black Sea and the Sea of Azov coasts under a possible greenhouse-induced global sea-level rise. In *Coastlines of the Black Sea*, ed. R. Kos'yan, American Society of Civil Engineers, New York, 1993, pp. 316-54.

10. Intergovernmental Panel on Climate Change, IPCC, WG II Third Assessment Report, Summary for Policymakers. Climate Change 2001: Impacts, Adaptation, Vulnerability, 2001; http://www.meto.gov.uk/sec5/CR_div/ipcc/wg1/WGII-SPM.pd.

11. Shuisky Y.D., The general characteristic of the Black Sea coasts. In *Coastlines of the Black Sea*, ed. R. Kos'yan, American Society of Civil Engineers, New York, 1993, pp. 25-49.

12. Jeftic L., Keckes S., and G. Sestini G. (eds), *Climatic Change and the Mediterranean*, vol. 1. Arnold, London, 1992.

13. Jeftic L., Keckes S., and Pernetta J.C., *Climatic Change and the Mediterranean*, vol. 2. London, Arnold, 1996.

14. Sapozhnikov V.V., Biohydrochemical Causes of the Changes of Black Sea Ecosystem and its present Condition. *GeoJournal*, **27**, 2, 1992, 149-58.

«When increasing a degree of eutrophication of the basin, the intensification of heterotrophic processes is expressed first in the growth of heretotrophic organic biomass, then hytoplankton and finally detritus. Such a prevalence of the heterotrophic processes over the autotrophic ones is caused, as a rule, by the flow of a large amount of allochthonic organic substance to their basin and, as possible, due to the small sizes of bacteria and accordingly the high correlation of the surface of these microorganisms to their volume compared with higher life».

15. United Nations Environment Programme. *State of the Mediterranean Environment*. MAP Technical Reports Series No. 28. UNEP, Athens, 1989.

16. Kos'yan R.D., Magoon O.T., Man on the Black Sea coast. In *Coastlines of the Black Sea*, ed. R. Kos'yan R., American Society of Civil Engineers, New York, 1993, pp. 1-13.

17. Mandych A.F., Shaporenko S.I., Influence of the Coastal Economic Activities upon the Coastal Waters of the Black Sea. *GeoJournal*, **27**, 2, 1992, 199-210.

18. Keondjian V.P., Kudin A.M., and Borisov A.S., Practical Ecology of Sea Regions—Concepts and Implementation. *GeoJournal*, **27**, 2, 1992, 159-68.

19. R. Kos'yan (ed), *Coastlines of the Black Sea*. Series of *Coastlines of the World*, series ed. O.T. Magoon, American Society of Civil Engineers, New York, 1993.

20. Aibulatov N.A., The history of the Black Sea coastal zone studies. In *Coastlines of the Black Sea*, ed. R. Kos'yan, American Society of Civil Engineers, New York, 1993, pp. 14-24.

21. Vigarié A., L'U.R.S.S. et sa participation à l'économie des océans. In *Marine Marchande 1976*, Journal de la Marine Marchande, Paris, 1976, pp. 107-18.

22. Tietze W., The Future Role of European Marginal Seas: The Pontus. *GeoJournal*, **27**, 2, 1992, 228.

23. Prescott J.V.R. (1985), *The Maritime Political Boundaries of the World*. Methuen, London & New York.

24. Scovazzi T., Management regimes and responsibility for international straits: with special reference to the Mediterranean Straits. *Marine Policy*, **19**, 2, 1995, 137-52.

25. UN, Office for Ocean Affairs and the Law of the Sea, Law of the Sea Bulletin

26. http://un.org

27. Gastescu P., The Danube Delta: Geographical Characteristics and Ecological Recovery. *GeoJournal*, **29**, 1, 1993, 57-67.

28. Drodzov V.A., Glezer O.B., Nefedova T.G., and Shabdurasulov I.V., Ecological and Geographical Characteristics of the Coastal Zone of the Black Sea. *GeoJournal*, **27**, 2, 1992, 169-78.

29. Vallega A., Regional level implementation of Chapter 17: the UNEP approach to the Mediterranean. *Ocean and Coastal Management*, **29**, 1-3, 1995, 251-78.

30. Vallega A, Geographical coverage and effectiveness of the UNEP Convention on the Mediterranean. *Ocean and Coastal Management,* **31**, 2-3, 1996, 199-218.

31. Pavasovic A., Integrated Coastal Management in the Mediterranean: Present State, Problems and Future. In *Regional seas towards sustainable development*, eds. S. Belfiore, M.G. Lucia, and E. Pesaro, Franco Angeli, Milano, 1996, 119-33.

32. http://www.unep.org/unep/program/natres/water/regseas

33. Vallega A., Integrated coastal area management in the framework of the UNEP Regional Seas Programme: The lesson from the Mediterranean. In *Ocean Yearbook*, vol. 13, eds E. Mann Borgese, A. Chircop, M. McConnell, and J. R. Morgan, Chicago University Press, Chicago, 1998, pp. 245-78.

34. Sorensen J., Integrated Coastal Zone Management: Lessons from the Black Sea and the Gulf. In *Regional seas towards sustainable development*, eds. S. Belfiore, M.G. Lucia, and E. Pesaro, Franco Angeli, Milano, 1996, pp. 134-56.

35. Meel. D., The Black Sea in Crisis: A Need for Concerted International Action. *Ambio*, **21**, 4, 1992, 278-86.

36. These investigations may be found in [12,13]

37. It has been provided by a special issue of *GeoJournal* (Black Sea environment 1992, 27, 2). Specifically concerned with physical features, contributions on the individual countries facing the Black Sea may be found in [19].

38. A framework of investigations carried out till the late 1980s on the physical context was provided by Aibulatov (1993). [20]

39. To calculate the LCP and CHP, only a part of the land surface and population of Russia, Spain and France has been considered. These estimates were tentative. The land surface and population of Turkey were divided into two equal parts, supposing that these are roughly equally gravitating on the Mediterranean and Black Seas.

40. Quoted by Sonrensen (1996: 138-9) [34]. See also Mee (1992). [35]

CHAPTER 13. MULTILATERAL MANAGEMENT OF NORTHEAST ASIAN SEAS: PROBLEMS AND PROGNOSIS

MARK J. VALENCIA

1. Introduction

Northeast Asia [1] is almost unique for its lack of regional institutions. Bilateralism dominates both political and economic relations. Indeed the region is remarkable for "its combination of several quite highly industrialized societies, with a regional international society so impoverished in its development that it compares poorly with even Africa and the Middle East." [2] This impoverishment reflects the conflicts among the governments in the region, particularly, between the divided countries, which create enormous obstacles to the establishment of regional regimes and institutions.

Further complicating the situation, recent extension of jurisdiction by China, [3] Japan, [4] and South Korea [5] has put maritime issues on the 'front burner' of international relations. However, just as the extension of jurisdiction creates an opportunity for the re-examination of national ocean management, it also presents an opportunity for re-examining a nation's relationship with its neighbors, with a view to moving toward a more co-operative structure of international relations. Technological change and increasing maritime use and user conflicts also makes the need for regional maritime cooperation even more obvious. [6] At the least, increased bilateral and multilateral consultations and a new degree of coordination are required to meet the challenge presented by changes in marine use patterns and concepts.

Moreover, the ratification and entry into force in 1994 of the 1982 United Nations Convention on the Law of the Sea [7] give further impetus to regional cooperation in ocean management. Indeed with the advent of the 1982 Convention, the venue for addressing issues of ocean law and policy has moved from the global to the regional and bilateral level. Most significant is the positive atmosphere created by China's joining with most of the world's nations in implementing this result of a truly multilateral process. [8] South Korea, Japan, and Russia have also ratified the Convention, thus legitimizing regional cooperation in maritime matters. [9] This should put pressure on North Korea to at least abide by the Convention's provisions.

The 1982 Convention serves as a framework within which nations can carry out their ocean management rights and responsibilities. [10] More specifically, the declaration of an EEZ confers certain responsibilities, e.g., for protection of the marine environment, [11] which may be fulfilled through multilateral cooperation. [12] Indeed, cooperation in the management of enclosed or semi-enclosed seas – like the Japan (East), Yellow (West) and East China Seas – is emphasized in Article 123 of the Convention, which holds that states bordering an enclosed or semi-enclosed sea should cooperate with each other as well as international organizations in the exercise of their rights and in the performance of their duties under this Convention – regarding management of living resources, preservation of the marine environment, and scientific research.

Even given these positive factors, there are still serious obstacles to marine regionalism in Northeast Asia. [13] Indeed, the immediate benefits of international functional arrangements may pale in the light of the immense political conflicts that still divide the region. In Northeast Asia there are four countries with six governments with little history or experience in multilateral cooperation. The participation of the major powers in the region is critical to a successful regime, but they may be reluctant to participate unless they can dominate. In general, most big powers prefer to avoid multilateral regimes in which the smaller nations can form blocks against them. It will thus be necessary to present a convincing argument that such major powers can gain more benefit from a multilateral regime than bilateral agreements which they can dominate.

Another difficulty for Northeast Asia is the isolation and non-participatory stance of North Korea. Since North Korea borders and claims continental shelf, 'security zones' and EEZs in the Japan and Yellow Seas, its *eventual* participation in functional marine policy regimes is important. As North Korea emerges from its isolation it should be invited and encouraged to join existing multilateral marine management dialogues and emerging regimes.

The current system of marine management and resource exploitation in the region is based mainly on national rights and obligations with some tentative first steps towards regional cooperation. It is ineffective because of overlapping claims to maritime jurisdiction, the lack of agreed maritime boundaries, and because countries continue to act largely in their own self interest – narrowly interpreted. In Northeast Asian waters in particular, a system of unilateral EEZs and sovereign resource rights is an obstacle to an effective regional system of marine management. Indeed, there is an unusually strong adherence to independence and sovereignty in the region. These countries are thus generally reluctant to agree to co-operative activities if they appear to compromise or qualify in any way their sovereignty or sovereign rights. This constraint to cooperation is particularly significant in the maritime domain given the extended jurisdiction allowed by the 1982 Convention the resulting and numerous overlapping claims to offshore areas, islands and reefs. And it is manifested by a lack of serious political commitment to regional seas programs and other co-operative activities as well as the failure to implement or comply with international instruments. [14]

The lack of funding and resources, particularly for capacity building and marine scientific research, is also an obstacle to implementing co-operative strategies. This is a major cause of the general failure to translate the co-operative rhetoric in International Maritime Organization treaties, the Law of the Sea and the Commission on Sustainable Development into operational activities, including a higher level of technical cooperation between developed and developing countries.

In sum, the general absence of multilateral maritime regimes in Northeast Asia reflects political calculations by the nation-states regarding the rewards/risks and losses/benefits of maintaining the *status quo* versus developing regimes acceptable and beneficial to all sides involved. Asian countries are simply not yet sufficiently aware of the seriousness of the need for a multilateral maritime regime that focuses on the management of fisheries resources and maritime environment production. Indeed, when countries in Asia think maritime, they think first and foremost about boundary disputes, not protection of the deteriorating marine environment or of management of dwindling fisheries. Perhaps most important there is no strong constituency in the region for ocean management. The region suffers from a lack of public knowledge and there are few NGO's active in this arena.

It is these perceptions and mechanisms that must change – and indeed they are changing. The end of the Cold War, the concomitant warming of relations in much of the region, the extension of maritime jurisdiction, and the coming into force of the Law of the Sea Convention provide a narrow window of opportunity to forge a new order for regional seas before resurgent nationalism further complicates these issues. As marine policy problems play an increasingly important role in the international relations of Northeast Asian states, the region's nations are being drawn slowly but surely into serial bilateral dialogues through which constructive and mutually beneficial marine policies will evolve. [15] Supporting this process is the growth of an epistemic community of maritime specialists. [16] This chapter will review current cooperative efforts in maritime management in Northeast Asia, delineate the need for, and problems preventing effective, multilateral cooperation, and provide a prognosis for successful multilateral maritime cooperation.

2. Ongoing and Potential Cooperation: Enhancing the Epistemic Community

There is actually a considerable base for regional cooperation involving Northeast Asian states in the maritime sphere, although most is bilateral or of a 'soft' character. [17]

2.1. SAFETY AT SEA

There are five important areas of ongoing cooperation in maritime safety [18] – regional nations' adherence to key safety conventions of the International Maritime Organization (IMO), the Tokyo Memorandum of Understanding (MOU) on Port State Control, [19] regional cooperation in Search and Rescue (SAR), cooperation in dealing with maritime disasters, and regional cooperation in training. The most important of IMO's safety conventions are the International Convention on Safety of Life at Sea (SOLAS 1974), the Convention on the International Regulations for Preventing Collisions at Sea (COLREGS 1972), the International Convention on Standards of Training, Certification and Watchkeeping for Seafarers (STCW 1978), and the International Convention on Maritime Search and Rescue (SAR 1979). All Northeast Asian nations except North Korea and Taiwan have ratified almost all these critical maritime safety agreements. This wide adherence provides an excellent basis for regional agreement on key maritime safety areas.

2.1.1. Law and Order at Sea

The key relevant IMO Convention regarding law and order at sea [20] is the Convention for the Suppression of Unlawful Acts Against the Safety of Maritime Navigation 1998 (Rome Convention) and the related protocol on offshore oil/gas platforms. The Rome Convention was originally a response to the *Achille Lauro* incident, but it also clearly applies to piracy incidents in which violence is used or the ship is seized. It obligates states either to extradite or prosecute persons who seize ships by force, commit acts of violence against persons onboard ships, or place destructive devices onboard ships. Clearly, regional cooperation in law and order at sea would benefit by greater adherence of regional countries to this basic international convention.

Piracy has been an increasing problem in recent years in Northeast Asia. Under international law (Article 101 of the 1982 United Nations Convention on the Law of Sea), piracy is defined as illegal acts of violence or detention committed for private ends on the high seas (i.e., outside the 12 nautical mile limit of territorial waters). But the broader definition of piracy of the International Maritime Bureau (IMB) of the International Chamber of Commerce, which includes such acts in territorial seas or even in ports, is more relevant to most Asian piracy, which generally occurs in such locations. In Northeast Asia in 1993, *ad hoc* agreements between the coast guard agencies of China and Japan, and unilateral naval patrols by Russia halted increasing piracy in the East China Sea.

Beyond piracy, transnational crime at sea – including illegal drugs, smuggling, and illegal migration – are of increasing concern in a globalized, modernizing and increasingly urbanized Northeast Asia. Almost all illegal drug intercepts/arrests of drug traffickers are in ports or territorial waters, and most regional cooperation is not public information, bilateral, and between law enforcement agencies. Illegal migration, intensified in times of economic crisis, poses particular problems for national maritime authorities and requires closer cooperation between neighboring states.

There are several potential initiatives dealing with transnational crime at sea. The United States Coast Guard has developed a Model Maritime Service Code which could assist some regional nations in improving their own legislative and operational framework for enforcement of law and order, maritime safety and environmental regulations. More region-wide information sharing on maritime smuggling and drug trafficking is needed, and could build on existing exchanges between cross-border criminal information databases. Unofficial cooperation in the Council on Security Cooperation in Asia-Pacific (CSCAP), which has International Working Groups on both Maritime Cooperation and Transnational Crime, has jointly developed specific recommendations on cooperation in law and order at sea issues for future official consideration.

Illegal fishing is a major area of concern to enforcement agencies. Indeed, fishing disputes complicate relations throughout Northeast Asia. But several recent initiatives have the potential to improve regional fisheries law enforcement. In recent years, Japan and China, Japan and South Korea, Japan and Russia, and China and South Korea have all concluded bilateral fisheries agreements. This web of bilateral agreement is a natural basis for a regional fisheries agreement including enforcement mechanisms. Other positive developments include an Asia-Pacific Economic Cooperation (APEC) Oceans Conference declaration which included agreement on enhanced cooperation and data sharing, and a comprehensive vessel registry. Also, the United States Coast Guard has an active program of cooperation with other regional nations, including China, to support the UN prohibition of large-scale high seas drift netting, and is also helping to develop a regional organization to implement the UN agreement on conservation and management of highly migratory species.

2.2 FISHERIES

2.2.1. Existing International Regimes and Their Inadequacies

There is presently a web of bilateral fisheries agreements in force in the region.[21] Each regional government is involved in at least one.[22] And several of these bilateral agreements have recently been renegotiated to respond to extended jurisdiction. Notable and innovative are the new agreements between China and Japan, South Korea and Japan[23] and China and South Korea.[24] These agreements provide a solid background of experience on which to build a multilateral fisheries regime.

China and Japan

The first to strike a bilateral fisheries agreement were China and Japan on 11 November, 1997, for the East China Sea. Although differences over the boundaries of the joint management zone and conditions of operation hindered its implementation, the agreement finally came into force on 1 June 2000.

China and Japan agreed to establish three different zones where different fisheries regimes apply:

- *exclusive fishing zones in their EEZs up to 52 miles from their respective baselines in the area between 27° N and 30° 40' N.* Here the principle of coastal-state jurisdiction applies. This is a change from the flag-state principle. If Japan or China does not have the capacity to harvest the entire allowable catch in their respective zones, as provided for in Article 62 of the 1982 United Nations Convention of the Law of the Sea, each state can allow the nationals and fishing vessels of other states to fish in its EEZ in accordance with this agreement and other related laws and regulations, based on the principle of mutual interest. Every year, Japan and China will determine the quotas of catch, the fishing area, and other terms of fishing for the nationals and fishing vessels of other signatory states who are allowed to fish in its EEZ, taking into account the condition of the living resources, their own capacity to harvest the living resources, traditional fishing operations, and other related matters. In doing so, however, each state is to observe the recommendations of the China-Japan joint fisheries commission. The agreement also provides that each party may take necessary measures in its EEZ in accordance with international law to ensure that the nationals and fishing vessels of the other state observe its conservation measures. To this end, each state is to immediately notify the other state of its measures for the conservation of the marine living resources and other terms provided for in its domestic laws and regulations.

- *joint regulation in the area beyond 52 nm from each state's baselines and between 27° N and 30° 40' N.* If one state observes violations of the measures of the China-Japan joint fisheries commission by the nationals and fishing vessels of the other state, it may bring the violations to their attention, and at the same time notify the flag state of this fact and related circumstances. The state of the alleged violator should inform the notifying state of the results after taking necessary action. Each state is to take appropriate measures to control catch in order to avoid over-exploitation in the joint management area, taking traditional fishing operations into consideration.

- *exclusion of the application of the fisheries agreement to the area south of 27° N.*

To summarise, each country will manage its fisheries within 52 nm of its baselines; beyond 52 nm and between 27° N and 30° 40' N boats of the two countries may fish without prior approval of the other's government.[25] The area south of 27° N including the area around the disputed Diaoyu/Senkaku islets remains unregulated high seas.

This agreement was not easily reached. China wanted a larger joint management area than Japan. Japan had wanted the eastern boundary of the joint zone set at 127° E whereas China had wanted it set at 128° E.[26] Moreover, the agreement authorizes Japanese authorities to inspect PRC boats within its EEZ and to seize them if they are found to be violating the agreement. This had not been permitted under the pre-existing 1975 China-Japan treaty. These terms satisfied the Japanese fishing industry and ruling Liberal Democratic Party members who had been upset by Chinese boats over fishing close to Japanese territorial waters and had demanded that the government scrap the current treaty. Boats from the two countries fishing for sea bream and squid in the area had often been involved in confrontations, such as trying to block each other's fishing operations.[27]

Further, the agreement adversely affected fishermen of both countries. More than 17,000 Chinese fishing boats will be unemployed affecting 170,000 fishermen and Chinese capture fish production was expected to be reduced by a million tons a year in Zhejiang alone.[28] Indeed, because of China's new bilateral agreements with Japan and South Korea, some one million people in the fishing and processing industries will lose their jobs.[29] Meanwhile Japan has set up a 6 billion yen fund to support its fishermen unemployed because of the new agreement.[30] The payments will last until fiscal 2003 and interest on fishermen's borrowings will also be paid. The government also set aside 2 billion yen to compensate for damage to gear caused by Chinese vessels fishing in Japan's EEZ.

The agreement is valid for five years and after this period will continue to remain in effect until it is terminated by a six-month advance notice of abrogation from either party. The agreement is also provisional pending boundary delimitation of the EEZ and the continental shelf. The two states have made a commitment to continue negotiating the boundary delimitation in good faith, so as to reach an agreement. However, both China and Japan have made it clear that the provisions of this agreement do not affect their positions on other legal matters, including the issue of disputed islands and boundary delimitation of their EEZs and the continental shelves.

Complicating matters is the fact that Taiwanese boats also fish in the East China Sea, particularly around the Diaoyu/Senkaku islets which Taiwan also claims. On 7 June 2001, a Taiwan fishing vessel capsized after colliding with a Japanese coast guard patrol boat off Yonagunijima. [31] According to Japan, the boat was illegally fishing in Japanese-claimed waters. Earlier, on 16 May 2001, a Taiwan fishing boat was intercepted by a Chinese naval vessel 32 nm northeast of Pengchiayu, a small islet off Taiwan's northeastern coast. [32] Taiwan dispatched coast guard vessels and a warship to lend assistance. Also, South Korea protested the agreement and demanded trilateral talks because the designated area overlaps its claimed EEZ.

Although this bilateral agreement is a step in the right direction, there remain several fundamental problems with negative implications for both the fisheries and international relations in the East China Sea. For example, the EEZ and continental shelf boundaries have not been agreed, leaving open the question of the relevance of Taiwan's claims and role. And although joint fishing areas have been declared *in lieu* of boundary delimitation, it is not clear whether the areas within the overlapping EEZ claims are high seas or EEZ. If they are high seas, a third state can fish there without restriction. If they are EEZs, then a third state would need permission from the coastal states claiming the area as their EEZ. Moreover there is no effective dispute settlement mechanism built into the agreements. They are provisional and specific quotas and conditions of operations must be negotiated every year and may be held hostage to the tenor of political relations between the parties. Nevertheless, this agreement was a significant step towards the building of at least a bilateral regime governing maritime activities in areas of overlapping or unclear jurisdiction.

Japan and South Korea [33]

Japan and South Korea have established a joint fishing area in the East Sea/Sea of Japan and a small provisional joint fishing area south of Cheju Island. They also agreed on the establishment of a South Korea-Japan joint fisheries commission to implement their cooperation for the conservation and management of the living resources in their joint fishing areas. The commission makes recommendations to states on the conditions of fishing operations, the maintenance of the fisheries order, the condition of the marine living resources, cooperation between the two states in the fisheries field, the conservation and management of the marine living resources in the joint fishing area, and other matters relating to the implementation of the agreement. The signatory states are supposed to respect the recommendations of the commission in their determination of the terms and conditions for allowing the fishing vessels of other states to fish in their EEZ.

Joint management is applied in the joint fishing areas. The absence of a clearly defined EEZ boundary should not present a problem of overlapping jurisdictions because in reality each state will most likely enforce the laws only on its own side of a hypothetical median line. Since the laws and regulations are to be determined by the joint commission, they will be the same for the entire joint fishing area.

Indicative of the need for trilateral coordinating the Japan-South Korea commission will be exercising prescriptive fisheries jurisdiction in an area south of Cheju which is also covered by the Japan-China agreement. The Japan-China agreement contains a clause that obligates them to take into account the habitual and present fishing operations of South Korea in the area north of the joint management area. This means that South Korea can continue its fishing operations in the joint management area and in the area north of 30° 40' N.

South Korea and China

In 1996, South Korea and China discussed the delimitation of their EEZ boundary. South Korea argued for the application of the median line, while China asserted the principle of equity, taking into consideration the size of its land mass, the length of its coastlines, and the existence of islands. The talks were unsuccessful.

In 1997, the two sides recognized that it would be difficult to reach an agreement on boundary delimitation of the EEZ and instead focused on the conclusion of a fisheries agreement of a provisional nature that could regulate fishing in the Yellow (West) Sea until a boundary could be agreed. However, the overlap of the two states' EEZ claims in much of the Yellow Sea complicated the negotiations and it became apparent that conflicts in the enforcement of laws and regulations would be unavoidable.

In these circumstances, South Korea and China agreed in principle to establish a joint fishing area as a buffer zone between their respective EEZs where they could exercise equal rights. Subsequent negotiations focused on determining the width of the joint fishing zone. Such an approach had already been used in the November 1997 China-Japan fisheries agreement and probably served as a precedent and model.

The most difficult problem was the lack of a clearly defined EEZ boundary. South Korea maintained that in the absence of a boundary, the coastal states could set hypothetical boundary lines in accordance with the principles of boundary delimitation in international law and that the establishment of a small joint fishing area around these hypothetical median lines would be "provisional." China argued that there would be no EEZ without a boundary and that therefore, no state could unilaterally enforce its EEZ laws and regulations unless a boundary could be agreed. It continued to reject South Korea's proposed hypothetical median line and instead insisted on the application of equitable principles in the boundary delimitation of the EEZ boundary.

There was also the issue of whether the coastal state or a joint management committee should be responsible for the management of the living resources. South Korea argued that it would be better for the coastal state to conserve and manage the living resources. On the other hand, China argued that joint management of the living resources would be more effective than unilateral management.

The two states also differed on whether a fisheries agreement should cover their entire EEZs. South Korea asserted that the entire EEZs of both states should be covered by the accord. It argued that both the Japan-China and South Korea-Japan fisheries agreement covered their entire EEZs and that there was no reason to confine the application of an accord only to the Yellow Sea. China maintained that the accord should cover only the EEZ between the opposite coasts of the two countries in the Yellow Sea, because the areas of EEZs of the two states were not equal.

Another important issue was the respective widths of the EEZ and the joint fishing area. Because many more Chinese fishing vessels were engaged in fishing in and near South Korean waters than South Korean vessels operating near the Chinese waters, China wanted a narrow EEZ and a wide joint fishing area. It argued that its larger territory and population warranted this arrangement. But South Korea argued that if the joint fishing area was too wide, the EEZ fisheries regime would be meaningless where the Yellow Sea is narrow. Thus, South Korea sought to establish a narrow joint fishing area so that the coastal states would have the major responsibility for conservation of the living resources in the Yellow Sea. These differences reflected more fundamental conflicting positions on boundary delimitation in general and their respective fisheries interests in particular.

The eventual agreement established three different zones between the two countries in the Yellow Sea: EEZs, joint, and transitional. The EEZ is the area where the coastal state exercises its sovereign rights; the transitional areas are zones of about 20 miles in width on both sides of the joint fishing area, where the nationals and fishing vessels of the two states are allowed to fish. The zones extend northward to 29° 45' latitude. These zones are to be incorporated into the respective EEZs after four years of joint management. The two countries agreed to gradually reduce their fishing so as to maintain balanced fishing in the transitional areas; take measures for the conservation and management of the living resources in accordance with the decisions of the joint fisheries commission; jointly conduct surveillance to ensure that the conditions for fishing are observed; and to exchange a list of fishing vessels for effectively carrying out these obligations. The establishment of the transitional area was a compromise between the positions of the two states, namely, South Korea's support of *de facto* coastal state EEZs and China's proposal for a wide joint fishing area.

Unresolved Issues and Inadequacies

Although these bilateral agreements are a step in the right direction, there remain several fundamental problems with negative implications for both the fisheries and international relations in the East China Sea.

- The EEZ and continental shelf boundaries have not been agreed, leaving open the question of the relevance of Taiwan's claims and role. [34]

- Joint fishing areas have been declared in lieu of boundary delimitation. Although innovative, it reflects the inability of the states to agree on the delimitation of their EEZs. Indeed, if they could have clearly delimited their boundaries, the contrived arrangement of the joint fishing areas would not be necessary. Moreover, the agreements leave open the question of whether these areas are within overlapping EEZ claims, high seas or national EEZs. If they are high seas, a third state can fish there without restriction. If they are EEZs, then, a third state needs permission from the coastal states claiming the area as their EEZ.

- There is almost no coordination between the China/Japan, China/ROK and Japan/ROK agreements, yet they include some of the same areas and some of the same stocks. China and Japan did not consult South Korea whose EEZ claim overlaps part of the area covered by their fisheries agreement. Article 311(3) of UNCLOS states that "the provisions of such agreements do not affect the enjoyment of other States Parties of their rights or the performance of their obligations under this Convention." Also Article 74(1) of the Convention provides that states with opposite or adjacent coasts should negotiate to determine the boundary delimitation if their EEZs are less than 200 nm. Moreover, migratory species move between the East China Sea, the Yellow Sea (West) Sea, and the Sea of Japan (East Sea). These facts suggest the practical need for China and Japan to negotiate with South Korea regarding their overlapping EEZs.

- The effectiveness of uncoordinated conservation, management and enforcement between the different joint fishing areas is questionable

- The area north of the China-Japan Provisional Measures Zone (PMZ) and south of the South Korea-China PMZ is open to fishing by any nation and may be over fished [35]

- There is no effective dispute settlement mechanism built into the agreements

- The agreements are provisional and specific quotas and conditions of operations must be negotiated every year and may be held hostage to the quality of political relations between the parties.

Thus the current fishery regime in Northeast Asia remains insufficient and fragmented. Although Northeast Asian countries have undertaken unilateral measures to regulate foreign fishing in their coastal waters, [36] none of the regulatory regimes includes all coastal or fishing nations nor is any state a party to all the agreements. Hence there is no forum wherein all Yellow Sea and/or East China Sea fishing nations can meet to discuss the condition of the stocks or the distribution of catches. Nor is there a clear dividing line for national fisheries areas. And geographically, none of the bilateral agreements takes into account all of the region. Some areas are not covered under any formal agreement, and in some areas agreements appear to overlap.

Under the present *de facto* regime, coastal states have attempted to reserve coastal fisheries for their own fishing interests. The bilateral agreements make some modest attempts to share catch within jointly regulated areas, or to limit efforts in order to stay within catch quotas for designated areas. But direct discussion of allocations would entail reviewing systematically the existing *de facto* allocations, and re-examining the legal regime in the area with respect to fishing rights. Discussion of shared stock questions in the Sea of Japan would also require involvement of North Korea.

This competitive environment constrains the sharing of scientific information. The bilateral fisheries commissions established under the agreements generally do not publish their decisions or the results of their scientific deliberations for peer evaluation or general public information. Without information on the basis for decisions, the necessity of, and rationale for the regulations cannot be fully understood, nor can their success be evaluated.

Despite some advantages, notably the lack of overt conflicts over fisheries until recently -- due mainly to self-restraint -- the present regime is fundamentally flawed in terms of fisheries management. Few species can be managed by only one country. Although the stocks are often transboundary in distribution involving two or more states, there is no corresponding multinational body to manage them. And the parties concerned often produce significantly different resource assessments. Japan is a fisheries hegemony; with a virtual monopoly of quality information -- although there remain significant knowledge gaps, for example, regarding stocks in the Seas of Japan and Okhotsk.

Though perhaps inequitable, this system, comprising an interlocking web of bilateral agreements dominated by one nation, could, in theory, successfully manage the region's fisheries, particularly if hidden factors serve to make the regime more equitable. But the advent of the 1982 Convention, the extension of jurisdiction by all of the states in the region, the development of China's offshore fishing capability, utilisation of unconventional species, and the full or overexploitation of most stocks indicate a need for regime evolution or expansion. The fact that many species are over-fished by itself indicates that the system is not working and underscores the need for multinational monitoring and regulation of this multispecies fishery, and ultimately, of fair allocation of the resource.

In Northeast Asia, fisheries policy must address the common problems of over fishing, management and allocation of shared stocks, scattered and fragmentary data, unresolved jurisdictional boundaries, difficult political relationships, and the inadequacies of the current fisheries management regimes. Historically, conflict rather than cooperation has been characteristic of fisheries relationships among Northeast Asian countries. Despite its outwardly tense stability, the current balance in the regime involves serious compromises in national positions, and for this reason it is unlikely to remain stable for long. In sum, regional agreements and organizations are needed to manage shared stocks, to allocate catch, and to obtain economies of scale in costly research, management and training, and in the capture, marketing and trade of fisheries products.

Fortunately, multilateral fisheries regimes for Northeast Asian seas are being officially "explored". In March 1993, Russia announced that it wanted to hold multinational negotiations with Japan, South Korea, China, Poland, and other countries on fishing in the Sea of Okhotsk. [37] And in what is probably the most notable regional fishery initiative to date, participants from Japan, Russia, China, and South Korea agreed in 1994 to establish an international committee on protection of the fishery resources in the Sea of Japan. The agreement came at a conference sponsored by the Hyogo prefectural government in western Japan. The meeting also agreed to urge North Korea to participate. [38] Movement toward a multilateral regime may be slow but it is inevitable and inexorable.

2.3. MARINE ENVIRONMENTAL PROTECTION

2.3.1. Overview

In a few other semi-enclosed seas are multilateral measures for marine pollution control as deficient as those in Northeast Asia. Indeed beyond coastal waters, most Northeast Asians seas are a '*mare nullius*' in terms of marine environmental protection. [39] Sensitive political relations and uncertain boundaries have not been conducive to information-sharing and cooperation on many matters, let alone the environment. This situation has made it difficult to evaluate the nature and extent of support for international environmental activities or even national positions thereon.

There is a general dearth of capacity and will to cooperatively monitor marine pollution. And there is no effective arrangement to provide the critical mass of international collaboration and cooperation in monitoring and research activities necessary to delineate the spatial distribution of a contaminant and its subsequent effects, and, in particular, whether it would cross national boundaries. The lack of such an arrangement prevents the development of well-coordinated cooperative baseline studies, and coordination in emergencies, such as a spill of oil or other toxic and hazardous materials.

Existing monitoring and research programmes are ineffective because they stop at artificial, politically-determined borders, rather than at a physical or chemical border. Among the countries bordering the Yellow and East China Seas, there are wide discrepancies in the level and effectiveness of marine pollution monitoring and research in support of regulation. Japan is clearly far superior to the others in terms of marine environmental knowledge and technology. Russia has the capability but neither the will nor the means to fully utilize it. China has carried out extensive surveys and research in its 'off-shore' areas since the early 1970s. But comprehensive marine research programs in South Korea and Taiwan have begun more recently. It is unknown, but doubtful, whether North Korea has undertaken such activities.

Except in response to occasional tanker accidents that have destroyed coastal fisheries and aquaculture areas, and to the severe effects of untreated industrial effluents on public health, there has been only minimal overt recognition by the Northeast Asian coastal state governments of the long-term effects of land-source, vessel, and other pollution on people and the marine environment. Scientific questions on factors affecting the health of marine species and ecosystems are poorly articulated, and the relevance of national laws and policies to regional environmental protection has not been seriously considered. A review of national legislation shows little evidence of laws and regulations being developed with specific relevance to natural features or processes that may affect pollutant transport, circulation, transformation, and dispersion. Laws and policies are couched in terms that separate legal justification and intent from the reality of people, ecosystems, and place. This problem exists in other regions as well but it is more important in Northeast Asia, because the apparent failure to relate law more directly to nature through improved scientific understanding supports a general disinterest in marine environmental issues.

Thus the degree of concern with marine pollution is quite varied, and practical policy towards it is even more diverse. Japan is clearly the leader in marine pollution policy and prevention in the Northeast Asian region. Marine pollution awareness and prevention are much more recent phenomena in China, South Korea, Taiwan, and Russia. Their laws and regulations are sufficiently strict, but a wide gap exists between the law and its implementation and enforcement. Although marine pollution is becoming a critical problem in these countries, industrial and economic growth remains the dominant national ethos. Moreover, uncertain jurisdiction due to the lack of agreed EEZ boundaries in Northeast Asian seas complicates any effort towards cooperation, and perhaps necessitates the involvement of an international organisation as an intermediary or coordinator.

In sum then, there is, except in some heavily polluted coastal areas, little public awareness of the importance of marine environmental protection. Governments still tend to see environmental problems as peripheral issues to be acknowledged but effectively ignored. Their attempts to draft regulations have invariably been hindered by the need to balance the interests of competing national and provincial level sectors, such as coastal and offshore shipping interests, fishing and fish processing enterprises, coastal inland development construction and water conservancy bureaucracies, port and harbour administrations, and agriculture and industrial ministries.

Thus the main constraints to regional cooperation in marine environmental protection remain poor political relationships and environmental ignorance and apathy. Transboundary pollution, coordination of regulations and their implementation, and a 'tragedy of the commons' [40] mentality are the main pressing issues. Two trends are apparent: increasing marine pollution with concomitant damage to living resources in semi-enclosed seas, especially in the Yellow Sea, and a growing environmental consciousness, which may spillover into the marine sphere. It is not clear that improving intra-regional relations and environmental consciousness will overtake and mitigate an environmentally damaging ethos before irreversible damage is done.

2.3.2. *Initiatives for Marine Environmental Protection and Their Deficiencies*

Despite the relatively poor record of the region's entities in joining or adhering to international conventions protecting the marine environment, the new wave of environmentalism has stimulated a proliferation of multilateral discussions and program proposals for environmental protection. But the motives and rationale for these new initiatives may be more broadly political than concern for the environment. By calling attention to politically benign but mutually threatening environmental issues, states sometimes can achieve broader political objectives such as opening channels of communication and building confidence.

Ongoing multilateral co-operative efforts relevant to marine environmental protection in Northeast Asia include the Working Group for the Western Pacific (WESTPAC) established under the auspices of UNESCO's Intergovernmental Oceanographic Commission 53; the UNDP/GEF Program on Prevention and Management of Marine Pollution in East Asian Seas; [41] the North Pacific Marine Science Organisation; [42] and the United Nations Environment Programme (UNEP) Northwest Pacific Action Plan (NOWPAP). The latter is the most advanced, although implementation has lagged well behind original expectations.

Indeed NOWPAP faces many problems. Some parties remain uncomfortable with the geographic definition of the region for cooperation. The present compromise definition is the Yellow Sea and the Sea of Japan (East Sea), with provision for including adjacent areas in the future. This excludes Taiwan and hence the China/Taiwan problem. But the Sea of Japan and the Yellow Sea are quite different oceanographically, and have different environmental problems. Moreover, South Korea and China prefer to focus initially on the Yellow Sea where they have already initiated a cooperative project. But Japan is more interested in the Sea of Japan where some cooperation exists because of the Russian dumping of nuclear waste there. In their discussions, different countries even use different names for the seas, underscoring the political difficulties for cooperative work.

Disagreement also continues over the priorities of projects for cooperation. The present compromise is regional assessment, establishment of a database, monitoring, and cooperation in emergencies. The differences in country priorities reflect different levels of economic development. China, North Korea and Russia all welcome, monitoring, but Japan is not as interested because it already has a well-established monitoring system.

Prevention of disposal of radioactive waste at sea was added as a NOWPAP task although Japan expressed reservations, presumably because it wishes to retain this disposal option. There are also differences regarding the means of implementation of the projects. In one instance, Russia proposed the use of a Russian ship and North Korea agreed, but Japan and China were not interested in financing it. China and Japan want to use work already done or being done by the coastal states to develop a standard method. But South Korea is more interested in a contingency plan for emergencies, particularly for the Yellow Sea.

Harmonisation of legislation and the ultimate goal of a regional convention also present technical and political problems. The coastal nations have different levels of economic development and different discharge standards. China emphasises its industrial development, rather than pollution control. North Korea is a reluctant and passive participant. Moreover, the industrial structure and technology differ from country to country, as does the preferred use of maritime areas. And even if the water quality and discharge standards were to be made similar, enforcement standards would differ widely.

Japan would have difficulty signing a formal convention, presumably because it has difficult relations with Russia and North Korea, and because it feels that ratifying a convention may obligate it to ratify all relevant international agreements. So Japan wants to separate an international legal agreement from the rest, and to make the regime *ad hoc*. For its part North Korea strongly suggested that such excuses should not be accepted as a means to avoid full participation in NOWPAP.

There is also disagreement over the all-important allocation of costs. Japan's Environment Agency does not have the resources to fund the program, which means that funds would have to come from the Ministries of Finance (MoF) and Foreign Affairs (MFA). Japan could provide funding as part of its ODA but Tokyo does not consider Russia a developing country, and it does not yet have political relations with North Korea. On top of all this, the MoF and MFA do not want to send representatives to UNEP meetings where they will be embarrassed by requests for money. Moreover from Japan's own point of view, Japanese have little expertise in foreign countries, do not generally speak foreign languages, and are familiar only with problems of an advanced country. North Korea's participation has been minimal in part because it is not able to contribute financially to the projects.

Despite these problems, NOWPAP has made some advances. It has educated regional elites on the need for a more comprehensive regional environmental policy. It has created or expanded domestic constituencies for environmental protection. It has enhanced knowledge of the quality of regional seas. And it has transferred marine science technology and knowledge to the developing countries, thereby increasing their capacity and confidence to participate in regional marine environmental protection regimes. It is this latter achievement that is likely to be most significant and long lasting.

With the assistance of the World Bank, both China and South Korea have developed an Action Plan for Monitoring and Protection of the Yellow Sea Large Marine Ecosystem (YSLME). The large marine ecosystem is a new concept linking exploitation, development, and administrative efforts. It sets aside national rights and interests to consider the whole ecosystem of the Yellow Sea. The goal is to establish bilateral cooperation to protect the ecosystem and make possible sustainable utilisation of the biologic resources of the Yellow Sea. North Korea recently accepted Global Environment Facility Funds for combating pollution in the Yellow Sea, thus enhancing the possibility of trilateral cooperation.

2.3.3 Inadequacies of Existing Regimes and Ways Forward

There is considerable redundancy of activities in the above programs. Although the programs could be complementary there is no coordination to achieve synergy. Despite the numerous initiatives, it will clearly be far easier to implement environmental assessment, legislation, and institutional arrangements than a regional management and financial structure. The concept of the EEZ is not yet ingrained in the psyche of policymakers. Besides, the more obvious problems and the initial effects of new ones are most likely to arise in waters close to land, and national attention is therefore concentrated on protecting the health of the coastal waters rather than the offshore, especially in enclosed and semi-enclosed seas. And countries generally resist the involvement of other nations in their coastal waters, no matter how well intentioned.

Despite efforts at national, regional, and international levels, the current sectoral and monodisciplinary approach to the multiple uses of marine and coastal resources will not provide an effective framework for achieving sustainability. Aside from physical and ecological degradation of the coastal and near-shore zones, and of course, nuclear waste dumping, pollution from land-based sources is at present the single most important threat to the Northeast Asian marine environment, contributing some 70 percent of the pollution load of the oceans.

Prospects for improved transnational cooperation in resource development and use may depend, therefore, upon better understanding of the causes and consequences of marine pollution in both coastal and open-sea areas. Indeed, increased knowledge is extremely important to the creation of regimes, and accounts for the expansion and strengthening of marine pollution regimes worldwide. [43] The most successful efforts to deal with marine environmental problems appear to have been carefully nurtured with simultaneous institution-building, scientific, and treaty-drafting activities at the regional level, but this can come about only with strong and sustained littoral state support and state or international organisational leadership. Environmental consciousness in the region must be raised further, new institutional arrangements developed, and new economic theory applied, incorporating environmental benefits and pollution costs. Laws must be harmonised and co-operative monitoring achieved, especially regarding future industrial development. Particular emphasis should be placed on ocean dumping, red tides, and the environmental hazards of nuclear waste.

Now that the 1982 United Nations Convention on the Law of the Sea is in force and ratified by all states in the region except North Korea, the region's countries must decide how to adjust their national initiatives so that they are compatible with emerging international legal and technical obligations or, conversely, the extent to which each of them wishes to ignore or deviate from international practice. The basic need to draft national regulations that reflect and incorporate the vaguely defined intent of Convention Articles 192 and 194 will require that choice to be made. These articles charge states with the 'duty to protect and preserve the environment' and obligate them 'to take all measures necessary to prevent, reduce, and control marine pollution and to ensure that activities under their jurisdiction or control do not cause pollution damage to other states or otherwise spread beyond the seas where they exercise sovereign rights.[44] Yet, there are no agreed scientific criteria to determine the precise meaning of such terms as 'prevent, reduce, and control.' It is also difficult to determine how to justify and enforce legal prescriptions, given the limitations of scientific and technical knowledge. And there is a large gap between acceptance of a vaguely defined legal framework, which moves from 'obligations of responsibility' to 'obligations of regulation and control,' and the willingness and ability of states to establish and enforce standards and rules.

An ideal regional marine environmental protection regime must satisfy many theoretical needs as well as national interests. Above all it should rectify existing inadequacies, and disparities in capacity. It should rationalise the redundancy of the existing and proposed international programs. It should provide the consultative channels, or infrastructure for cooperation needed for synchronic monitoring, coordinated baseline studies, and prevention and cleanup of transnational pollution. It should coordinate policies and regulations for national zones and tailor them to fit natural features and processes, such as current systems and ecological zones, whether near shore, offshore, temperate, or boreal. It should foster coordination and sharing of the results of research in individual zones. It should serve to educate the public and policymakers as to the causes and consequences of marine pollution, thereby evening up the degree of knowledge and concern among countries, particularly for offshore living resources and ecosystems.

Perhaps most important, it must provide opportunities to upgrade the capacities of North Korea and others to assess, monitor, prevent, control, and combat marine pollution. Without this assistance, North Korea and perhaps China may not be able to participate effectively in negotiations and ensure their concerns are reflected in policy formulation and would thus be less likely to comply with resultant agreements. Indeed such assistance may be the major incentive for North Korea, China, and perhaps Russia to participate in the regime. And Japan and South Korea may be motivated by the opportunity to establish a diplomatic relationship with North Korea. Moreover, the increased scientific knowledge that would accrue could reduce ecological and economic uncertainty regarding the distribution, effects, and costs of pollution. This could, as in the Regional Seas Programme for the Mediterranean, mitigate the common fear of developing countries that developed countries are trying to make them pay for pollution the developed countries caused, and to make them less competitive by diverting resources from economic development.[45]

Initially each government could manage its own jurisdictional areas according to agreed standards, perhaps using those of centrally located South Korea as a base, but with the monitoring capability of Japan, the most developed country in the region. Decisions should be by consensus and implemented by voluntary acquiescence to the rules. Compliance would be achieved through detection, publicity, and persuasion.

The benefits of an ideal regime would be positive but varied for each participant. Although all participants will lose the ability to treat the sea as a free waste dump, all will clearly benefit from cleaner seas. Though considerable, these benefits are also unquantifiable because of the long-term nature of the impacts of an environment, which is less polluted than it might have been, and the uncertainty regarding the causal relationships between pollution and ecosystem damage.

Perhaps the most important benefit would be the evening up of the levels of marine environmental technology and expertise throughout the region. The overall objective of the arrangement is to manage the marine environment of Northeast Asian seas. The most important spin off may be the provision of greater equity: equity in the sense of increased national responsibility to control pollution with potential transnational effects, and equity in the sense of a transfer of technology and knowledge from the rich for the benefit of all. In short, the major trade-off would be the benefit to Japan and South Korea of adherence by China, North Korea, and Russia to a predictable regime with common minimum standards of discharge, in exchange for training and technical assistance from Japan and South Korea.

2.4. MARINE SCIENTIFIC RESEARCH

"Illegal" scientific research in another country's EEZ is increasingly considered a security issue when the countries in question are rivals for power and leadership in Northeast Asia. Since 1998, Chinese marine scientific research and navy ships have been frequenting the area claimed by both China and Japan in the East China Sea, causing great concern and consternation in Japan. Indeed, Japan claims that activities carried out by Chinese ships in Japan's claimed EEZ over the past five years include a collection of data for military purposes as well as exploration of natural resources - both in violation of the 1982 United Nations Convention on the Law of the Sea. [46] Chinese research vessels were sighted on 16 occasions in 1998, 30 times in 1999 and 24 times in 2000 operating within Japan's claimed EEZ in the East China Sea. [47] In 1999 four sightings occurred within the 12 nm territorial waters of the disputed Diaoyu/Senkaku islands. Although Japan's Maritime Safety Agency asked the vessels to leave the area and to cease the research, they refused. The vessels appeared to be collecting oceanographic data necessary for naval operations and or to delineate oil and gas potential in the area. Some of the Chinese activities were concentrated near the Amami islands and some involved magnetic and seismic exploration for hydrocarbons.

Japan suspected that the increasing activities of Chinese marine research vessels on the Japanese side of the Japan-China equidistant line are designed to make such activities a *fait accompli* that China can use to its advantage in negotiating the boundary of the EEZ and continental shelf. [48] It is also concerned because the major sealane for tankers importing its vital oil runs through the East China Sea.

These "intrusions" eventually raised domestic political hackles in Japan and forced then Japanese Foreign Minister Yohei Kono to urge China to curb its ship operations in Japan's claimed EEZ. [49] Indeed, Japanese lawmakers threatened to postpone a US $161 million loan to China because of concern among ruling party lawmakers with Chinese "spy" ships. In talks with Chinese Foreign Minister Tang Jiaxuan, the two nations agreed on 31 August 2000 to negotiate an agreement for advance notification of such "surveys" by either party. [50] China argued that the problem was the lack of an agreed boundary in the East China Sea and wanted to negotiate a demarcation line. But Japan preferred to negotiate an advance notification system first.

On 15 September 2000, the first working level meeting was held in Beijing to set up a framework for dealing with Chinese vessel entry into Japan's claimed EEZ. Japan proposed that the area which Japan claims as its EEZ be that to the east of a median line between undisputed Japanese and PRC territory, but the PRC rejected the proposal. [51] China argued that it was conducting research in its claimed EEZ and on its claimed continental shelf, which is allowed by the 1982 Convention. It also said that any prior notification agreement would not constitute recognition of Japan's claims. However, on 8 October 2000, China's Prime Minister Zhu Rongji vowed to reduce the amount of PRC research activities in seas around Japan and to establish a mutual prior notification system. [52]

On 13 February 2001, Japan and China agreed on such a mutual prior notification system. [53] The agreement cleverly avoids specifying any line beyond which advance notification is required. It simply says that China is to give Japan at least two months notice when its research ships plan to enter waters "near Japan and in which Japan takes interest" and that similarly, Japan is to inform China before its vessels enter waters "near" China. The notification must include the name of the organisation conducting the research, the name and type of vessels involved, the responsible individual, the details of the research, such as its purpose and equipment to be used, the planned length of the survey, and the areas to be surveyed.

In April 2001, Chinese marine research ships resumed their activities in Japan's claimed EEZ. [54] China notified Japan in advance of the operations of seven ships in accordance with the agreement. The research focused on seabed exploration for minerals and hydrocarbons northeast of Okinawa with one ship undertaking drilling. From 10 to 25 July 2001 the Chinese Navy conducted detailed marine research in a wide area extending off southern Kyushu to near Iwojima. [55] But in late December 2001, citing several violations of the agreement, Japan repeated its request to China to abide by the mutual prior notification agreements. [56]

2.5 NAVAL COOPERATION

There is already a gossamer web of bilateral naval arrangements being spun.[57] Incidents at Sea (INCSEA) or similar agreements exist between the United Sates and Russia, the United Sates and China, Japan and Russia, Japan and South Korea, and South Korea and Russia. [58] Further, Russian and Japanese naval forces staged a first-ever joint marine rescue drill in the Sea of Japan in 1998. And Japanese and South Korean naval vessels staged a path-breaking joint search and rescue operation in the extreme northern East China Sea in early August 1999. [59] South Korea has proposed joint maritime search and rescue to China as well as an exchange of visits by naval ships; [60] it also plans a joint naval exercise with Russia. Recently Russia has proposed that it Japan, and the United States stage joint search and rescue drills. Meanwhile, the United States has proposed that the U.S.-Malaysia joint military search and rescue training mission be expanded to include China and Japan. And even Japan and China have resume their security dialogue, while Russia and North Korea have signed a new treaty on friendship and cooperation. In what could presage an emerging security relationship, Russia and China held joint naval manoeuvres in October 1999. More important however is the convergence of proposals by China, South Korea and Russia for a multilateral security forum for Northeast Asia.

Initially, a sub-regional approach might be best, at least for specific maritime confidence building measures, albeit under the overall umbrella of a regional security forum. Given the existence of this network of agreements and arrangements in Northeast Asia, a multilateral agreement could be based on these standards.

But there are still formidable obstacles to a multilateral arrangement. How can China, Taiwan and North Korea be persuaded to join? North Korea has heretofore shown little desire to participate in multilateral discussions on security issues that would be necessary for a sub-regional INCSEA agreement. However, North Korea's participation in the ASEAN Regional Forum may bode well for its eventual participation in maritime cooperation. But to attempt to include both China and Taiwan in an official agreement would be pure folly, unless the arrangements were considered "informal."

There are also other serious political as well as practical obstacles of a military nature to strengthening navy-to-navy cooperation in Northeast Asia. These practical obstacles stem from the fundamental political fact that each views the others as potential enemies. Thus there are no 'natural' naval 'partners' in the region. The practical problems overlaid on these deep-seated political sensitivities include tight operating budgets; lack of common doctrine, language and interoperability of equipment; widely varying stages of technological development and the reluctance of less advanced navies to reveal their technological weaknesses; the possibility that naval cooperation may be used to gain intelligence about the capabilities of potential adversaries; and the confined maritime geography of Northeast Asia and the sensitivities about foreign naval vessels operating in areas of overlapping EEZs or near features whose sovereignty is contested. However, the U.S. – led war on global terrorism may forge a new level of multilateral maritime security cooperation in Northeast Asia. Indeed the United States is in the process of negotiating a series of bilateral treaties giving U.S. forces the right to pursue and board vessels in international waters as well as in other countries waters. [61] Nevertheless, in the long term progress on the harder maritime security issues – such as a military security – may well depend on successful development of a softer, essentially civil, maritime safety regime. Asian specialists list similar maritime problem areas for greater cooperation: piracy, smuggling, illegal immigration, transnational oil spills, incidents at sea, search and rescue, navigational safety, exchange of maritime information, illegal fishing, and management of resources in areas of overlapping claims. These issues are all maritime safety problems of a civil, as opposed to a military, nature. Proposals for maritime cooperation can be formulated against common problems of 'terrorism', transnational crime, terrorism human depredation, pollution and natural disaster, rather than a single adversary. Further out to sea, in time and space, an international naval or "self-defence" force might ensure ocean peacekeeping, including safety of navigation. This joint force could focus on areas outside national jurisdiction and emphasise protection of fisheries, air-sea rescue and environmental monitoring. [62] In this context, regional oceans management may be the most significant of all the current proposed maritime confidence building measures.

2.6 THE NEXT STEP: A REGIME REGULATING THE USE OF FORCE AND MILITARY AND INTELLIGENCE GATHERING ACTIVITIES IN THE EEZ

The U.S. EP3 incident over China's EEZ and the 'North Korean' spy boat incident in Japan's EEZ have brought to the fore important questions regarding military and intelligence gathering activities in EEZs. On April 1, 2001 a U.S. spy plane collided with a Chinese jet fighter over China's EEZ about 70 miles Southeast of Hainan. In the ensuing controversy over who was to blame, the United States argued that the aircraft was enjoying freedom of navigation over 'international waters.' However, China claimed that such freedoms are not absolute and that such foreign aircraft flying over its EEZ should abide by China's laws, and 'refrain from activities which endanger the sovereignty, security and national interests' of the coastal country. China further demanded that all such spy flights cease. [63]

In late December 2001 Japanese coast guard vessels pursued and fired at a 100-ton suspect ship of unknown origin, which then sank in China's EEZ. [64] Although Japan later said it was a North Korean smuggling or spy ship, and that it sunk itself, [65] North Korea denied any link to the ship and argued that the attack and sinking were 'piracy' and 'unpardonable terrorism.' [66] Surprisingly, given China's inadvertent use of force in the EP3 incident, China also condemned Japan's use of force in the 'spy' boat incident calling it 'rash' and 'indiscreet.' [67] Japan responded that it was within its rights to use force to stop a vessel that refused to comply with its orders to stop for inspection, and that its sinking was an act of "self-defence".

Meanwhile, the activities of China's navy intelligence vessels in Japan's EEZ have been increasing. On 14 May 1999, 12 Chinese naval vessels, including a Jianghu I-class frigate were observed in Japan's claimed EEZ about 110 km north of the Diaoyu/Senkakus. Again on 15 July 1999, 10 Chinese naval vessels including three Luda I-class destroyers were observed in Japan's claimed EEZ, 130-260 km north of the disputed islets. In May through June 2000, *Haibing*-723, a Chinese icebreaker/intelligence gathering ship, circumnavigated Japan on a suspected intelligence gathering mission. The ship, after carrying out a series of activities in the sea area near the Tsushima Strait, sailed north through the Sea of Japan, crossed the Tsugaru Strait three times back and forth, sailed south along the seashore of Japan bordering on the Pacific, past the Boso Peninsular, Shikoku and Amami Oshima. Japan also alleged that the *Dongtiao*-232, a Chinese missile range instrumentation ship, had engaged in intelligence gathering activities in July 2000 in sea areas off Irako-misaki, Aichi Prefecture, and in sea areas south of the Kii Peninsula.

Such military and intelligence gathering activities in the EEZ were a controversial issue in the negotiations of the text of the 1982 Convention and continue to be in state practice. [68] Indeed, some coastal states, such as India, Malaysia, Brazil, Cape Verde, and Uruguay hold that other states cannot carry out military exercises or manoeuvres in or over their EEZ without their consent. [69] Their concern is that such uninvited military activities could threaten their national security or undermine their resource sovereignty. Another unresolved but relevant question sharply dividing nations is whether some of the intelligence-gathering activities carried out by maritime powers in the EEZs of other coastal nations could be considered "scientific research." If so, according to the 1982 Convention, scientific research can be carried out only with the consent of the coastal state. However, maritime powers like the United States insist on freedom of military activities in the EEZ out of concern that their naval and air access and mobility could be severely restricted by the global EEZ 'enclosure' movement. The 1982 Convention does not address these issues directly and thus leaves many questions unanswered, and the regime must be negotiated between the parties concerned.

2.6.1 The Way Forward

Negotiating various confidence-building measures, such as limitations on the number and scale of naval exercises in specific areas of Northeast Asia and exchange of information on naval matters, would be first steps. These could be followed by prohibition of naval exercises and weapons tests in certain areas, especially near busy sea-lanes, and expansion of such areas into 'peace zones' where no military activity would be permissible.

A specific objective could be bilateral conflict avoidance agreements along the lines of the 1972 "Incidents at Sea" agreement (INCSEA) between the United States and Soviet Union, which has proven effective in regulating the interaction of their fleets on the high seas. [70] The value of an INCSEA agreement lies in the obligation of the countries to consult regularly on safety. The purpose of an INCSEA would be to prevent incidents at sea and in the air that could affect relations among the countries; to minimise the chance of accidents resulting from normal activities; and to develop more predictable standard operating procedures at sea. The annual Japan-China security talks and the recent agreement to exchange naval visits is a basis for such a step, between these prime powers in the region. [71]

INCSEA-type arrangements for the seas of Northeast Asia could take different forms and operate at different levels. Initially, their contents could be simple and later, more comprehensive. The development of a mutual prior-notification regime for surface and air reconnaissance missions might be a first step. Moreover, INCSEA-type arrangements might be bilateral first and multilateral later. An immature multilateral approach could quickly become bogged down by particularistic interests. China, for example, is unlikely to accept a regional solution to the problems of incidents at sea as it has a marked preference for bilateral agreements that escape the influence of third parties.

In the longer term, a comprehensive regional INCSEA for Northeast Asia together with supervisory mechanisms, would be needed to prevent or minimize conflict escalation and the likelihood of potentially dangerous naval confrontations. Its contents might include the following: [72]

- Definition and clarification of the extent of the jurisdiction of coastal states in the EEZ, vis-à-vis the freedom of navigation in the EEZ

- Procedures governing notification of the intent to conduct marine scientific research in the EEZ

- Procedures governing notification of the intent to conduct naval and air exercises in the EEZ

- Procedures to prevent the collision of submarines with other undersea vessels and with surface vessels

- Regulations governing the tracking of submarines

- Procedures to prevent the escalation of conflicts that result from accidental or unintended weapons use, or unauthorised weapons use by subordinates, and implementation of a reporting process for such incidents

- The means for increasing military transparency, especially with regard to naval build-ups and naval strategy

- Emergency consultation systems regarding unexpected events

- A regime governing air-to-air and air-to-sea encounters, including methods of preventing aircraft collisions and rules governing the interception of reconnaissance aircraft and vessels and of engagement at sea

- A regime governing interdiction and arrest of suspect vessels in the EEZ

- Anti-piracy measures and systems for, or coordinated joint patrols

- Cooperation in humanitarian assistance, search and rescue, mine countermeasures, and cooperation in combating drug trafficking and illegal migration.

3. Conclusions

Integrative forces for multilateral maritime regime building in the region include the transnationality of marine ecosystems and their living resources, pollution, and shipping; advances in technology and marine use patterns and concepts; growing conflicts among uses and users; extension of maritime jurisdiction and the responsibilities therein prescribed by the 1982 United Nations Convention on the Law of Sea, especially its emphasis on regional management of semi-enclosed seas; and the growth of an epistemic community of maritime affairs specialists. Disintegrative forces include the remaining isolation of North Korea and the tense relations between North and South Korea, between China and Taiwan, and to a lesser degree between Japan and Russia. An unstable Russia, disputes over islands and maritime boundaries, and the conceptual dichotomy in the 1982 Conventions between cooperation and sovereignty would also have disintegrative influence.

For China, multilateral regime participation could lead to technology transfer from Japan, South Korea and Taiwan, and confidence building in itself as a member of the international community. But China would have to limit its flexibility in the marine sphere and commit scarce resources to fulfil its regime responsibilities. Furthermore Beijing prefers bilateral relationships which it can dominate. For Taiwan, regime participation would expand its channels for discussions with China, enable its maritime issues to be addressed, and would enhance its status vis-vis China. On the other hand, Taiwan might have to share its technological know-how and possibly sensitive data with the other participants, including China. And it would probably have to pay more than its share for the implementation of the regime. More fundamentally, if China and Taiwan are to be included in a regime, it must be non-governmental -- and the East China Sea -- should initially be excluded.

For Japan, participation in such a maritime regime is favoured by its economic and technological dominance, its knowledge and experience; and its web of bilateral maritime agreements. Benefits include protection of fisheries resources and the environment; elimination of the transaction costs of annual bilateral fisheries negotiations; and enhancement of its status in the region. But a prominent role for Japan in a multilateral regime is inhibited by its preference for bilateral relationships which it can dominate; the Kuriles, Tok Do and Diaoyu disputes with Russia, South Korea and China; the bitterness among its prospective partners regarding Japan's wartime behaviour; Japan's priority on immediate national economic gain; and its bureaucratic conservativeness and generally reactive posture regarding international affairs.

North Korea could use participation in a regime as a 'coming out' into the international community; to feel out potential regional partners; to increase its financial, technical and knowledge capacity; and to gain a cleaner environment. But for North Korea, opening of its society to foreigners and their cultures and practices could help undermine internal control. Moreover, scarce resources would have to be diverted to fulfil its obligations to the regime.

For South Korea, participation in a multilateral maritime regime could help enhance conservation and management of fishery resources in general and the management of transnational fish stocks in particular; possibly improve its access to neighbour's stocks; improve its international stature; and expand its points of diplomatic contact with North Korea.

South Korea is the only state to border all three Northeast Asian seas and recognises that environmental degradation, depletion of fishery resources, maritime anarchy and political conflict are not in its own, or any other nation's, interest. Indeed, South Korea supports a multilateral marine environmental protection regime and might even be willing to exercise leadership thereof. However, South Korea is the only state to border all three Seas, it puts more emphasis on the Yellow Sea because it is more polluted, and because Korea's western coast is more developed than its eastern shore. Moreover, a regime for the Yellow Sea could employ China as a go-between to help broaden its contacts with North Korea.

Russia has much to gain and little to lose politically and economically by participating in a regional maritime regime, especially for the Sea of Japan. It could gain technological and financial assistance as well as diminished poaching by foreigners. But fisheries cooperation as well as cooperation in environmental protection is hampered by bureaucratic confusion, ineffectiveness, lack of infrastructure, economic malaise, and a relative lack of interest in maritime affairs, particularly in Northeast Asian seas. Nevertheless, Russia does have considerable experience in fisheries arrangements and international cooperation in marine environmental protection, a legislative base upon which to build, and considerable fisheries and oceanographic expertise. And it is interested in fulfilling its proper role as a member of the international community. The rise of a domestic Green movement in the Russian Far East and pressure from aid organisations could promote its participation in a regional marine environmental protection regime. However, Russia's participation in regional maritime regimes will likely be determined by the progress of Russian market reform, political relations between Russia and Northeast Asia, particularly Japan, and the impact of incipient non-governmental organisations.

Taken together, the array of positive and negative factors in Northeast Asia argues strongly for an *ad hoc,* issue-specific evolutionary process for multilateral maritime regime building. To move the process forward, the countries might agree to begin discussing objectives and principles for multilateral maritime regimes in Northeast Asia. *Ad hoc* agreement on a 'core' area for a fisheries regime is necessary, perhaps beginning with two separate regimes for the Yellow Sea/East China Sea and the Sea of Japan. Trade-offs will also be necessary, such as transfer of technology, training, information and fees from Japan and perhaps South Korea to China, North Korea and Russia in exchange for responsible access to their fish stocks. Governments might begin experimentally with a variety of relatively low-risk initiatives with decentralised power and authority such as academic networks and meetings; co-operative research on the effectiveness of existing regional fisheries commissions; intergovernmental task forces; discussions of possible regional fisheries regimes for Northeast Asia; and establishment of an informal intergovernmental forum to facilitate harmonisation of national policies and practices. Meanwhile, the agreements for bilateral access to fisheries in overlapping claim areas around Tok Do and north of the Senkakus will hopefully ameliorate tensions, and make such dialogue possible.

There appears to be a confluence of incentives for such a dialogue. Russian foreign ministry officials are concerned that the United States may engage in "adventures" regarding North Korea and are exploring ways and means to enhance multilateral dialogue in the region, specifically including maritime cooperation. Thus, Russia would certainly support such an initiative. South Korea is also supportive of such a dialogue.

And even China appears to be willing to consider limited multilateral approaches on a sea-by-sea basis. Although Japan would be reluctant to lead such an effort, it would probably participate if it is initiated by another state or United Nations agency. Thus the time seems to be right for such exploration of *ad hoc* multilateral maritime cooperation in Northeast Asia.

In Northeast Asia, there is thus considerable basis for regional maritime cooperation. But what is less certain is the extent to which that cooperation will have real or lasting effect. Most existing cooperation is only bilateral and the multilateral cooperative activities so far have been "talk shops" that have led to little action or implementation of the ideas were discussed. Nevertheless, "second track" forums have a particular role to play in spreading awareness of the problems and potentially identifying solutions that may be too sensitive or embryonic for consideration at a "first track" level. And they do help build an all-important epistemic community supportive of cooperation in the marine sphere.

References and Notes

[1] Barry Buzan, "The Post Cold War Asia-Pacific Security Order: Conflict or Cooperation?", Paper presented at the Conference on Economic and Security Cooperation in the Asia-Pacific: Agenda for the 1990s, Canberra, 28-30 July 1993, 16.

[2] For the purposes of this chapter Northeast Asia includes China, Japan, North Korea, South Korea, Taiwan and Russia. The marine regions covered in this analysis include the Yellow Sea, the East China Sea and the Sea of Japan (East Sea). Part of this introduction is from Mark J. Valencia, "Asia, the Law of the Sea and International relations," *International Affairs*, v.73 (1977), 263-282.

[3] "Asian Reaction Swift To China's Maritime Expansion." *Korea Times*, 20 May 1996, 1.

[4] "Japan Approves Sea Zone Including Disputed Isles," *Japan Times*, 21 February 1996, 1.

[5] "Seoul to Proclaim 200-nautical mile EEZ," *Korea Herald*, 21 February 1996, 1.

[6] Edward L. Miles, "Concept, Approaches, and Applications in Sea Use Planning and Management," in *Ocean Development and International Law*, vol. 20 (1989), 213-238.

[7] United Nations, Convention on the Law of the Sea, Dec. 10, 1982, U.N. Doc. A/CONF.62/122, reprinted in *International Legal Materials* 21 (1982): 1261 [hereinafter 1982 Convention]. See *Law of the Sea United Nations Convention of the Law of the Sea with Index and Final Act of the Third United Nations Conference on the Law of the Sea, U.N. Doc. A/CONF.62/122, U.N. Sales No. E.83.V.5*

[8] Mark J. Valencia, "Northeast Asia: Navigating Neptune's Neighborhood," in Benjamin L. Self and Yuki Tatsumi, eds., *Confidence Building Measures in Northeast Asia*, The Henry L. Stimson Center, Report no. 33, February 2000, 1-35.

[9] *Ibid.*

[10] United Nations, *supra* n.7, Article 122.

[11] *Ibid.*, Part XII.

[12] *Ibid.*, Part XII, e.g., Section 2, 4 and 9, in particular Article 235 "States are responsible for fulfillment of their international obligations concerning the protection and preservation of the marine environment. They shall be liable in accordance with international law."

[13] Valencia, *supra* n.8.

[14] The International Convention on Maritime Search and Rescue, 1979 has a relatively low level of acceptance possibly because of the costs involved in establishing a search and rescue infrastructure and a reluctance to allow searching ships or aircraft of another country access to sovereign waters or territory.

[15] An East-West Center Conference series coordinated by the author has focused on these issues – *International Conference on East Asian Seas: Cooperative Solutions to Transnational Issues*, Seoul, 21-23 September 1992; *The Soviet Far East and the North Pacific Region: Emerging Issues in International Relations*, Honolulu, 20-23 May 1991; *East China Sea: Transnational Marine Policy Issues and Possibilities of Cooperation*, Dalian, China, 27-29 June 1991; *International Conference on the Japan and Okhotsk Seas*, Vladivostok, Russia, September 1989; *International Conference on the Sea of Japan*, Niigata, Japan, 11-14 October 1988; *International Conference on the Yellow Sea*, Honolulu, 23-27 June 1987. Also see: "Japan to Seek Regional Meeting to Look at Water Pollution, Other Problems," in *International Environment Reporter*, 4 December 1991; Northeast Asian Conference on Environmental Cooperation, 13-16 October 1992, Environment Agency of Japan and Niigata Perfecture. Indeed, Japan has established a Center to elaborate the concept of regional cooperation and to prepare specific proposals for cooperation around the Sea of Japan (*Russia in Asia Report* No. 15, July 1993, p. 44).

[16] The above conferences were attended by policy makers in their personal capacities from all the coastal states – North Korea, South Korea, Japan and Russia – as well as China, Taiwan, and international organizations. The topics discussed ranged from scientific assessments of the resources to transnational fishery management and conflicting navigation regimes.

[17] Stanley B. Weeks, "Overview of Regional Cooperation Activities," CSCAP Maritime Cooperation Working Group, Kuala Lumpur, 17-18 November 1998. Stanley B. Weeks, "Strengthening Maritime Cooperation in the Asia-Pacific," a paper presented at the 13th Asia-Pacific Roundtable, Kuala Lumpur, 30 May-2 June, 1999.

[18] Valencia, *supra* n.8.

[19] The Tokyo MOU on Port State Control of 1993 is one of several recent regional agreements encouraged by the IMO and designed to establish regional systems of reciprocal cooperation in inspecting and surveying ships to verify their compliance with international safety standards.

[20] Valencia, *supra*, n.8.

[21] Parts of this section are derived from Mark J. Valencia and Yoshihisa Amae, "Regime Building in the East China Sea", *Ocean Development and International Law Journal*, in press.

[22] Mark J. Valencia, *International Conference on the Sea of Japan*, East-West Environment and Policy Institute Occasional Paper No. 10, 1989, 77-85.

[23] "Seoul, Tokyo Agree on Joint Fishing Zone," *The Korea Times*, 26 September 1988; "Lessons From Fisheries Now," *Korea Herald*, 23 March 1999.

[24] Valencia, *supra* n.8.

[25] Hiroyuki Seiguyama, "Japan, China Conclude Bilateral Fisheries Treaty," *The Daily Yomiuri* 28 February 2000.

[26] Chen Yaobang, "Japan-China Fisheries Agreement To Take Effect in June," *BBC Monitoring Asia Pacific-Political*, 27 February 2000.

[27] "Japanese Official Comments on Fisheries Agreement with China", *BBC Monitoring Asia Pacific-Political*, 28 February 2000.

[28] "Enforcement of Fishery Pact on the Horizon, *China Daily* 31 May 2000.

[29] *China Daily*, 9 December 2000.

[30] "Japan: Fund Planned to Aid Fishermen Affected By Pact with China", *BBC Monitoring Asia Pacific-Political*, 18 October 2000.

[31] "Taiwan Vessel Capsizes After Collision with Japanese Patrol Boat", *BBC Monitoring Asia Pacific-Political*, 7 June 2000.

[32] "Taiwan Fishing Boat Intercepted by Chinese Naval Ship", *BBC Monitoring Asia Pacific-Political*, 16 May 2001; "Taiwan Islet Said to be New Cross-Strait Smuggling Hub", *BBC Monitoring Asia Pacific-Political*, 8 August 2001.

[33] Chi Young Pak, "Resettlement of the Fisheries Order in Northeast Asia Resulting from the New Fisheries Agreements among Korea, Japan and China", unpublished manuscript.

[34] For example, Taiwan boats not only fish in the East China Sea but also buy catch from Chinese fishing boats. The Taiwan coast guard also arrests Chinese fishing boats poaching near Pengchiayu an islet located in the East China Sea 55 nm north of Keelung. *BBC Monitoring Asia Pacific-Political, supra* n.33.

[35] Jong-hiva Choi, "A New Subregional Fisheries Cooperation System in Northeast Asia", paper presented at SEAPOL Inter-Regional Conference on Ocean Governance and Sustainable Development with East and Southeast Asian Seas: Challenges in the New Millennium, Bangkok, Thailand.

[36] Valencia, *supra* n.22.

[37] *BIS-EAS-93-044*, 9 March 1993, 24.

[38] *FBIS-EAS-92-175*, 9 November 1992, 10.

[39] This section is excerpted from Mark J. Valencia, <u>A Maritime Regime for Northeast Asia</u>, Oxford University Press, Hong Kong, 1996, 175-244.

[40] Garrett Hardin, "The Tragedy of the Commons," *Science*, 162 (1968), 1243-8.

[41] Peter Hayes and Lyuba Zarsky, "Regional Cooperation and Environmental Issues in Northeast Asia," *Nautilus Pacific Research*, 1 October 1993, 9. Chua Thia Eng, Robert Cordover, Miles Hayes, Celso Roque, David Shirley, Gurpreet Singhota and Philip Tortelli, "Prevention and Management of Marine Pollution in East Asian Seas," Formulation Mission Report Prepared for the United Nations Development Programme Division, Regional Bureau for Asia and the Pacific, April 1993, 2.11-2.27; *The Nation*, 16 February 1993.

[42] United Nations Environment Programme, *Third Meeting of Experts and National Focal Points on the Development of the North-West Pacific Action Plan*, Report of the Meeting, UNEP (OCA)/NOWP.WG3/6, 10-12 November 1993.

[43] Boleslaw A. Boczek, "Concept of Regime and the Protection and Conservation of the Marine Environment", in Elisabeth Mann Borgese and Norton Ginsburg, eds., *Ocean Yearbook* 6, Chicago: University of Chicago Press, 1986 288-292, 294.

[44] Peter M. Hass, "Protecting the Baltic and North Seas," in Peter M. Haas, Robert O. Keohane, and Marc A. Levy eds., *Institutions for the Earth: Sources of Effective Environmental Protection*, Cambridge, Massachusetts: The MIT Press, 1993, 133-81; United Nations, *supra* n.7, Article 194 (2).

45 John Birger Skjaerseth, "The 'Effectiveness' of the Mediterranean Action Plan," *International Environmental Affairs,* 5, 4 (Winter 1993), 313-34.

46 "China Said Conducting Drills in Japan's EEZ", *The Daily Yomiuri,* 26 July 2001.

47 The National Institute for Defense Studies, Japan, East Asian Strategic Review 2000, 104-105, East Asian Strategic Review 2001, 200-203.

48 *Ibid.*

49 "Japan, China Agreement on Maritime Notice System Detailed", *BBC Monitoring Asia Pacific-Political,* 13 February 2001.

50 Hu Quihua, "Dispute Over East China Sea Research May Be Ending," *China Daily,* 29 August 2000, 1.

51 Yoshihisa Komori, "Japan and PRC Agree to Swiftly Promote Co-operation in Dealing with Maritime Activities," *Sankei Shimbun,* 15 September 2000.

52 Sugiyama Hiroyuki, "China Will Stop Goading Japanese Over Wartime Past", *Daily Yomiuri,* 9 October 2000.

53 "Japan, China Agreement on Maritime Notice System Detailed", *BBC Monitoring Asia Pacific-Political,* 13 February 2001.

54 "Chinese Ships Resume Research in Japan's Economic Zone", *Kyodo News Service,* 23 April 2001.

55 Buzan, *supra* n.2.

56 "Japan Asks China to Abide by Maritime Notification", *BBC Monitoring Asia Pacific Political,* 26 December, 2001.

57 Valencia, *supra* n.8.

58 Pauline Kerr, "Maritime Security in the 1990's: Achievements and Prospects," in Andy Mack, ed., *A Peaceful Ocean?: Maritime Security in the Pacific in the Post-Cold War Era,* Canberra: Allen and Unwin, 1993, 186-198.

59 "Japan, South Korea to Hold First Joint Naval Exercises," *Japan Times,* 8 August 1999, 2.

60 "Seoul, Beijing Begin to Build Military Confidence," *The Korea Herald,* 25 August 1999.

61 "Broader Sea Powers Sought", *Honolulu Advertiser,* 10 August 2002, A3; Thomas E. Ricks, "Aggressive New Tactics Proposed for Terror War: Covert Acts Eyed as Rumsfeld Seeks to Revitalize U.S. Effort", *The Washington Post,* 3 August 2002, A1.

62 Mark J. Valencia, "Road to Asian Security May Begin at Sea", *Honolulu Star Bulletin,* 11 April 2001, A2.

63 "China; USA Violates International Law" in Plane Collision", Xinhua, *BBC Monitoring Asia Pacific-Political,* 4 April 2001.

64 "Japan All Out to Rescue Crewman, Identify Vessel", *Jiji Press English News Service,* 23 December 2001.

65 Teruaki Ueno, "North Korea Suspected of Stepping Up Drug Smuggling," *Reuters,* 10 January 2002.

66 "Clash at Sea Sparks Verbal Shots - - North Korea Calls Encounter 'Brutal Piracy, Pardonable Terrorism'", *Asian Wall Street Journal,* 27 December 2001.

[67] "Japan Asks China to Abide by Maritime Notification", *BBC Monitoring Asia Pacific-Political,* 26 December 2001.

[68] Francisco Orego Vicuna, "The Excusive Economic Zone: Regime and Legal Nature Under International Law", Cambridge University Press, 1989.

[69] Stephan V. Molodtsov, "The Exclusive Economic Zone: Legal Status and Regime of Navigation" in E. M. Borgese and Norton Ginsburg, eds., *Ocean Yearbook 6,* Chicago: University of Chicago Press, 1986, 203-216; Mark J. Valencia, "Law of the Sea in Transition: Navigational Nightmare for the Maritime Powers?", *Journal of Maritime Law and Commerce*, 418, 4 1987, 541-54.

[70] Ji Guoxing, "Rough Waters in the South China Sea: Navigation Issues and Confidence–Building Measures", *Asia Pacific Issues 53*, Honolulu: East-West Center, August 2001.

[71] "China, Japan Warships to Exchange Port Visits", *BBC Monitoring Asia Pacific-Political,* 20 March 2002.

[72] Ji, *supra* n.70.

CHAPTER 14. THE ARCTIC OCEAN

Yu.G. MIKHAYLICHENKO

1. Introduction

A stock phrase "growing importance of so-and-so in the future century" can be to the largest extent applied to the Arctic Ocean. It essentially stems from its extreme geographical location, resulting from it harsh natural conditions and, subsequently, relatively low status of studies, development and involvement in economic activity. Its vast resources remain unused and even undiscovered.

A growing importance of the Arctic region seems an objective and inevitable process. Throughout the history of mankind the priority values were space and related to it resources. The Arctic Ocean and surrounding shoreland constitute a resource, spatial and environmental reserves of not only the countries located here, but essentially of global nature.

2. Resources

With a high degree of certainty the Arctic region will for a long time remain primarily a resource region.

2.1. MINERAL RAW MATERIAL RESOURCES

2.1.1. Hydrocarbon raw materials
To-date in the world the major prospects of exploring for new oil and natural gas provinces have to be related to submarine continental margins. Meanwhile the largest resources of hydrocarbon (HC) raw materials offshore in the world ocean are concentrated in the Arctic. According to expert estimates the Arctic offshore zone contains about 180 billion metric tons of HC in oil equivalent. A greater part – 66 billion metric tons (37%) is presumably located in Asian part of the Arctic over the area of 2 million sq. km, 60 billion metric tons of HC (33%) - in European part of the Arctic over the area of 1.8 million sq. km, and 54 billion metric tons (30%) of HC - in polar offshore regions of North America over the area of 2.2 million sq. km.

The first production of oil in the Arctic offshore began in 1987 at Andycott field located in the Beaufort Sea. Different areas of the Arctic shelf have highly variable natural conditions. At some of them HC prospecting and extraction present no difficulties since the necessary equipment and technology are available. But substantial and most prospective part of Arctic offshore regions are located within frozen water areas with a severe ice regime and harsh natural and climate conditions. For these reasons the activities in these areas may be launched in future upon development of principally new means and technologies.

Concentration of some HC reserves in giant deposits considerably facilitates their development, while geographic location within the Arctic offshore creates favourable conditions for HC transportation to industrially developed countries. All these factors make continental margins at the Arctic Ocean the most important reserve of HC raw materials in the 21st century.

In the future a key prospective non-traditional source of methane are gas hydrates. Their reserves in the ocean are immense, including the Arctic Ocean coasts and offshore regions where research exploration started in 1998 (Mackenzie Delta, Canada).

2.1.2. Minerals
The Arctic shelf zone mineral resource potential is characterised by a wide range of minerals. The major placer minerals on shoreland and water areas are gold and tin. The major bedrock minerals on islands within shelf basins are metallic including manganese, basic metals and gold and non-metallic including fluorite, coal and lignite. The zone features large and unique deposits of both the placer and vein types.

The combination of a large petroleum and gas potential, large resources and variety of types of minerals means that the shelf zone of the Arctic region is among world leading resource regions.

2.2. SPATIAL RESOURCE

Similar to some other water areas the Arctic Ocean is not free from problems of delimitation of jurisdiction zones between the countries. The issue of boundaries between Norwegian and Russian zones in the Barents Sea remains unsolved. The agreement between the USSR and USA on delimitation of marine space, including the Chukchi Sea, of 1 June 1990 has not been ratified by Russia. Norwegian innovations concerning interpretation of the Svalbard Treaty of 1925 are often misunderstood by other countries.

A problem of the near future will be definition and validation of the outer limits of the continental shelf beyond the 200 nautical mile exclusive economic zones of countries within legal frames of the 1982 United Nations Convention on the Law of the Sea (UNCLOS). Apart from respective rights on resources within its juridical continental shelf, a country has jurisdiction regarding such important activities as scientific research, creating artificial islands and installations, drilling operations for any purpose and identifying the routes for underwater cables and pipelines.

The Arctic shelf is the largest continental shelf on the planet and simultaneously a marine boundary between five Arctic states – Canada, Denmark, Norway, Russia and USA. In view of a high forecast evaluation of oil and natural gas and mineral raw materials reserves in continental margins of the Arctic Ocean, the active desire of Arctic states to develop these resources, and also the remaining defence importance of the Arctic region, the problem of validating the outer limits of their continental shelves acquires extreme actuality for the countries of the region.

Based on several geological and geomorphological parameters the Arctic basin continental shelf is not limited by a continental slope but extends far into the Arctic Ocean. In accordance with UNCLOS defining extensions to offshore areas beyond 200 nautical miles is based on bathymetric and geological and geophysical criteria. A littoral state may ultimately establish the outer limits of its continental shelf only upon the boundary and substantiating scientific and technical data have been considered by the UN Commission on the Limits of the Continental Shelf. Meanwhile UNCLOS stipulates a definite period within which a littoral state must determine the outer limits of its continental shelf – ten years after UNCLOS enters in to force for a given state.

For instance Russia, which ratified UNCLOS in February 1997, has currently processed the existing mass of geological, geophysical and hydrographic information, including the data acquired by an expedition on board the scientific vessel "*Academician Fedorov*" to eliminate "white spots" within the region of the Mendeleev ridge in summer and autumn of 2000. This has facilitated the knowledge of bathymetric and morphological aspects of sea bottom, in-depth structure of the crust, structure and thickness of sediment layer of the Arctic basin within the Russian sector of the Arctic Ocean. The results obtained allow the compilation of a draft map showing location of the outer limits of Russian juridical continental shelf within the Arctic basin (Fig. 1). Its eastern part should be defined during relevant negotiations with Canada and USA.

According to preliminary estimates the establishment of the outer limits of the juridical continental shelf as per UNCLOS will allow increasing the Russian offshore areas beyond 200 nautical miles in the Arctic by almost 1.2 million sq. km with oil and natural gas potential of about 4.9 billion metric tons of conventional fuel.

Considering a more remote perspective, one could forecast a joint desire of the Arctic states to extend to a certain degree its influence over the central parts of the Arctic basin related to the high seas. The future will show to what extent they succeed in substantiating it.

2.3. TRANSPORTATION RESOURCE

2.3.1. Present status and close prospects

The Arctic seas were traditionally used primarily for coastal or short-range international transportation. A special place in terms of transportation is occupied by the Russian so-called Northern Sea Route (NSR). Geographically it stretches from north-western national boundaries (Murmansk and Arkhangelsk seaports) up to the Bering Strait (about 3000 nautical miles) (Fig. 2).

The NSR is the unique latitudinal main line connecting all Arctic and sub-Arctic regions in Russia rich with mineral, energy and biological resources, and of influencing the development of territories several hundred kilometres southward the Arctic coast, primarily along the rivers. The NSR operation demands non-traditional approaches. It is due to the complex environmental conditions of the region, primarily ice cover along navigation routes and extreme vulnerability of the Arctic environment, that Russia has spent vast resources and efforts over several generations for its development and created infrastructure, powerful ice-breaking and transport fleet, systems for pilotage, hydrographic and hydrometeorological support of navigation.

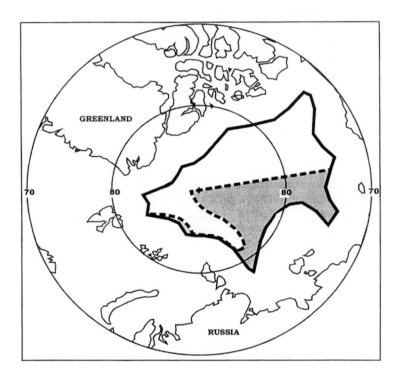

200 nautical mile limits of Exclusive Economic Zones (solid curve). Preliminary delineation of the Russian juridical continental shelf (dashed curve).

Figure. 1. Draft map of the outer limits of Russian juridical continental shelf within the Arctic basin.

The volume of transportation along the NSR reached 6.5 million metric tons in 1987. The consequences of economic transformation in Russia have reduced this indicator almost four times. The country's economic revival should no doubt result in restoration and further growth of the importance of the NSR. The principal factor of the expected growth of economic activity within its zone is development of oil and natural gas resources in coastal and offshore regions in the Barents, Pechora (south-east of the

Barents Sea) and Kara Seas. Potential removal of hydrocarbon raw materials from the above regions by sea is estimated at up to 50 million metric tons per year in the first stage.

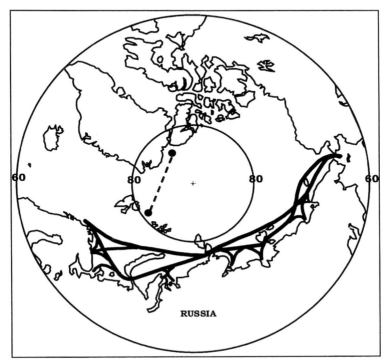

Path of acoustic measurements of ocean temperature (dashed line).

Figure 2. The Northern Sea Route (solid curve).

To date long-term navigation practice has demonstrated the possibility of all-year-round navigation along the NSR of large tonnage vessels using both traditional and high latitude routes. Demilitarising the Russian Arctic will facilitate the international use of the NSR.

2.3.2. Potential resource
The point in question is using the NSR for transit between North-western Europe and the Pacific region - Japan, China, USA and Canada. As early as the 18th century the great Russian scientist Mikhail Lomonosov substantiated the idea that using ice breakers would make the NSR the shortest route to the Pacific Ocean. For instance, the Hamburg-Yokohama route (11400 nautical miles via the Suez Canal) is reduced 1.7 times using NSR.

The Northern Sea Route User Conference held in November 1999 in Oslo, Norway, summarised the results of the International Northern Sea Route Program (INSROP) fulfilled in 1993-1998 by scientists from Russia, Norway, Japan and other countries. Russian and foreign experts optimistically evaluated the competitiveness of transit use of the NSR compared to the existing operational southern alternative, navigation reliability and safety along this route. According to expert assessment appropriate investments in the NSR infrastructure and creation of a new generation of ice breaker transport vessels would help increase cargo flow along the NSR up to 10 million metric tons per year in 2010 and 50 million metric tons per year in 2020. Considering higher requirements to safe navigation under ice conditions and particular vulnerability of the Arctic environment, transportation via the NSR should be carried out under strict government control without contradicting the UNCLOS under which the states with waters covered by ice over six months a year have a right to establish their own rules of marine transportation.

Thus in the coming century the Arctic Ocean will have all chances to compete with the Indian Ocean in the transport scheme connecting two centres of the world economic system – Euro-Atlantic and Asia-Pacific. However, considering the totality of economic, environmental and technical problems it will hardly be achieved in the near future.

2.4. BIORESOURCES

Fisheries and marine mammals hunting are traditional activities of a small number of indigenous peoples of the North. But only fishing in the Barents Sea has national and international importance. Specific hydrological factors resulting from inflow of warm North Atlantic waters facilitates high productivity of the sea and rich ichthyofauna compared to other Arctic seas. The Barents Sea fishery basis is constituted of cod, haddock and polar cod; after a seven year interval commercial catches of capelin started in 2000.

The stock of commercial species in the Barents Sea region is estimated at 20 to 22 million metric tons on the average including 16 to 18 million metric tons of fish. The average productivity is 11 to 13 metric tons per sq. km. Cod is the most common species of benthic fish in the Barents Sea and adjacent water bodies where homogeneous fish populations are caught. This factor highlights its importance for the economy of northern countries. Its share is estimated at 30% of the total fish catch in the region. Following World War II cod stocks tended to reduce. There are some cyclic fluctuations induced by natural factors, but the main reason is high intensity of commercial catches not of cod alone, but also its main food species (capelin, shrimp). To compare, in the 1950s – 1970s the average cod stock was estimated at 2.8 million metric tons, and total annual catches about 800 thousand metric tons. In 1980 - 1986 these figures reduced to 1 million and 370 thousand metric tons respectively. In 2001 the total allowable catch (TAC) was approved at 435 thousand metric tons (e.g. the Russian quota was 183 thousand metric tons). A serious problem in recent years was unregulated catches of cod, primarily by Icelandic trawlers in the open part of the Barents Sea beyond the jurisdiction of Russia and Norway.

Management of the Barents Sea bioresources is based on the USSR - Norway intergovernmental agreements on co-operation in fisheries (1975) and mutual relations (1976). It is implemented through decisions of a Joint Russia-Norway Commission on Fisheries. The basis for regulating catches is fixing the TAC for each stock for the coming year. This measure is based on assessment of commercial stocks by scientists from both countries in line with joint and national research programmes. In addition the Joint Commission has elaborated and adopted such important decisions as transition of catches of most fish species to strictly limited catch with sharing into national quotas, introducing minimum species size during catches of the key species, and a number of scientifically based limitations for using particular fishing equipment (e.g. regions, mesh size and others).

An important role in protecting the stock and optimum management of catches of marine fish resources is still played by such international organisations as the International Council on Exploration of the Sea and some others. Although their decisions are of recommendatory nature, the littoral states try to follow them in their practical activities.

Life has prompted a necessity of searching for and introducing in fisheries practice of techniques to ensure more effective management of fisheries, including the so-called precautionary approach to identifying TAC, balanced fisheries with due account for ecosystem links and others.

Under shrinking reserves emphasis on cod mono catches is apparently unwise. In recent years shrimp catches in the region have grown, particularly those by Norway. Russia and Norway have started on a par basis experimental catches of king (Kamchatka) crab that was successfully acclimatised to the Barents Sea. Its growing number and expanding area with every year indicate that commercial catches could possibly start in the near future under strict regulation. The search is underway for new prospective targets for commercial catches, e.g. sea-hen and others.

It seems that fisheries development in the region in the near future will be characterised by:
a) fluctuations in the number of the most common targets for commercial catches related to climate changes;
b) regulating fisheries in high sea regions and imposing international control over commercial catches;
c) accelerated development of aquaculture, particularly in the coastal zone and economic zones of the states.

A serious challenge to the Barents Sea fisheries will be the deployment of offshore oil and natural gas extraction. The vulnerability of Arctic ecosystems would necessitate validating a new environmental protection strategy based on identifying pollution impacts on primary production, vital activity of key systematic groups of marine life, commercial species stock and reproduction. Special attention should be paid to promoting research activities related to EIA procedures of oil and natural gas deposits planned for development.

Obviously, oil and natural gas operations within the offshore zone will cause certain damage to marine fisheries. To compensate the damage to marine bioresources it would be worthy to consider organising polar mariculture, reproduction of valuable commercial fish species and creating marine natural reserve zones. The scope of such activities should correspond to the scope of development of offshore oil and natural gas extraction.

2.5. TOURISM

Natural and ethnographic features of the Arctic region predetermine the development of such types of tourism as environmental, exotic (including hunting and fishing), and extreme. The ways for further development of indigenous peoples of the North are also connected to a large extent with promoting of tourism. Understanding the interrelationship between tourism and environmental protection has been broadly acknowledged lately.

The concept of linking tourism and conservation in the Arctic originated from the 1995 Second International Symposium on Polar Tourism in St. Petersburg, Russia. Since that time, a series of workshops have developed Principles and Codes of Conduct for Arctic Tourism and a mechanism for their practical implementation. Because tourism differs significantly in Russia from other parts of the North, a separate effort will be required to introduce the principles and implement mechanisms there. The challenge will be to adapt lessons learned elsewhere for conditions in Russia - the largest and least disturbed parts of the circumpolar Arctic.

It has become quite common in recent years for tourists from USA, Japan, Canada and other countries to take cruises on board Russian ice breakers to the North Pole, Franz Joseph Land or from Providence Bay, near the Bering Strait, along the NSR. A major prerequisite for such kind of tourism is creating a specialised fleet and coastal tourist infrastructure.

3. Environmental Status

Harsh climatic conditions are typical of the Arctic region, with extreme fluctuations of light and air temperature, short summers and snowy winters, ice cover of the ocean and permafrost on islands and coasts – all this makes Arctic ecosystems extremely vulnerable. The Arctic flora and fauna have adapted to such conditions but this adaptation has resulted in their extreme sensitivity to anthropogenic impact. Due to extreme conditions it is extremely difficult to remedy the emerging environmental situations.

Anthropogenic pollutants are transported into Arctic ecosystems both along water primarily by river runoff, and atmospheric pathways, including transboundary transport.

The most dangerous pollution factors are persistent toxic organic pollutants, heavy metals and radionuclides. The radiation situation in the Arctic is predetermined by two basic sources: global fall-out as a result of previous nuclear weapon tests in the atmosphere and subsequent collection from vast territories and transport by rivers, and also discharge of liquid radioactive wastes by West European nuclear fuel recycling facilities and subsequent transport by marine currents into the Arctic Ocean. A potential threat is created by the USSR nuclear burial sites in the Barents and Kara Seas and decommissioned Russian nuclear powered submarines awaiting utilisation.

A comparative analysis of the status of ecosystems in the Arctic Seas with that of other regions of the World Ocean leads to the conclusion that the Arctic Seas are relatively clean, and the status of pelagic, open sea, ecosystems is generally stable, excluding a noticeable overcatch of commercial fish species. Yet some offshore regions in the Arctic Seas and some coastal zones are seriously polluted, and the status of ecosystems in some bays, gulfs and estuaries can be evaluated as critical and even disastrous.

Among the key problems of the Arctic environment are:
a) specifying the amount of incoming pollutants and importance of various pathways;
b) studies of assimilation capacity of specific regions of the ocean in relation to the most hazardous substances and evaluation of their critical impacts on the biotic system;
c) creating and implementing a system of biological and chemical techniques to assess the cumulative impact of different anthropogenic factors;
d) elaborating measures for reducing the impacts on man and the living environment of the existing pollution levels with due account for a delayed nature of self-purification and rehabilitation of Arctic ecosystems, and
e) assessment of climatic change impact on ecosystems.

Special importance should be paid to the need to expand the specially protected areas and water areas. According to several scientists their optimum share should amount to 25% of the total Arctic area.

Successful resolution of environmental problems depends largely on environmental management arrangements within the framework of each Arctic state. The content of the environmental management strategy of nature conservation should be creation of highly effective advanced compensatory measures, which prevent negative anthropogenic transformations and reduce environmental risk on local, regional and global scales.

4. Ethnic Problems

The means of survival and development of indigenous populations constitute possibly the most complex and painful problems without explicit decisions as yet. The major contradiction in evolution of life of the Arctic aborigines – selecting the development pattern – remains unsolved.

So far there exists no definite answer to the question: to what extent the problems of peoples of the North result from costs of the general course of civilisation; and to what extent these result from drawbacks of the existing social and economic structure or intentional activities by the authorities. In any case, vital problems of aborigines and means for their solution are quite similar, not only in all parts of the North and Arctic, but also in other regions with small-scale ancient ethnic groups.

While economic development of the Arctic intensifies, the problem will inevitably get worse: may it be considered as a universal principle that aborigines – pioneers of a territory who played a relatively modest role in exploring and developing lands and who now constitute an overwhelming minority, are granted exclusive property rights over lands and natural resources in the region?

It could be stated that the crucial issues of aborigines have nowhere been resolved. But this just increases general human responsibility for their situation. Rescue of these peoples is of universal humanitarian importance. They are the carriers of a culture perfectly adapted to extreme conditions.

5. International Environmental and Scientific Co-operation

5.1. INFRASTRUCTURE

The strategic location of the Arctic Ocean for a long time hampered full scale international co-operation in the region. Tangible changes in the political situation during the last fifteen years have resulted in a more rapid than in any other region of the Earth, growth of co-operation between eight circumpolar nations, Canada, Denmark, Finland, Iceland, Norway, Russia, Sweden, USA, primarily in political, environmental and scientific spheres (Table 1).

Bilateral intergovernmental agreements on co-operation in the Arctic were concluded. Numerous projects and programs are implemented by UNEP, UNDP, WB, GEF, and other UN system organisations. Arctic programs are launched by WWF, IUCN, Advisory Committee on Protection of the Sea (ACOPS), and other international NGOs. The Land-Ocean Interaction in the Russian Arctic (LOIRA) is being fulfilled. Large scale complex bilateral Russia-Norway, -Sweden, -Germany, -USA scientific environmental and socio-economic expeditions are organised; Russia-France expedition is scheduled for 2003.

5.2. COOPERATION IN GLOBAL CHANGE STUDIES

The results of long-term studies indicate that despite the relatively small area and volume of the Arctic Ocean (3.6% of the area and less then 1.5% of the volume of the World Ocean) and adjacent seas, they have serious impact on the Earth's climate status and play a crucial role in various global processes.

At the same time the Arctic environment is characterised by a unique combination of such components as the polar ocean, ice and snow cover, glaciers, tundra, permafrost, boreal forests and wetlands each of which serves a sensitive indicator of global change. All of these are subject to considerable modifications even under weak impacts manifested as variations of incoming solar radiation, underlying surface temperature, transfer of heat by the ocean, chemical composition of the atmosphere and ocean, content and properties of aerosol in the atmosphere and others. Subsequently the dynamics of processes in polar regions may serve as the most reliable indicator of global environmental change.

Table 1. Developmental Landmarks

Year	
1987	Famous speech of Soviet leader Mikhail Gorbachev in Murmansk
1989-91	Preparation and launching of the Arctic Environmental Protection Strategy (AEPS) at the Ministerial Meeting, Rovaniemi, Finland, starting the Arctic Monitoring and Assessment Program (AMAP);
1990	International Arctic Science Committee formed, the Northern Forum established;
1992	Inaugural meeting of the AEPS Program for the Conservation of Arctic Flora and Fauna (CAFF);
1993	Kirkenes Declaration on Barents Euro-Arctic Council, Norway;
1994	First meetings of the AEPS Protection of the Arctic Marine Environment (PAME) and Emergency Prevention, Preparedness and Response (EPPR) working groups
1996	Third AEPS Ministerial in Inuvik, Canada, established a Working Group on Sustainable Development, Arctic Council inaugurated in Ottawa, Canada, to continue the work of the AEPS;
1998	First Arctic Council Ministerial in Iqaluit, Canada, adopted the Regional Programme of Action for the Protection of the Arctic Marine Environment from Land-Based activities (RPA);
2000	Second Arctic Council Ministerial in Barrow, USA, adopted the Arctic Council Action Plan to Eliminate Pollution of the Arctic (ACAP) and the Arctic Climate Impact Assessment (ACIA) Program

Studies of the climate-forming processes in the Arctic whose main feature is predetermined by an ocean isolated from the atmosphere by the ice cover, dictate some constraints to traditional approaches such as seagoing studies and satellite observations, and demand application of new approaches.

A successful example of international co-operation in developing and applying such approaches is the Russian-US experiment on acoustic thermometry of the climatic status of the Arctic Ocean. Acoustic thermometry is remote observation of large-scale, including climate variability of the ocean temperature based on almost linear dependence between the rate of sound propagation in water and its temperature. Measuring the time of acoustic signals propagation between two fixed points in the ocean permits evaluating the integral variations of water temperature along the path of this propagation. Besides that, attenuation of sound signals during propagation in the acoustic channel is largely related to thickness and roughness of the ice cover. Subsequently, by measuring the amplitude of the received signals it is possible to trace variations in the average ice thickness, which combined with satellite data on the ice cover area makes it possible to obtain unique evaluations of one of the most important climate characteristics – floating ice volume.

In October 1998, within the framework of the above experiment, Russian scientists mounted the first stationary autonomous acoustic emitter at the northern point of the Franz Victoria Strait. Simultaneously, at a distance of 1250 km in the Lincoln Sea, the US scientists mounted a receiving array. The emitter - receiving array formed a path crossing the Arctic basin directly northward of the Framm Strait (see Fig. 2). The experiment was finished in May 2001 when the autonomous receiving array was removed. Now scientists are analysing the signals and data registered by the array and recorded in the control memory over 2.5 years.

To study and reliably forecast large-scale climate change in the Arctic Ocean similar to that currently observed, including considerable warming of water in some regions of the Arctic Ocean, and shrinking areas under ice cover it is expected to create an autonomous system for remote observations over variations in the aquatic medium and ice cover of the ocean. The system enables data collection using new technologies and transmits the data in real time.

6. Perspective

6.1. ECOSYSTEM RESOURCE

The problem of preserving the human environment occupies a crucial place when searching for a paradigm of global development. Awareness is growing that the strategies for development of individual countries and regions should not always be oriented towards industrial and resource development of new territories and water areas. Possibly, this is particularly true of the regions with extreme climate and conditions for human life.

The world system of humankind's existence relies largely on preserving the global sustainability of the biosphere ensured by functioning of natural ecosystems. A significant trend in the end of the past century were the first attempts in history to formulate approaches to realising the principle of economic value of ecosystem services, in other words, at least partial accounting for economic costs of the normal process of natural ecosystems functioning. These attempts include the implementation of schemes for restructuring the external debts of developing countries and countries in transition upon condition that they implement environmental programmes, manifestation by the developed countries' middle class (mass consumers) of conscientious preference when buying products of environmentally oriented responsible companies-producers, and advancement towards practical realisation of the Kyoto Protocol of the UN Framework Convention on Climate Change.

6.2. THE RUSSIAN FACTOR

Russia's Arctic zone embraces almost one half of the Earth's circumference in these latitudes. Harmonious resolution of the Russian Arctic problems is possible only in the context of solving the crucial social, economic, environmental, political and other problems the whole country is faced with that intensified or emerged during the transition period. There is an urgent need for a substantial revision of the forms and mechanisms for state protectionism that is essential in the Arctic region under market conditions as well.

Structural, technological lagging of the country's economy has exacerbated the problem of orientation towards extensive, raw material patterns of development when over half of the country's budget is due to extracting fuel and energy, with the North and Arctic being the main regions for advancement.

Reliance on the country's strengths including personnel, scientific and technical innovation and production potential, particularly a diversified potential accumulated in the military-industrial complex; and attraction of large-scale and correctly oriented foreign and local investments – would all facilitate attaining the goals of restructuring the economy and export potential from raw materials to competitive mass end products of science-based production.

Attaining this goal could be facilitated by environmentally-oriented restructuring of aggregate debts, namely, both Russia's debt to developed countries and developing countries' debts to Russia, almost equivalent to USD 150 billion each, for instance, as partial payment for ecosystems services rendered by these countries to the world community.

7. Conclusion

Scenarios for development of the Arctic Ocean and its future potential role would largely depend on the selected model of development and successful transformations in Russia, to what extent it would succeed in combining the universal principles of market economy and democracy with the country's reality.

The emerging positive shifts in the Russian economy, obviously manifested recently, displayed Russia's intentions to participate in resolving the global problems jointly with the world community, the harshness of the Arctic, and also the fact that the leading actors in this region besides Russia are the developed countries with growing environmental awareness of their people, give the Arctic a chance to become, in due course, the first globally important region with nature including natural ecosystems and biodiversity becoming its acknowledged and most valuable renewable resource in the market for ecosystem services, with appropriate compensation to the local population for these services.

Main Sources

Arctic Pollution Issues: A State of the Arctic Environment Report (1997) AMAP, Oslo.

Azizov, Ya.M., Shpachenkov, Yu.A., and Studenetskiy, S.A. (2000) *Russian Fisheries on the Edge of Centuries,* Goskomrybolovstvo Rossii, Moscow (in Russian).

P.J. Cook and C.M. Carleton (eds.), (2000) *Continental Shelf Limits (The Scientific and Legal Interface)* Oxford University Press, Oxford.

I.S. Gramberg and N.P. Laverov (eds.), (2000) *The Arctic on the Threshold of the Third Millennium (Resources Potential and Ecological Problems)* Nauka, St.Petersburg (in Russian).

Kotlyakov, V.M. and Agranat, G.A. (1999) Russian North – land of great opportunity, *Herald of the Russian Academy of Sciences,* **69**, No. 1, 3-8.

Mikhaylichenko Yu.G. (ed.), (1998) Management of the Russian Arctic Seas *Journal of Ocean & Coastal Management,* Special Issue, **41**, No. 2-3, 123-280.

WWF Arctic Bulletin (1998) No. 4.

(1997) *Russia: strategy of development in 21st century* Noosfera, Moscow, v. 1, 2 (in Russian).

CHAPTER 15. THE SOUTHERN OCEAN
Environmental and Resource Management under International Law

DONALD R. ROTHWELL

1. The Southern Ocean Area

The Southern Ocean is a unique body of water which surrounds Antarctica. Unlike its polar counterpart the Arctic Ocean, the Southern Ocean is not ice bound for much of the year. Because the Southern Ocean is relatively open to the effects of ocean currents and prevailing winds, both pack ice and icebergs have a large area within which they can disperse. In total, 85 per cent of Antarctic pack ice melts each summer. Accordingly through a combination of these factors and the improvement in technology, there has been increasing commercial interests in Southern Ocean living and non-living resource exploitation.

Defining the extent of the Southern Ocean is a difficult task as there are no northern geographical limits. The Southern Ocean has been estimated at being 37.5 M km^2 in size, [1] however these estimates vary because its exact limits are difficult to define. Some maps and charts do not even acknowledge the existence of the Southern Ocean. Rather, they indicate that the Atlantic, Pacific and Indian Oceans extend as far south as Antarctica. This ignores the close relationship between the Southern Ocean and the Antarctic continent. One commonly accepted definition of the Southern Ocean is the area to which the 1959 Antarctic Treaty [39] applies. Under Article VI, the Treaty extends to the area south of 60°S. This area includes within it the complete Antarctic continent and many sub-Antarctic islands, especially those which cluster around the Antarctic Peninsula. However, some major islands such as South Georgia and Kerguelen are not included within this area. Importantly, a boundary extending to 60°S does not include any other continent: the most southern tip of South America extends to 55°S at Tierra del Fuego where the Drake Passage creates a natural barrier to Antarctica. While a boundary set at 60°S is convenient for administrative purposes under the Antarctic Treaty, it has not proven to be as adequate when defining the extent of the Antarctic marine environment. Accordingly, the 1980 Convention on the Conservation of Antarctic Marine Living Resources [40] adopted a boundary which while respecting the limits of the Antarctic Treaty extends further north to the 'Antarctic convergence' (CCAMLR, Art I). The Antarctic convergence marks the boundary between cold Antarctic surface water and warmer subantarctic water. It represents a major marine ecological boundary which distinguishes the Antarctic ecosystem from those to the north. CCAMLR defines the outer limits of the area of the Convention's application by way of a set of co-ordinates that closely approximates the outer limits of the convergence. As a result of the increasing importance that has been placed by the

Antarctic Treaty System upon the protection of the environment it is therefore appropriate to adopt the northern boundary of CCAMLR as the outer limits of Antarctica for the purposes of this study (Figure 1).

The management of the Southern Ocean raises a vast array of legal and political issues, many of which have evolved from the development of the Antarctic Treaty System, which is based on the Antarctic Treaty. This chapter will explore these issues in detail, however first it is necessary to understand the Southern Ocean environment and its ecosystem.

1.1. THE SOUTHERN OCEAN ENVIRONMENT

More than one third of Antarctica's coastline is covered with ice shelves. This accounts for about 10 per cent of the continent's area. [2] Three types of sea ice are common in Antarctica: pack ice, fast ice, and icebergs. [3] However, ice shelves have dual characteristics. They are thick bodies of ice which at the coastal edge extend out over the water so as to float atop the water or ground themselves on the seabed. Pack ice usually takes the form of large bodies of frozen sea ice. In Antarctica, pack ice is also present for much of the year but expands considerably during the winter. [2] Fast ice is characteristically attached to land and is frozen sea water which clings to the shores of rocks, inlets, and bays. Icebergs are formed through ice shelves calving. Because there are a greater number of ice shelves in Antarctica, icebergs are more common in the Southern Ocean. The size of icebergs can vary widely with some of enormous proportions having been reported.

The Antarctic ecosystem is both complex and diverse. This is partly a result of the very short summer season, lack of daylight during the winter, and the impact of the vast maritime area. In Antarctica there are no terrestrial animals. Rather bird and marine life dominate with large populations of penguins and seals found around the coastline of the continent and sub-Antarctic islands. The marine life in the Southern Ocean is extensive. Whales and other cetaceans exist in the polar oceans and seas, though not to the extent they were found previously due to extensive whaling operations in the nineteenth and early twentieth centuries.

An issue of emerging concern throughout the twentieth century has been environmental impact. It has gradually been recognised that because of the structure of polar ecosystems an inherent characteristic is that they are threatened by any activity which has an environmental impact. The principal reason for this is the fragility of the ecosystem. An activity which may be sustainable in temperate or tropical regions, is more difficult to sustain where the ecosystem is continually under threat because of the harshness of the environment. In Antarctica, local environmental impact has not been as substantial, principally due to a lack of major resource-based industries. The exception to this has been nineteenth century commercial whaling and sealing. [4] In the 1980s grave concerns were expressed over the potential impact upon the Antarctic marine ecosystem from the development of a substantial Southern Ocean fishing industry. The particular focus of this concern was the overexploitation of krill, a small crustacean which plays a pivotal role in the Antarctic food chain, [4] however the exploitation of finfish and toothfish have also been a concern. A further impact upon the marine environment has been pollution resulting from normal operating activities of vessels in the Southern Ocean, and also shipping incidents.

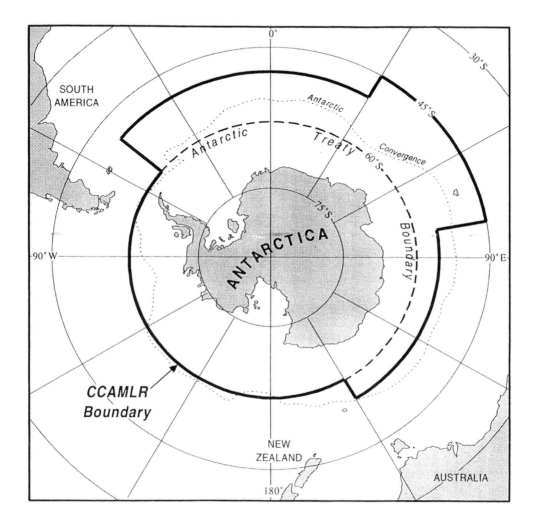

Figure 1. The limits of the Antarctic Treaty, the Convention on the Conservation of Antarctic Marine Living Resources (CCAMLR) and the Antarctic convergence

1.2. SOUTHERN OCEAN MARINE LIVING RESOURCES

Marine living resources have been the principal Antarctic resource subject to exploitation. [5] For many years these resources went unexploited because their existence was unknown and vessels had been unable to penetrate the pack ice. This changed during Captain James Cook's circumnavigation of the continent in 1772-75 when he not only succeeded in sailing into and through the pack ice (though never so far as actually to sight the continent) but also sighted abundant numbers of seals on sub-Antarctic islands. This proved the catalyst for an influx of sealers eager to exploit the newly discovered resource and it was not long before whalers followed. As a result, Antarctic seal and whale stocks were placed under considerable pressure during the nineteenth century and have only just begun to recover as a result of international management plus the lack of significant commercial interest in the resource during the latter part of the twentieth century. The Southern Fur seal and the Southern Elephant seal were the most exploited of the seal species, while Blue, Fin, Sei and Minke whales have all been subject to heavy exploitation. [5] Commercial exploitation of penguins also occurred on some sub-Antarctic islands in the nineteenth and early twentieth century but ceased prior to World War II. [6] The fish stocks of the Southern Ocean are limited, being largely found in areas surrounding the continental shelves of the Antarctic Peninsula and sub-Antarctic islands. [4] While there is evidence that attempts to exploit the Southern Ocean fishery began as far back as 1904 in and around South Georgia, scientific research into the potential of such a fishery did not seriously begin until the 1950s. [7] The Soviet Union and other communist countries began commercial fishing activities throughout parts of the Southern Ocean in the 1960s. These activities extended into the 1970s, but peaked by the end of the 1970s and have not returned to those levels. Krill are found in abundant quantities in the Southern Ocean, however it has proven difficult to accurately assess the size of krill stocks because of a lack of scientific knowledge about the species. This has in turn made it difficult to assess the potential impact of krill exploitation upon the Antarctic marine ecosystem. While krill catches were reported as far back as 1964, significant commercial exploitation did not commence until 1976. Today, management of krill exploitation is dealt with under the 1980 Convention on the Conservation of Antarctic Marine Living Resources (CCAMLR).

Considerable efforts have also been expended in the search for oil and gas in the Southern Ocean. Some important discoveries were made in the 1970s and it is now known that varying reserves lie scattered throughout parts of the Southern Ocean. [8] While there is some debate amongst scientists as to the actual minerals potential of the region, it does seem widely accepted that few of these resources have any current commercial value. [9]

2. The Southern Ocean and the Antarctic Treaty Regime

The 1959 Antarctic Treaty developed in response to a number of political, legal and scientific concerns over the future of Antarctica. The Treaty, which became operative in 1961, has proven particularly robust and effective in dealing with these issues during its 40 years of operation. Part of the key to its success has been its capacity to evolve through the adoption of new legal instruments which has expanded the scope of the original Treaty, the most prominent example of which is the 1991 Protocol on Environment Protection (Madrid Protocol). [41] The Treaty contains a number of pivotal provisions which has allowed it to succeed. One of these is Article IV dealing with Antarctic sovereignty, which provides that pre-existing and new Antarctic claims are not recognised for the duration of the Treaty. [10,11] However, as the Treaty's area of application extends north to 60°S, this also has implications for the Southern Ocean, especially with respect to the capacity to assert maritime claims. The Antarctic Treaty regime therefore provides a legal regime not only for the continent but also for the Southern Ocean, which through the extended reach of CCAMLR applies beyond 60°S.

3. Southern Ocean Maritime Zones

3.1. THE NEED FOR A COASTAL STATE

Only coastal states may assert maritime claims. It is therefore necessary to determine whether any state has a valid claim over a coastal area before it is eligible to assert a maritime claim. In Antarctica this raises particular problems where Article IV of the Antarctic Treaty dominates any discussion on the ability of the territorial claimants to assert maritime claims. Given the uncertain status in international law of the validity of the Antarctic claims, a threshold question is whether there exists any states in Antarctica whose territorial claims are recognised so as to allow them to assert a maritime claim. This question can only be answered by a long historical review of each of the territorial claims, [10] which is beyond the scope of this work. It is appropriate, however, to consider whether the Antarctic Treaty's provisions allow for the assertion of Antarctic maritime claims. In the lead-up to the negotiation of the Treaty in 1959 the resolution of disputes over the status of Antarctic territorial claims was considered one of the pivotal issues. The provisions of Article IV(1) of the Treaty were considered the answer to this problem. Article IV(2) also prohibits the making of any additional or new sovereignty claims while the Treaty is in force. How Article IV(2) is interpreted can have a significant impact upon the ability of Antarctic territorial claimants to assert maritime claims. [12]

3.2. TERRITORIAL SEA

In Antarctica there has been variable practice concerning territorial sea claims. In some instances claims have been made in conjunction with territorial claims. In other instances, separate proclamations have been made over territorial sea zones. In the case of New Zealand's territorial sea off the Ross Dependency it seems that there was no formal proclamation as such but rather a gradual recognition that through the application of previous UK legislation a territorial sea did indeed exist. [13] Despite this variable practice, the limitations on claims in Article IV(2) of the Treaty can not be interpreted as inhibiting territorial claimants from asserting territorial sea claims. This follows because the territorial sea is considered an inherent right of coastal states recognised in customary international law and conventions prior to the entry into force of the Antarctic Treaty. [14,15] A more difficult question arises over whether territorial sea claims can be enlarged without breaching Article IV(2). If customary international law now recognises that coastal states are entitled to a 12 mile territorial sea it could be argued that the enlargement of an Antarctic territorial sea claim from 3 to 12 miles is not an enlargement for the purposes of Article IV(2) of the Treaty, but merely an act adopting a current coastal state entitlement recognised by international law. Nevertheless, it must be conceded that the enlargement of a pre-existing claim to a territorial sea offshore Antarctica will result in the assertion of sovereignty over a greater maritime area. [16]

3.3. CONTINENTAL SHELF

Continental shelf claims have also been asserted by the Antarctic claimants in the Southern Ocean, however, the practice has been variable. Both Australia and Chile made claims to the continental shelf adjacent their Antarctic claims prior to the entry into force of the Antarctic Treaty. [17] In 1966 Argentina claimed sovereignty over its adjacent Antarctic continental shelf out to the edge of the 200 mile limit or the limit of exploitation. [17] No other claimant has asserted a continental shelf claim, though France has asserted claims with respect to the adjacent continental shelves of its sub-Antarctic islands. [17] The validity of Antarctic continental shelf claims, whether asserted prior to or after the entry into force of the Antarctic Treaty, raises similar issues to the validity of territorial sea claims. Coastal state rights to a continental shelf had been recognised under both customary international law and convention prior to 1961. [14,15] While the outer limits of the continental shelf have been extended by the 1982 United Nations Convention on the Law of the Sea (UNCLOS), [42] the extent of coastal state rights to a continental shelf did not change. The law of the sea recognises that coastal states have inherent sovereign rights over a continental shelf which need not be actively proclaimed. As such, it can be argued that an Antarctic continental shelf claim is an inherent right of every claimant and does not represent the assertion of a new claim. Consequently under the terms of the Antarctic Treaty it would seem that coastal states are entitled to assert continental shelf claims in Antarctica. However, as in the case of the territorial sea, an issue can arise over whether the enlargement of a previously asserted continental shelf claim in order to meet the new standards set by UNCLOS breaches the terms of Article IV(2). [18]

3.4. EXCLUSIVE ECONOMIC ZONE

In Antarctica, Exclusive Economic Zone (EEZ) or fisheries zones have only been claimed by Argentina, Australia and Chile. [17] However, varying practices have been adopted by the claimant states towards their mainland claims and those of offshore and sub-Antarctic islands. More than in the case of either the territorial sea or continental shelf, EEZ claims raise difficulties for the Antarctic territorial claimants under Article IV(2). This follows because the EEZ concept was not recognised in international law prior to 1961, [14,19] and unlike the territorial sea or continental shelf it is not recognised as an inherent sovereign right of a coastal state. [20] It follows that EEZ claims must be proclaimed if they are to be asserted. Under these circumstances, the conclusion seems inescapable that the declaration of an EEZ in Antarctica seeking to assert resource sovereignty and jurisdiction is either an enlargement of an existing claim or assertion of a new claim and thereby infringes Article IV (2). [15,18]

3.5. THE SOUTHERN OCEAN DEEP SEA-BED

One of the major developments in the law of the sea during the past 20 years has been the development of a regime for the deep seabed which is found in Part XI of UNCLOS. Following the resolution of a political dispute concerning some of these provisions, Part XI is now actively in the process of being implemented. Part XI seeks to give effect to the principle of the "common heritage of mankind, and establishes the International Seabed Authority (ISA) to oversee its implementation. Any consideration as to whether parts of the Southern Ocean could be classified as part of the 'Area' under UNCLOS raises several difficult issues. [21] First, because only three Antarctic claimants have actually asserted continental shelf claims, the question arises whether for the purposes of determining the area beyond the limits of national jurisdiction in the Southern Ocean account is to be taken of potential maritime zones or only those which have actually been asserted. An issue here is whether territorial claimants who failed to assert offshore claims prior to the entry into force of the Antarctic Treaty are now precluded from doing so under Article IV (2) or whether rights to a continental shelf are inherent in every coastal state. A second issue is whether the ISA would have authority over the seabed offshore the Antarctic unclaimed sector. As there is no territorial claimant no maritime zones have been proclaimed offshore this sector. As a result, it could be argued that the ISA would be entitled to exploit the seabed up to the edge of the low-water mark of the unclaimed sector. [2] The third, and perhaps most substantial question is whether Part XI of UNCLOS can even be said to apply in Antarctica. [22] The question of Antarctica was expressly excluded during the negotiation of UNCLOS, which, given the tentative discussions then taking place between the Antarctic Treaty parties over the implementation of a mining regime in Antarctica, can be seen as significant. In this context, the provisions of the now abandoned 1988 Convention for the Regulation of Antarctic Mineral Resource Activities (CRAMRA) [43] are important. Article 5 of CRAMRA adopted the same area of application as the Antarctic Treaty. It specified that "Antarctic mineral resource activities which take place on the continent of Antarctica and all Antarctic islands, including all ice shelves, south of 60^0south latitude and in the seabed and subsoil of adjacent offshore areas up to the deep seabed" were regulated by the Convention (CRAMRA, Art. 5 (2)). However, for the purposes of CRAMRA, the "deep seabed" was defined as the seabed and subsoil beyond the geographic extent of the continental shelf (CRAMRA, Art 5 (3)). The CRAMRA regime therefore implied that irrespective of UNCLOS, the Antarctic Treaty parties sought to exercise some jurisdiction over mining activities in the Southern Ocean. The Protocol also affects seabed mining because all minerals activities are prohibited under Article 7. Unlike CRAMRA, the Protocol applies to the same area as the Treaty. This implies that the Protocol's area of application includes both the continental shelf and deep seabed up to 60^0S. It follows that by adopting the same area of application as the Antarctic Treaty, the Protocol seeks to regulate all mineral resource activities that may take place in the deep seabed within the Antarctic Treaty area. Irrespective of what constitutes the 'Area' in the Southern Ocean, the Protocol is an indicator that the Antarctic Treaty parties do not believe there is any scope for the application of the UNCLOS deep seabed mineral's regime in Antarctic waters. [11]

4. Antarctic Environmental Protection: The 1991 Madrid Protocol

The 1991 Madrid Protocol to the Antarctic Treaty is perhaps the most significant addition to the Treaty since its inception. The Protocol, which entered into force in 1998, provides for a range of new environmental mechanisms applicable to both the continent and the Southern Ocean. [23] Annex IV deals with the 'Prevention of Marine Pollution' and makes applicable within the Treaty area provisions similar to those found in the 1973/78 International Convention for the Prevention of Pollution from Ships (MARPOL). [44] In particular the following polluting activities are prohibited: the discharge of oil except in cases permitted under Annex I of MARPOL (Protocol, Annex IV, Art. 3), the discharge into the sea of noxious liquid substances (Protocol, Annex IV, Art 4), the disposal of garbage at sea (Protocol, Annex IV, Art 5), and the discharge into the sea of sewage (Protocol, Annex IV, Art 6). While the adoption of these measures under the Protocol is a positive step, it would be inaccurate to argue that they are an improvement on the MARPOL regime. As noted by Bush: "While attempts were made to repeat here the phraseology of the MARPOL annexes, the language necessarily diverges. This is generally because the present annex encompasses in one relatively short annex what is dealt with in five lengthy annexes under MARPOL" (Bush 1990+:Bin 1, Pt AT91C, 129). Few attempts are made in Annex IV to go beyond the reach of the MARPOL provisions. Where there is an exception to the MARPOL regime, as in the case of Article 9 (1) which imposes obligations upon parties to ensure that vessels supporting its Antarctic operations are equipped with adequate retention capacity, the overriding impact of Article 14 providing that nothing in the Annex is to derogate from rights and obligations under MARPOL seems to defeat the effort.

5. Southern Ocean Resource Management

5.1. CCAMLR

The most comprehensive management mechanism which exists in the polar regions for any resource is the 1980 Convention for the Conservation of Antarctic Marine Living Resources (CCAMLR). The Convention was adopted as a result of concerns which developed during the 1970s that unregulated exploitation of fishery stocks in the Southern Ocean could develop into a major commercial activity. Given what had occurred with unregulated whaling in the nineteenth and early twentieth centuries there was a desire to deal with resource management issues before major problems arose. [24] CCAMLR represents an attempt to adopt the 'precautionary approach' to resource management in the Antarctic and this is recognised in its Preamble. This is encapsulated in Article II which sets down a series of objectives for Antarctic marine living resource conservation. These objectives outline the 'ecosystem approach' to resource management. [24] For the purposes of the Convention this approach can be summarised as involving adherence to conservation principles which allow for the rational use of existing resources providing that:

(a) harvested populations do not fall below a level which ensures their stable recruitment;
(b) the ecological relationship between harvested and non-harvested dependent species is maintained; and,
(c) irreversible changes in the marine ecosystem are prevented or minimised.

Unlike some of the other institutions established under the Antarctic Treaty System (ATS), CCAMLR is supported by a permanent Commission administered by a small Secretariat in Hobart, Australia. The Commission's functions are wide-ranging. Its primary aim is to "give effect to the objectives and principles set out in Article II" (CCAMLR, Art IX (1)). Perhaps the most important role it has is the adoption of Conservation Measures to manage Antarctic marine living resource harvesting. The Commission makes its decisions by consensus (CCAMLR, Art XII (1)), and is assisted by a Scientific Committee which acts as a consultative body to the Commission (CCAMLR, Art XIV). The Scientific Committee reviews data concerning Southern Ocean marine living resources, but does not have the ability to conduct its own research (CCAMLR, Art XV). In this respect, both the operation of the Commission and the Scientific Committee is dependent on the annual provision of statistical, biological and other data by Commission members (CCAMLR, Art XX). Membership of both the Commission and the Scientific Committee is open to original contracting parties and other parties which engage in research and harvesting activities in the Convention area (CCAMLR, Arts. VII, XIV).

Following the speedy entry into force of CCAMLR in 1982 there were high hopes that it would be able to deal with the emerging problem of Antarctic marine living resource management and to right any difficulties which had been identified following years of unregulated access to the resource. [25,26,27] However, these expectations were not met, due to a combination of institutional and political problems. One immediate problem, identified at the First Meeting of the Commission, was the consensus decision-making structure. This approach had been adopted because of its success at Antarctic Treaty meetings. However, even before the first meeting of the Commission it was recognised that this decision-making procedure had the potential to impact upon the strength of the Conservation Measures adopted. This has since been reflected in practice, with the ability of a single state to 'veto' any consensus impairing the Commission's ability to adopt strict measures affecting fishing states. [25] Another problem was the lack of adequate data upon which the Scientific Committee or the Commission could assess how an ecosystem approach to resource management could be best implemented. [28] Much of the data initially used was provided by commercial sources and proved to be inadequate. As a result, it was very difficult to implement sensitive management of fishing and harvesting activities. [29] It therefore became vital for the success or otherwise of CCAMLR that the parties supplied information to the Commission and Scientific Committee on a regular basis. A further area of difficulty which emerged early on was that the Convention did not adequately determine the nature of the relationship between the Commission and the Scientific Committee.

CCAMLR has focussed on regulating two fisheries in particular. The krill fishery has received considerable attention, primarily because it was one of the driving factors behind the development of CCAMLR due to the concerns over unregulated harvesting of a stock which plays a key role in the Antarctic food chain. Catch limits have progressively been placed on krill in the South Atlantic and South Indian Ocean sectors of the Southern Ocean. One difficulty which has been encountered in adopting an ecosystem approach in the management of the resource is that scientists still have an incomplete understanding of krill. [30] Despite criticism over the delay in the CCAMLR Commission adopting effective conservation measures for krill, [31] there is no evidence to suggest detrimental impact on Southern Ocean krill stocks. More recently there has been much attention given to the management of Patagonian toothfish (*Dissostichus eleginoides*). The management of this fishery has proved to be particularly challenging due to stocks being found both outside the CCAMLR area of operation and also adjacent to the maritime zones of sub-Antarctic islands over which sovereignty is not disputed, but which do fall within the CCAMLR area. In addition, there have been widespread reports of illegal fishing for the toothfish by vessels from non-CCAMLR parties, which has further highlighted the difficulty in applying CCAMLR to non-parties. [32] The enforcement of fisheries laws in the Southern Ocean is particularly difficult due to the vast distances patrol vessels are forced to operate from their bases and the operating conditions which exist in the Southern Ocean. Australia has successfully arrested three illegal fishing vessels offshore the Heard and McDonald Islands, and the French also have been operating regular patrols in the vicinity of the Kerguelen Islands. [32] While this matter has been extensively discussed at meetings of the CCAMLR Commission in 1998 and 1999, there has been no cessation of the illegal fishing.

5.2. ANTARCTIC SEALING

A considerable degree of unregulated Antarctic sealing took place in the nineteenth century. The sealers were mostly from America, Britain, Canada and South America. The consequences were devastating with fur seal populations on certain islands completely wiped out. Britain did take some action to regulate sealing in the late nineteenth century when Ordinances were proclaimed to regulate sealing activities in and around the Falkland Island Dependencies. However, regulation and enforcement has been variable. Under the Antarctic Treaty, the protection of seals was first raised in 1961 at a meeting of the Treaty Parties. Recommendation I-VIII encouraged the Treaty parties to implement measures to conserve the living resources of the Treaty area. More particular action was taken to combat sealing activities in the 1964 Agreed Measures. [17] Article VI provided that certain mammals could be listed as 'Specially Protected Species' under which permits could be issued for their taking providing there was a "compelling scientific purpose". However, the limitations in the Agreed Measures with respect to sealing were quickly realised when a 1964 sealing expedition revived concerns that full scale commercial sealing could recommence. [11] Various interim Recommendations were adopted by the ATCPs until such time as a special Conference could be convened to discuss a Convention. The Conference on the Conservation of Antarctic Seals convened in 1972 and resulted in the adoption of the Convention for the Conservation of Antarctic Seals (CCAS). [45] CCAS was the first substantive attempt by the Antarctic Treaty Parties at resource management in the Antarctic Treaty area. While it would be incorrect to say that it became a prototype for subsequent efforts to regulate marine living resources and mineral resources, it did provide the Treaty parties with experience in regulating a resource activity and its institutional implications.

CCAS has been praised as the first example in which a legal framework was created for the management of a resource prior to any commercial exploitation of the resource occurring.[11] In this respect CCAS can be said to represent the 'precautionary principle' at work. While these assessments are correct in the post-Antarctic Treaty context, they disregard the uncontrolled exploitation of seals which occurred in the nineteenth century. Nevertheless, the Convention has contributed to effective management of seals in the Southern Ocean. CCAS places any potential commercial exploiter on notice that not only does the Convention place catch limits on certain seal species, but also that once commercial sealing commences any Party may request a meeting to consider the establishment of further measures to ensure an "effective system of control" of sealing activities (CCAS, Art 6).

5.3. SOUTHERN OCEAN MINING

When the Antarctic Treaty was negotiated there was no consideration given to the question of mining being carried out on the continent or the Southern Ocean. A number of reasons existed for this. First, the Treaty negotiations were dominated by concerns over the resolution of sovereignty. If the possible exploitation of minerals had been discussed there was the potential for the consensus over the Article IV sovereignty solution to break down. Second, Antarctica's resource potential at the time was unknown and while scientists had discovered a wide range of minerals throughout the continent, none were then thought to have much commercial potential. Third, up till that time there had been very little mineral exploration in the Southern Ocean and this in conjunction with the relative world-wide abundance of oil and gas reserves meant that little interest was shown in Antarctica's non-renewable resource potential. As a result, it comes as no surprise that the Antarctic Treaty made no reference at all to minerals and that the only provision in the Treaty dealing with resource management was the reference in Article IX(1)(f) to living resources.

A change began to occur in this attitude during the 1970s. In 1973 the research vessel *Glomar Challenger* discovered ethane and methane gas in the Ross Sea area and this raised expectations that commercially recoverable reserves could exist in the Southern Ocean. There had also been, in the period since the Treaty was negotiated, a substantial development in mining technology which had been applied in the Arctic and which made mining operations in Antarctica more feasible. The final factor which saw commercial mining become a possibility was the global concern which existed in the 1970s over the scarcity of energy resources and the need to find alternate sources, especially given the political instability which then existed in the Middle East oil fields. As a result of these developments the exploitation of Antarctic mineral resources became an agenda item for the ATCPs.

The 1991 Madrid Protocol was not designed as a supplement to CRAMRA but rather as an alternative instrument. It establishes a total prohibition on mineral resource activity in Antarctica. The only exception to this is where scientific research is being conducted (Protocol, Art 7). The Protocol does not extend to the mining of ice, which is a surprising gap given the studies which have demonstrated the potential commercial viability of ice and iceberg harvesting/mining. [33,34,35] This does not imply that ice harvesting is unregulated as the other provisions of the Protocol are considered to apply to any such activity. Under these circumstances an environmental impact assessment would have to take place before such an activity could occur. The Protocol's prohibition on mining can also be subject to modification. Two methods are open to the parties. The first adopts the same procedure found in Article XII (1) of the Antarctic Treaty requiring unanimous agreement of all the parties (Protocol, Art 25 (1)). The other adopts a Review Conference mechanism which may be convened 50 years after entry into force of the Protocol (Protocol, Art 25 (2)). In this event, special provision is made for any renunciation of the Protocol's mining prohibition. Before any change to the mining prohibition can occur, there must be "in force a binding legal regime on Antarctic mineral resource activities that includes an agreed means for determining whether, and, if so, under what conditions, any such activities would be acceptable" (Protocol, Art 25 (5a)). This provision seems to imply that a resource management mechanism similar to CRAMRA may be required before any mineral activities could occur in Antarctica if the Protocol's mining prohibition was one day lifted.

6. Other International Regimes operating in the Southern Ocean

While the legal frameworks established under the ATS are dominant in the Southern Ocean, they are not the only legal regime of relevance. In the case of whaling, the 1946 International Convention for the Regulation of Whaling [46] is particularly important because of the large numbers of whales which frequent the Southern Ocean and the history of Antarctic whaling, The Convention, which has been signed by most whaling nations and the principal parties to the Antarctic Treaty, creates a regulatory regime for the catching of whales in all the world's oceans. Administered by the International Whaling Commission (IWC), state parties through the IWC forum can set catch quotas, designate protected species, and regulate whaling methods. While the Convention seeks to prevent the over-exploitation of whales it could not be claimed that it is protectionist. Rather it seeks to ensure sustainable whaling. [27] Given the serious depletion of whale stocks that had occurred before the Convention's entry into force, the IWC has always sought to closely monitor and regulate any whaling activities taking place in the Southern Ocean. In 1982 all commercial whaling was prohibited in Antarctic waters, and elsewhere from 1986. [36] However, 'scientific whaling' still occurs for research purposes, principally by Japan. This has been the subject of controversy as claims have been made that Japanese whalers have used this as a loophole to engage in commercial whaling in the Southern Ocean. [37] The issue has caused some concern for Antarctic territorial claimants as Japanese vessels have engaged in whaling offshore the continent within some Fishing Zones. [38]

Of the specific global marine pollution regimes, MARPOL, and the 1972 London Convention [47] apply in the Southern Ocean. [2] However, while these Conventions are generally considered to have been successful in combating vessel-sourced oil pollution, dumping, and other forms of vessel-sourced discharges which occur at sea, neither deal with any of the specialised pollution problems which can arise in polar waters. This was highlighted by the fact that it was only in 1991 that the Southern Ocean was declared a 'special area' under MARPOL. This had the effect that all operational discharges were absolutely prohibited from vessels except under cases of extreme peril. A further amendment was proposed in 1992 so that the Southern Ocean south of 60^0S would also be included as a Special Area under Annex II. With this exception, the major conventions dealing with vessel-source pollution do not take into account the navigational and operation difficulties which vessels can experience in polar waters, or recognise that the polar marine environment is particularly susceptible to damage from pollution. The 1991 Madrid Protocol, which incorporates some MARPOL provisions in Annex IV, has now partially dealt with this gap in the legal regime.

7. Conclusion

The Southern Ocean despite its relative remoteness has increasingly become the focus of international attention. Initially through the operation of the Antarctic Treaty and more recently under CCAMLR and the 1991 Madrid Protocol the legal and management regime for the Ocean has developed. Resource management and conservation in the Southern Ocean has increasingly become subject to greater regulation as concerns have grown over unregulated fishing and mining. The protection of the marine environment has also received some attention, though this is by no means comprehensive. The challenge which remains for environmental and resource management under international law in the Southern Ocean is to not only develop a comprehensive regime, but one which is capable of effective enforcement. This will require a great deal of good will on the part of the Antarctic Treaty parties, and global commitment to an ocean which is not subject to traditional claims but is more akin to being part of the common heritage of all peoples.

References

1. Stonehouse, B. (1989*) Polar Ecology*, Blackie, Glasgow

2. Joyner, C.C. (1992) *Antarctica and the Law of the Sea*, Martinus Nijhof, Dordrecht

3. Armstrong, T., Roberts, B. and Swithinbank, C. (1973) *Illustrated Glossary of Snow and Ice* 2nd, Scott Polar Research Institute, Cambridge

4. Gulland, J.A. (1988) The Management Regime for Living Resources, in C.C. Joyner and S.K. Chopra (eds) *The Antarctic Legal Regime*, Martinus Nijhoff Publishers, Dordrecht, pp. 219-240

5. Knox, G.A. (1983) The living resources of the Southern Ocean: A scientific overview, in F.O. Vicuna (ed) *Antarctic resources policy: Scientific, Legal and Political Issues*, Cambridge University Press, Cambridge, pp. 21-60

6. Chittleborough, G (1984) Nature, extent and management of Antarctic living resources, in S. Harris (ed) *Australia's Antarctic policy options*, Centre for Resource and Environmental Studies, Australian National University, Canberra, pp. 135-161

7. Hureau, J.-C. and Slosarczyk, W. (1990) Exploitation and Conservation of Antarctic Fishes and Recent Ichthyological Research in the Southern Ocean, O. Gon and P.C. Heemstra (eds) *Fishes of the Southern Ocean*, Institute of Ichthyology, Grahamstown, South Africa, pp. 52

8. Gjelsvik, T. (1983) The Mineral Resources of Antarctica: Progress in Their Identification, in F.O. Vicuna (ed) *Antarctic Resources Policy*, Cambridge University Press, Cambridge, pp. 61

9. Larminie, G. (1991) The Mineral Potential of Antarctica: The State of the Art, in A.Jorgensen-Dahl and W. Ostreng (eds) The Antarctic Treaty System in World Politics, Macmillan, London, pp. 79

10. Auburn, F.M. (1982) *Antarctic Law and Politics*, C.Hurst & Co, London

11. Watts, A. (1992) *International Law and the Antarctic Treaty System*, Grotius Publications, Cambridge

12. Vicuna, F. O. (1988) *Antarctic Mineral Exploitation*, Cambridge University Press, Cambridge

13. Bush, W.M. (1988) *Antarctica and International Law: A Collection of Inter-State and National Documents, Vol III & IV*, Oceana, London

14. O'Connell, D.P. (1982) *The International Law of the Sea*, Clarendon Press, Oxford

15. Churchill R.R. and Lowe, A.V. (1988) *The law of the sea*, Rev ed, Manchester University Press, Manchester

16. Opeskin, B.R. and Rothwell, D.R. (1991) Australia's Territorial Sea: International and Federal Implications of its Extension to 12 Miles, *Ocean Development and International Law* **22**: 395-431

17. Bush, W.M. (1982) *Antarctica and International Law: A Collection of Inter-State and National Documents, Vol I & II*, Oceana, London

18. Crawford, J. and Rothwell, D.R. (1992) Legal Issues Confronting Australia's Antarctica, *Australian Yearbook of International Law* **13**: 53-88

19. Attard, D.J. (1987) *The Exclusive Economic Zone in International Law*, Clarendon Press, Oxford

20. Kwiatkowska, B. (1989) *The 200 Mile Exclusive Economic Zone in the New Law of the Sea*, Martinus Nijhoff, Dordrecht

21. Vidas, D (1999) The Relationship between the Environmental Protocol and the Law of the Sea Convention regarding the Southern Ocean Seabed, *Antarctic Project Reports* **7/99**

22. Infante, M.T. (1983) The continental shelf of Antarctica: legal implications for a regime on minerals resources, in F.O. Vicuna (ed) *Antarctic resources policy: Scientific, Legal and Political Issues*, Cambridge University Press, Cambridge, pp. 253

23. Vicuna, F.O. (1996) The effectiveness of the Protocol on Environmental Protection to the Antarctic Treaty, in O.S. Stokke and D. Vidas (eds) *Governing the Antarctic*, Cambridge University Press, Cambridge

24. Edwards D.M. and Heap, J.A. (1981) Convention on the Conservation of Antarctic Marine Living Resources: A Commentary, *Polar Record* **20**: 353-362

25. Lagoni, R. (1984) Convention on the Conservation of Marine Living Resources: A Model for the Use of a Common Good, in R. Wolfrum (ed) *Antarctic Challenge*, Duncker & Humblot, Berlin, pp. 93

26. Zegers, F. (1983) The Canberra Convention: objectives and political aspects of its negotiation, in F.O. Vicuna (ed) *Antarctic resources policy: Scientific, Legal and Political Issues*, Cambridge University Press, Cambridge, pp. 149

27. Lyster, S. (1985) *International Wildlife Law*, Grotius Publications, Cambridge

28. Howard, M. (1989) The Convention on the Conservation of Antarctic Marine Living Resources: A Five Year Review, *International and Comparative Law Quarterly* **38**: pp. 104-149

29. Nicol, S. (1991) CCAMLR and its approaches to management of the krill fishery, *Polar Record* **27**: 229-236

30. Nicol S. and de la Mare, W. (1993) Ecosystem Management and Antarctic Krill, *American Scientist* **81 (1)**, pp. 36-47

31. Nicol, S (1992) Management of the krill fishery: was CCAMLR slow to act?, *Polar Record* **28**: 155-157

32. Bateman, S. and Rothwell, D.R. (eds) (1998) *Southern Ocean Fishing: Policy Challenges for Australia*, Centre for Maritime Policy, University of Wollongong, Wollongong

33. Schwerdtfeger, P. (1986) Antarctic Icebergs as Potential Sources of Water and Energy, in R. Wolfrum (ed) *Antarctic Challenge II*, Duncker & Humblot, Berlin, pp.377-389.

34. Lindquist, T.R. (1977) The Iceberg Cometh? International Law Relating to Antarctic Iceberg Exploitation, *Natural Resources Journal* 17:1-41

35. Carroll, J.E. (1983) Of Icebergs, Oil Wells and Treaties: Hydrocarbon Exploitation Offshore Antarctica, *Stanford Journal of International Law* 19: pp207-227.

36. Birnie, P (ed) (1985) *International Regulation of Whaling*, Oceana, New York

37. Ellis, S.L. (1988) Japanese Whaling in the Antarctic: Science or Subterfuge? *Oceanus*: **31(2)**: pp. 68-69

38. Shevlin, J. (1992) Japanese whaling activity observed near Davis station, *ANARE News* **69**: 18-19.

39. 402 UNTS 71.

40. (1980) 19 ILM 841.

41. (1991) 30 ILM 1455.

42. (1982) 21 ILM 1261.

43. (1988) 27 *ILM* 868.

44. (1973) 12 ILM 1319; (1978) 17 ILM 546.

45. (1972) 11 ILM 251.

46. 161 UNTS 74.

47. 1046 *UNTS* 120.

INDEX

A

abandonment, 135, 144, 184
access, 49, 71, 77, 79, 80, 81, 84, 102, 122, 126, 128, 150, 155, 156, 164, 167, 169, 185, 201, 212, 222, 223, 229, 274, 276, 277, 279, 305
Ad hoc Expert Panels under Fish Agreement, 40
Agreement, 26, 27, 28, 29, 38, 40, 42, 46, 56, 59, 76, 156, 157, 202, 225, 227, 279, 280, 281
American Revolution, 220
Antarctic ecosystem, 297, 298
ANTARCTIC SEALING, 306
Antarctic Treaty, 4, 297, 298, 299, 300, 301, 302, 303, 304, 305, 306, 307, 308, 309, 310
Antarctic Treaty System, 298
Antarctica, 297, 298, 300, 301, 302, 303, 307, 308, 309, 310, 311
Anthropogenic pollutants, 290
antifouling paints, 210
aquaculture, 9, 11, 13, 153, 154, 265, 289
archipelagic baselines, 28, 71, 73
archipelagic state, 73
Arctic Ocean, 216, 225, 283, 284, 285, 288, 291, 292, 294, 295, 297
ARCTIC OCEAN, 283
Assembly, 26, 28, 29, 30, 31, 32, 34, 35, 42, 45, 46, 51, 67, 156, 192, 230
Awareness, 97, 210, 211, 294

B

ballast water, 186
Barcelona Convention, 234, 245, 246
Barcelona Convention on the Protection of the Mediterranean Sea against Pollution, 234
Barents Sea, 70, 189, 284, 287, 288, 289
BATHYMETRY, 216
beach, 8, 9, 164, 169, 170, 181, 186, 188
beaches, 164, 165, 172, 186, 188, 229
Behavioural approaches, 7
Best Practicable Environmental Option, 187, 190
bilateral agreements, 238, 239, 254, 257, 258, 261, 262, 263, 274

BIODIVERSITY, 206
Biodiversity Convention, 13
Black Sea, 88, 148, 185, 231, 232, 233, 234, 235, 236, 237, 238, 240, 241, 242, 243, 244, 247, 248, 250, 251, 252
BLACK SEAS, 231
boundaries, 2, 3, 19, 63, 66, 68, 71, 74, 133, 136, 161, 201, 203, 257, 259, 262, 263, 264, 265, 284, 285
Boundary Delimitation, 68
BPEO, 187
bulk carriers, 84, 91, 95, 98, 242

C

Camouco, 52, 58, 59, 61
capacity, 40, 48, 49, 85, 86, 92, 95, 96, 97, 98, 100, 101, 102, 103, 116, 117, 127, 137, 142, 146, 151, 152, 154, 156, 173, 181, 190, 207, 211, 248, 254, 257, 264, 267, 269, 276, 291, 300, 304
catches, 145, 147, 148, 149, 150, 152, 154, 155, 204, 223, 225, 228, 262, 288, 289, 300
CBD, 206
CCAMLR, 297, 299, 300, 301, 304, 305, 306, 309, 311
CFP, 224
Chaisiri Reefer 2, 52, 60, 61
Chamber for Fisheries Disputes, 37
Chamber for Marine Environment Disputes, 37
Chemical tankers, 99
choke points, 87
CLCS, 32, 63, 74
Club of Rome, 178, 192
coastal defences, 200, 205
coastal lagoons, 200
coastal settlement, 14
COASTAL STATE, 301
coastal state rights, 64, 302
coastal states, 63, 64, 65, 66, 67, 73, 74, 156, 189, 222, 224, 229, 238, 247, 259, 260, 261, 262, 263, 266, 274, 275, 279, 301, 302
coastal water quality, 181, 182, 185, 188
coastal zone management, 228
co-management, 158
commercial shipping, 11
Commercial Shipping, 227

Commission on Sustainable Development, 34, 245, 249, 254
Commission on the Limits of the Continental Shelf, 31, 32, 33, 34, 63, 74, 285
common inheritance of humankind, 69
COMMON MARITIME ZONES, 69
common zone, 69
Compulsory Procedures Entailing Binding Decisions, 39, 43
conservation, 2, 3, 7, 11, 13, 14, 16, 21, 26, 27, 37, 40, 48, 57, 59, 66, 74, 155, 156, 158, 162, 163, 169, 170, 173, 192, 193, 201, 202, 203, 207, 208, 211, 212, 213, 222, 223, 224, 225, 226, 243, 257, 258, 259, 261, 262, 276, 290, 291, 304, 306, 309
Conservation and Management of Straddling Stocks and Highly Migratory Fish Stocks, 27, 38
Conservation and Sustainable Exploitation of Swordfish Stocks in the South-Eastern Pacific Ocean, 37, 52, 59, 61, 62
Container Sector, 101
container ships, 86, 92, 95, 100, 236
Container traffic, 242
containership fleets, 237
continental shelf, 28, 29, 31, 32, 40, 41, 49, 50, 63, 66, 67, 71, 74, 118, 135, 136, 144, 216, 225, 239, 240, 244, 254, 259, 262, 270, 271, 284, 285, 286, 302, 303, 310
Continental Shelf, 2, 29, 32, 50, 66, 178, 249, 296
CONTINENTAL SHELF, 66, 73, 302
continental shelves, 12, 14, 73, 185, 229, 232, 233, 238, 244, 259, 285, 300, 302
Convention for the Conservation of Antarctic Marine Living Resources, 304
Convention for the Regulation of Antarctic Mineral Resource Activities, 303
Convention on Biological Diversity, 202, 206
Convention on the Conservation of Antarctic Marine Living Resources, 297, 299, 300, 310, 311
Conventional Freight Sector, 100
Corfu Channel case, 50
Council, 6, 18, 29, 30, 31, 34, 40, 42, 43, 46, 49, 193, 200, 204, 212, 214, 230, 256, 289, 293
cradle-to-grave, 179

CRAMRA, 303, 308
crisis, 87, 95, 102, 107, 128, 132, 141, 145, 158, 192, 196, 256
cruise industry, 103, 182
cruise shipping, 3
cruise ships, 95, 103
CS, 63, 66, 67, 68, 69, 239, 249

D

DCs, 83
decentralisation, 158
decommissioning, 134, 136, 144, 185, 189
deep sea mining, 12
DEEP SEA-BED, 303
deep-ocean and mid-ocean routes, 242
DELIMITATION PRINCIPLES, 68
demand, 3, 104, 105, 106, 107, 108, 109, 110, 113, 114, 117, 119, 125, 127, 134, 136, 138, 142, 144, 150, 153, 154, 164, 165, 166, 167, 169, 170, 234, 293
Deregulation, 157, 158
Developing Countries, 83
diesel engine, 94, 95
dilution, 176, 181
Directive on the discharge of certain dangerous substances into the aquatic environment, 181
discarding, 148, 204
Dispute Settlement Institutions, 29, 35
Dispute Settlement Mechanisms, 46
distribution chains, 152
dredging spoil, 180, 185
Drift nets, 203
drilling technology, 127
dry bulk, 91, 96, 97, 98, 227
Dutch, 5, 8, 9, 15, 22, 64, 101, 218, 219

E

East China Sea, 256, 257, 259, 261, 262, 270, 271, 276, 277, 278, 279, 280, 281
Ecological Conditions, 234
ecological stress, 235
Economic Planning Commission, 29, 30
ecosystem approach, 201, 202, 204, 212, 213, 304, 305, 306
ecosystems, 168, 185, 192, 201, 202, 203, 205, 208, 234, 242, 265, 269, 289, 290, 291, 294, 295, 296, 298

INDEX 315

EEZ, 40, 41, 43, 48, 49, 53, 54, 55, 58, 59, 60, 63, 65, 66, 68, 69, 74, 85, 86, 155, 222, 223, 230, 239, 249, 253, 257, 258, 259, 260, 261, 262, 265, 268, 270, 271, 273, 275, 278, 281, 302
EFZ, 155, 239, 249
endocrine disruptors, 208
Enterprise, 29, 31, 67, 100, 101
Environment and Safety Measures., 99
environmental impact, 30, 133, 175, 202, 206, 298, 308
Environmental Management Systems, 182
Environmental Status, 290
equidistance, 68, 71
equitability, 68
Ethnic Problems, 291
Eutrophication, 234
EVOLUTION AND STRUCTURE, 5
Evolution of the World Fleet, 91
EXCLUSIVE ECONOMIC ZONE, 65, 302
exclusive economic zones, 63, 155, 156, 238, 284
EXTERNAL RELATIONS, 78

F

Fast ice, 298
fast ships, 94
ferries, 95, 97
Finance Committee, 29, 30
finfish, 299
fisheries, 2, 6, 7, 9, 10, 11, 13, 18, 20, 21, 39, 40, 41, 43, 48, 49, 64, 70, 71, 145, 146, 147, 148, 154, 155, 156, 157, 158, 169, 181, 189, 190, 199, 202, 203, 204, 208, 212, 216, 218, 219, 222, 223, 224, 225, 226, 230, 255, 256, 257, 258, 259, 260, 261, 262, 263, 264, 265, 273, 276, 277, 289, 290, 302, 306
Fisheries, 11, 18, 19, 37, 50, 70, 146, 155, 157, 159, 160, 204, 213, 224, 230, 243, 257, 279, 280, 288, 289, 296
fisheries management, 18, 154, 263
fishing effort, 154, 204
Fishing fleets, 85
Fishing Free' zones, 203
fishing industry, 145, 150, 153, 228, 258, 298
fishing resources, 28, 146, 148, 155, 156, 158
fishing zones, 66, 86, 155, 156, 257
flag of convenience., 92

flags of convenience, 92, 157
food, 9, 11, 16, 78, 79, 87, 88, 99, 100, 104, 145, 148, 153, 163, 185, 199, 203, 208, 210, 217, 221, 226, 228, 288, 298, 306
Framework Convention on Climate Change, 13, 295
freedom of navigation, 53, 64, 65, 73, 273, 275

G

gas hydrates, 284
GDP, 241, 249
general cargo, 84, 95, 100, 101
general management, 18
geo-economy, 79, 81
GEOGRAPHY OF MARINE WASTE, 177
geography of the sea, 1, 5, 6, 7, 8, 16, 18, 19, 20
geopolicy, 81
GEO-POLICY, 79
Geopolitical Context, 235
geopolitical stress, 236, 238, 242, 244
geostrategic evolution, 11
geo-strategy, 77, 78, 81, 82, 86, 87, 88, 89, 92, 103
gill nets, 203
Global Ballast Water Management Programme, 186
Global Programme of Action (GPA) for the Protection of the Marine Environment from Land-Based Activities, 185
GloBallast, 186
GPS, 71, 166
Grand Prince, 52, 59, 60, 61
Greenpeace, 135, 183, 194
gross domestic product, 241

H

Hanseatic League, 219
harmful aquatic organisms, 186
HDI, 241, 249
heavy metals, 208, 291
high seas, 59, 156, 193, 203, 220, 223, 224, 225, 229, 256, 257, 258, 259, 262, 274, 285
horizontal drilling technology, 127
hormone-mimicking pollutants, 186
hovercraft, 94
human development indicator, 241

Hydrocarbon raw materials, 283
hydrocarbon reserves, 119
hydrofoils, 94
hypoxia, 235

I

ICCAT, 226
ice shelves, 298, 303
Icebergs, 298, 311
ICJ, 39, 44, 45, 48, 49, 50, 51, 61, 62, 68, 71, 72
Illegal fishing, 256
IMB, 256
IMO, 35, 100, 186, 194, 209, 227, 228, 255, 256, 279
implementation, 12, 13, 26, 77, 147, 158, 187, 188, 207, 209, 210, 212, 222, 235, 242, 244, 245, 251, 257, 259, 265, 266, 275, 276, 278, 290, 294, 303
Implementation Agreement, 26, 27, 29, 30, 31, 42, 203
indigenous populations, 291
Individual Transferable Quotas, 158
industrialisation of the world ocean, 11, 18
INSTITUTIONALISATION OF WASTE MANAGEMENT, 182
International Baltic Sea Fishery Commission, 224
International Chamber of Commerce, 256
International Commission for the Conservation of Atlantic Tunas, 226
International Convention for the Prevention of Pollution from Ships, 304
International Convention for the Regulation of Whaling, 308
International Council for the Exploration of the Sea, 6, 18, 200
International Council on Exploration of the Sea, 289
International Court of Justice, 38, 39, 44, 45, 46, 49, 51, 68, 88, 133, 240
International Geographical Congresses, 6
International Maritime Bureau, 256
International Maritime Organisation, 100, 186, 193, 194, 195, 209, 227
International Safety Management, 100
INTERNATIONAL SEABED, 67
International Seabed Authority, 2, 29, 67, 74, 303
international straits, 216, 251

International Tribunal for the Law of the Sea, 29, 31, 35, 36, 37, 38, 39, 42, 45, 49, 51, 52, 54, 56, 59
International Whaling Commission, 225, 230, 308
intertidal wetlands, 200
island groups, 216
islands, 9, 68, 70, 71, 73, 87, 99, 133, 165, 207, 217, 220, 230, 233, 254, 259, 260, 270, 276, 284, 290, 297, 298, 300, 302, 303, 306
Islands, 70
ISM, 100
ITLOS, 31, 35, 36, 37, 44, 46, 48, 49, 51, 53, 54, 56, 57, 58, 59, 60, 61, 62
ITOs, 158
IWC, 225, 308

J

Japan, 15, 18, 52, 56, 57, 58, 67, 70, 81, 83, 92, 93, 94, 96, 107, 133, 150, 151, 153, 179, 215, 221, 225, 253, 254, 256, 257, 258, 259, 260, 261, 262, 263, 264, 265, 266, 267, 269, 270, 271, 273, 274, 276, 277, 278, 279, 280, 281, 282, 287, 288, 290, 308
Joint Offshore Zones, 70
Jurisdiction, 50, 222
Jurisdiction cases, 50
jurisprudence, 43, 44, 62, 68
Jurisprudence, 43

K

krill, 298, 300, 306, 311
Kyoto Protocol, 134, 295

L

LAND BASED MARINE POLLUTION, 208
land bridges, 229
large marine ecosystems, 3, 4, 7, 18
Law of the Sea Convention, 1, 12, 13, 37, 67, 69, 230, 255, 310
LDC, 177
LDCs, 83, 177, 184, 189
Legal and Technical Commission, 29, 30
leisure, 2, 3, 7, 11, 13, 16, 20, 23, 134, 165, 166, 173

INDEX

Less Developed Countries, 83, 177
Liberal industrialised countries, 82
Libya-Malta, 68
Libya-Tunisia, 68
life cycle analysis, 179
liquid bulk, 97
Living Marine Resources, 222
LIVING MARINE RESOURCES, 217
LNG, 107, 114
London Convention, 182, 184, 186, 188, 189, 190, 194, 309
London Dumping Convention, 176, 182, 183, 184, 194, 227
long waves, 1, 12, 15

M

M/V "Saiga, 51, 52, 53, 54, 61
Madrid Protocol, 301, 304, 308, 309
management, 2, 3, 4, 6, 7, 9, 12, 13, 14, 16, 18, 19, 20, 21, 22, 23, 26, 28, 37, 40, 57, 74, 100, 108, 130, 135, 145, 146, 147, 154, 155, 156, 157, 158, 159, 167, 169, 171, 172, 173, 175, 176, 177, 178, 179, 180, 182, 184, 186, 187, 188, 189, 190, 192, 193, 194, 199, 201, 202, 203, 204, 205, 206, 207, 210, 211, 212, 213, 222, 224, 225, 226, 228, 234, 246, 247, 248, 251, 253, 254, 255, 257, 258, 259, 260, 261, 262, 263, 268, 272, 275, 276, 279, 289, 291, 298, 300, 304, 305, 306, 307, 308, 309, 310, 311
Management, 178
manganese nodule deposits, 67
MAP, 234, 244, 245, 246, 249, 250
Mare Liberum, 219
Mariculture, 226
marine biodiversity, 199, 208
marine conservation, 163, 208, 212
marine ecosystems, 19, 21, 162, 167, 172, 275
Marine Environmental High Risk Areas, 210, 214
marine environments, 164, 166, 169, 173
Marine Geography, 5, 16
marine life, 163, 289, 298
marine litter, 186
Marine Nature Conservation, 206
marine parks, 163, 169, 170
marine policy, 254, 255
Marine Pollution, 21, 191, 193, 194, 195, 196, 208, 226, 245, 249, 266, 280, 304

marine recreation, 163, 164, 165, 168, 169, 170, 171
Marine recreation, 168
MARINE RECREATION, 164
Marine scientific research, 13
MARINE SITE PROTECTION, 207
marine turtles, 204
marine waste disposal, 175
marine waste management, 175, 179, 182, 187
marine wildlife, 3, 135, 199, 203, 206, 207, 208, 209, 210, 211, 212
Maritime Biodiversity Action Plans, 206
maritime boundaries, 1, 2, 6, 11, 19, 43, 63, 70, 71, 72, 74, 75, 81, 254, 276
MARITIME BOUNDARIES, 63
maritime boundary agreements, 66, 68, 71
maritime boundary delimitation, 12, 64, 68, 69, 70
Maritime boundary delimitation, 63, 68, 74
Maritime boundary disputes, 72
maritime policy, 81, 82, 83, 84, 86
maritime power, 77, 82, 216, 220
maritime powers, 11, 219, 220, 274
maritime trade links, 5
Maritime transportation, 242
MARITIMISATION, 81
Market and Price Trends, 105
MARPOL, 99, 182, 187, 242, 249, 304, 309
maximum sustainable yield, 222, 223
maximum yield, 146, 148
MDC, 177
Mediterranean, 3, 4, 22, 88, 96, 102, 140, 148, 200, 216, 227, 231, 232, 233, 234, 235, 236, 237, 238, 239, 240, 241, 242, 243, 244, 245, 246, 247, 248, 249, 250, 251, 252, 269, 281
MEDITERRANEAN, 231
Mediterranean Action Plan, 234, 244
Mediterranean surface water, 232
Meetings of States Parties, 29, 33
MEHRAs, 210, 214
merchant fleets, 86
merchant navy, 83, 85, 92
methane carriers, 99
Methodological Approach, 81
migratory species, 203
mineral and energy resources, 7, 11, 12
mineral and energy use, 12
mineral production, 85
Minerals, 30, 32, 284
Monte Confurco, 52, 58, 59, 61

More Developed Countries, 177
Mox Plant Case, 52
MSY, 154, 155, 160, 223
mudflats, 200
municipal waste, 235

N

NAFO, 224
Natura 2000, 207
natural gas, 87, 107, 109, 110, 111, 112, 113, 116, 119, 120, 121, 123, 124, 125, 127, 132, 134, 137, 138, 283, 284, 285, 286, 289, 290
natural prolongation, 66, 68
Naval Fleet, 86
naval power, 5, 13, 21, 86
naval topics, 7
near shore environments, 168
New Industrial Countries, 83
New sea powers, 92
newly industrialised countries, 83, 92
No Take Zones, 203
North Atlantic, 3, 4, 7, 14, 21, 70, 84, 133, 158, 159, 160, 194, 215, 216, 217, 218, 219, 220, 221, 222, 224, 225, 226, 227, 228, 229, 249, 288
North Korea, 253, 254, 255, 263, 264, 266, 267, 268, 269, 270, 272, 273, 276, 277, 278, 279, 281
North Sea Continental Shelf cases, 50
North Sea Inter-ministerial conferences, 184
NORTHEAST ASIAN SEAS, 253
Northeast Atlantic, 184, 202, 224, 225
Northern Sea Route, 285, 287, 288
Northwest Atlantic, 224
Northwest Pacific Action Plan, 266
NOWPAP, 266, 267
NSR, 285, 286, 287, 288, 290
Nuclear Tests cases, 50

O

Ocean and Coastal Management, 16, 140, 144, 159, 173, 191, 251
OCEAN CIRCULATION, 217
oceanography, 6, 15, 18, 19, 21, 220, 228
Offshore Exploration and Development, 113
offshore jurisdiction, 69

Offshore oil and gas, 23, 244
OFFSHORE OIL AND GAS, 105, 218
offshore oil and gas industry, 105, 133, 205, 238
Offshore Production, 110
oil and gas, 3, 12, 66, 69, 74, 105, 109, 110, 125, 128, 129, 130, 132, 133, 136, 139, 140, 141, 142, 143, 144, 205, 210, 217, 218, 229, 236, 238, 242, 244, 270, 300, 307
oil price, 107, 108, 109, 113, 114, 116, 117, 127, 128, 130, 137, 143
oil tanker spills, 209
oil tankers, 82, 84, 91, 95, 98, 99, 209
Oil tankers, 98
OPEC, 79, 107, 108, 109, 117, 128, 129, 138, 142, 144
orange juice tankers, 99
organic pollutants, 208, 291
organotins. See . See
OSPARCOM, 184, 194
outer limits of the continental shelf, 284, 302
Overfishing, 145, 147, 226
overlapping claims, 71, 254, 272

P

pack ice, 297, 298, 300
PARADIGMS, 6
Paris and Oslo Conventions, 182
Particular category chambers, 37
Particularly Sensitive Sea Area, 210, 214
Partnerships, 210, 212
passengers, 91, 95, 97, 99, 100, 102, 103
physical geography of the sea, 7
Piracy, 256, 281
Points of entry to environment, 181
political uncertainty, 131, 136
Pollutant sources, 184
Polluter Pays, 179
pollution, 3, 7, 12, 14, 23, 48, 52, 60, 61, 70, 100, 135, 148, 170, 175, 176, 177, 178, 179, 180, 181, 182, 184, 185, 186, 187, 188, 189, 190, 191, 199, 200, 201, 202, 208, 209, 210, 211, 226, 227, 229, 235, 264, 265, 267, 268, 269, 270, 272, 275, 289, 291, 299, 309
population growth, 12, 13, 138, 153, 177, 178, 189
port development, 11, 205

ports, 6, 7, 10, 12, 83, 84, 88, 92, 100, 101, 102, 103, 180, 182, 185, 186, 193, 219, 227, 229, 256
precautionary approach, 3, 184, 204, 209, 289, 304
PRECAUTIONARY APPROACH, 183
Precautionary Principle, 175, 179, 184
Preparatory Commission, 29, 44
PREPCOM, 29
privatisation, 3, 83, 113, 128, 137, 158
production, 6, 11, 12, 13, 42, 66, 78, 79, 80, 81, 85, 101, 105, 107, 108, 109, 110, 111, 112, 113, 114, 117, 118, 119, 120, 122, 125, 126, 127, 128, 129, 131, 134, 135, 137, 138, 139, 140, 141, 146, 147, 148, 149, 150, 151, 153, 154, 167, 178, 179, 183, 218, 223, 241, 255, 258, 283, 289, 295
property rights, 154, 155, 292
Proportionality, 68
PSSA, 210
PSSAs, 210, 214
Public awareness, 187

R

radionuclides, 291
recreation, 7, 13, 161, 162, 163, 164, 165, 166, 167, 168, 169, 170, 171, 172, 181, 201, 205, 228
reflagging, 156
refrigerated fleet, 101
regional conflicts, 237
Regional Co-operation, 244
Regional Seas Programme, 179, 192, 244, 247, 249, 251, 269
renewable energy, 12
Research Perspectives, 87
Reserves, 119
Roll-on Roll-off, 97, 100
Royal Commission on Environmental Pollution, 190, 196
runoff, 168, 185, 226, 234, 235, 290
Russia, 47, 48, 49, 70, 81, 84, 88, 93, 108, 123, 124, 132, 142, 215, 221, 229, 236, 238, 240, 252, 253, 256, 264, 265, 266, 267, 269, 270, 271, 276, 277, 278, 279, 284, 285, 286, 288, 289, 290, 292, 295, 296

S

SAC, 207
sailing liners, 103
saltmarsh, 200
SCUBA, 166
sea ice, 298
Sea of Japan, 70, 259, 262, 264, 266, 271, 273, 278, 279
sea order, 84
sea transport, 7, 20, 91, 92
Sea Use Development, 240
Seabed Authority, 27, 34
Sea-Bed Disputes Chamber, 29, 31, 35, 36, 42
Seabed Institutions, 29
seal, 300, 306, 307
Secretariat, 29, 31, 245, 305
Self-reporting, 182
shipping, 2, 6, 7, 10, 11, 13, 65, 83, 88, 92, 95, 97, 98, 100, 101, 102, 103, 178, 189, 192, 204, 209, 210, 219, 220, 222, 227, 229, 265, 275, 299
SHIPPING, 209
shipping and strategy, 11
shipping geography, 7
short sea passenger traffic, 102
smuggling, 256, 272, 273
socialist countries, 82, 92
South Korea, 15, 18, 70, 92, 93, 94, 96, 98, 107, 253, 256, 257, 258, 259, 260, 261, 262, 264, 265, 266, 267, 269, 270, 271, 276, 277, 278, 279, 281
Southern Bluefin Tuna, 52, 56, 57, 58, 60, 61
Southern Ocean, 4, 297, 298, 300, 301, 302, 303, 304, 305, 306, 307, 308, 309, 310, 311
SOUTHERN OCEAN ENVIRONMENT, 298
SOUTHERN OCEAN MINING, 307
Soviet Union, 64, 105, 107, 114, 116, 117, 120, 123, 124, 125, 128, 129, 132, 134, 136, 215, 221, 236, 237, 238, 274, 300
SPAs, 207
spatial analysis, 7, 20
SPATIAL RESOURCE, 284
Special Area, 187, 210, 309
Special Areas of Conservation, 207
Special Chambers, 36
Special Protection Areas, 207

species, 135, 146, 147, 148, 149, 154, 156, 161, 170, 171, 186, 199, 200, 201, 202, 203, 206, 207, 209, 214, 217, 218, 223, 224, 225, 226, 228, 235, 243, 257, 262, 263, 265, 288, 289, 290, 291, 300, 305, 307, 308
state practice, 68, 69, 71, 274
State Practice in Resolving Disputes, 49
stocks, 26, 40, 57, 145, 146, 147, 148, 154, 155, 156, 157, 199, 200, 201, 202, 203, 204, 210, 217, 222, 223, 224, 225, 226, 228, 262, 263, 276, 277, 288, 289, 300, 304, 306, 308
straddling stocks, 156, 225
straight baselines, 50, 71, 73
STRAITS, 65
strategic interests, 7
subsea completions, 127
Suez Canal, 216, 231, 236, 237, 242
sulphur tankers, 99
Summary Procedure Chamber, 36
supply, 54, 78, 87, 99, 107, 108, 110, 117, 134, 139, 144, 153, 163, 169, 170, 212, 217, 220
sustainability, 16, 157, 162, 163, 207, 268, 294
Sustainability, 16, 18, 155, 230
sustainable development, 20, 22, 159, 179, 192, 196, 246, 251
systems approaches, 7

T

TAC, 223, 224, 230, 288, 289
technical management, 18
TECHNOLOGICAL DEVELOPMENTS, 126
TERRITORIAL SEA, 64, 301
TEU, 95, 96, 101, 102
The High Seas, 156, 225
The NATURE OF THE MILLENNIUM, 14
the UN Conference on Environment and Development, 179
Theoretical Foundations, 78
toothfish, 299, 306
total allowable catch, 222, 223, 224, 288
Total Quality Management, 182
tourism, 20, 23, 134, 161, 162, 163, 164, 165, 166, 167, 168, 169, 170, 171, 172, 173, 186, 189, 200, 244, 290
TOURISM, 161, 164, 167, 290

Tourism and recreational uses, 244
trade, 1, 2, 5, 8, 9, 10, 11, 78, 80, 82, 84, 85, 87, 89, 96, 97, 100, 103, 107, 153, 211, 215, 219, 220, 221, 236, 237, 263, 270
TRADITIONAL SOCIETY, 8
trampships, 95
transhipment routes, 242
transit routes, 242
transnational crime at sea, 256
transportation, 137, 236, 242, 243, 284, 285, 286, 288
typology of nations, 82

U

ULCCs, 95, 98
Ultra Large Crude Carriers, 95
UN Conference on Environment and Development, 202
UN Conference on the Law of the Sea, 29, 63, 64
UN Continental Shelf Convention, 66
UN Development Programme, 241
UNCED, 157, 179, 182, 186, 202, 245, 249
UNCLOS, 2, 25, 26, 27, 28, 29, 30, 31, 32, 33, 34, 35, 36, 38, 39, 40, 42, 43, 46, 49, 51, 57, 58, 59, 60, 61, 62, 64, 65, 67, 68, 73, 156, 182, 188, 202, 239, 262, 284, 285, 288, 302, 303
UNDP, 84, 241, 249, 266, 292
UNEP, 35, 179, 192, 193, 233, 234, 244, 247, 249, 250, 251, 266, 267, 280, 292
UNICPO, 34, 35
United Nations Conference on Environment and Development, 245, 249
United Nations Conference on the Human Environment, 178
United Nations Environment Programme, 135, 179, 192, 196, 234, 249, 250, 266, 280
United Nations Open-ended Informal Consultative Process on Oceans and the Law of the Sea, 34
United States, 15, 26, 27, 36, 46, 64, 66, 67, 69, 73, 76, 78, 80, 84, 87, 92, 129, 143, 164, 166, 189, 194, 220, 221, 224, 225, 226, 228, 229, 256, 257, 271, 272, 273, 274, 277
unmanned structures, 127
urban areas, 103, 164, 168, 177
URBAN INDUSTRIAL SOCIETY, 11

Urban Wastewater Directive, 181
Urban Waste-Water Directive, 180
US Ocean Dumping Act, 189

V

VLCCs, 95, 98

W

WASTE CHARACTERISATION, 176
waste disposal, 7, 11, 12, 13, 20, 23, 175, 176, 177, 178, 179, 180, 181, 182, 183, 184, 187, 188, 189, 190, 194, 196, 201
waste management, 175, 178, 179, 182, 189, 190
Waste recycling, recovery and reuse, 180
Waste reduction and minimisation, 179
Waste treatment and disposal, 180

WESTPAC, 266
wilderness experiences, 168, 169
wine tankers, 99
Working Group for the Western Pacific, 266
World Bank, 135, 159, 196, 247, 248, 267
World Commission on Environment and Development, 179, 192
World Conservation Strategy, 179, 192
World Oceans Day, 211
World Tourism Organisation, 161, 164, 172
World War I, 220
World War II, 66, 95, 221, 224, 288, 300

Y

Yellow Sea, 260, 261, 262, 265, 266, 267, 277, 278, 279

The GeoJournal Library

1. B. Currey and G. Hugo (eds.): *Famine as Geographical Phenomenon.* 1984
ISBN 90-277-1762-1
2. S.H.U. Bowie, F.R.S. and I. Thornton (eds.): *Environmental Geochemistry and Health.* Report of the Royal Society's British National Committee for Problems of the Environment. 1985
ISBN 90-277-1879-2
3. L.A. Kosiński and K.M. Elahi (eds.): *Population Redistribution and Development in South Asia.* 1985
ISBN 90-277-1938-1
4. Y. Gradus (ed.): *Desert Development.* Man and Technology in Sparselands. 1985
ISBN 90-277-2043-6
5. F.J. Calzonetti and B.D. Solomon (eds.): *Geographical Dimensions of Energy.* 1985
ISBN 90-277-2061-4
6. J. Lundqvist, U. Lohm and M. Falkenmark (eds.): *Strategies for River Basin Management.* Environmental Integration of Land and Water in River Basin. 1985
ISBN 90-277-2111-4
7. A. Rogers and F.J. Willekens (eds.): *Migration and Settlement.* A Multiregional Comparative Study. 1986
ISBN 90-277-2119-X
8. R. Laulajainen: *Spatial Strategies in Retailing.* 1987
ISBN 90-277-2595-0
9. T.H. Lee, H.R. Linden, D.A. Dreyfus and T. Vasko (eds.): *The Methane Age.* 1988
ISBN 90-277-2745-7
10. H.J. Walker (ed.): *Artificial Structures and Shorelines.* 1988
ISBN 90-277-2746-5
11. A. Kellerman: *Time, Space, and Society.* Geographical Societal Perspectives. 1989
ISBN 0-7923-0123-4
12. P. Fabbri (ed.): *Recreational Uses of Coastal Areas.* A Research Project of the Commission on the Coastal Environment, International Geographical Union. 1990
ISBN 0-7923-0279-6
13. L.M. Brush, M.G. Wolman and Huang Bing-Wei (eds.): *Taming the Yellow River: Silt and Floods.* Proceedings of a Bilateral Seminar on Problems in the Lower Reaches of the Yellow River, China. 1989
ISBN 0-7923-0416-0
14. J. Stillwell and H.J. Scholten (eds.): *Contemporary Research in Population Geography.* A Comparison of the United Kingdom and the Netherlands. 1990
ISBN 0-7923-0431-4
15. M.S. Kenzer (ed.): *Applied Geography.* Issues, Questions, and Concerns. 1989
ISBN 0-7923-0438-1
16. D. Nir: *Region as a Socio-environmental System.* An Introduction to a Systemic Regional Geography. 1990
ISBN 0-7923-0516-7
17. H.J. Scholten and J.C.H. Stillwell (eds.): *Geographical Information Systems for Urban and Regional Planning.* 1990
ISBN 0-7923-0793-3
18. F.M. Brouwer, A.J. Thomas and M.J. Chadwick (eds.): *Land Use Changes in Europe.* Processes of Change, Environmental Transformations and Future Patterns. 1991
ISBN 0-7923-1099-3

The GeoJournal Library

19. C.J. Campbell: *The Golden Century of Oil 1950–2050*. The Depletion of a Resource. 1991   ISBN 0-7923-1442-5
20. F.M. Dieleman and S. Musterd (eds.): *The Randstad: A Research and Policy Laboratory*. 1992   ISBN 0-7923-1649-5
21. V.I. Ilyichev and V.V. Anikiev (eds.): *Oceanic and Anthropogenic Controls of Life in the Pacific Ocean*. 1992   ISBN 0-7923-1854-4
22. A.K. Dutt and F.J. Costa (eds.): *Perspectives on Planning and Urban Development in Belgium*. 1992   ISBN 0-7923-1885-4
23. J. Portugali: *Implicate Relations*. Society and Space in the Israeli-Palestinian Conflict. 1993   ISBN 0-7923-1886-2
24. M.J.C. de Lepper, H.J. Scholten and R.M. Stern (eds.): *The Added Value of Geographical Information Systems in Public and Environmental Health*. 1995
ISBN 0-7923-1887-0
25. J.P. Dorian, P.A. Minakir and V.T. Borisovich (eds.): *CIS Energy and Minerals Development*. Prospects, Problems and Opportunities for International Cooperation. 1993   ISBN 0-7923-2323-8
26. P.P. Wong (ed.): *Tourism vs Environment: The Case for Coastal Areas*. 1993
ISBN 0-7923-2404-8
27. G.B. Benko and U. Strohmayer (eds.): *Geography, History and Social Sciences*. 1995
ISBN 0-7923-2543-5
28. A. Faludi and A. der Valk: *Rule and Order. Dutch Planning Doctrine in the Twentieth Century*. 1994   ISBN 0-7923-2619-9
29. B.C. Hewitson and R.G. Crane (eds.): *Neural Nets: Applications in Geography*. 1994
ISBN 0-7923-2746-2
30. A.K. Dutt, F.J. Costa, S. Aggarwal and A.G. Noble (eds.): *The Asian City: Processes of Development, Characteristics and Planning*. 1994   ISBN 0-7923-3135-4
31. R. Laulajainen and H.A. Stafford: *Corporate Geography*. Business Location Principles and Cases. 1995   ISBN 0-7923-3326-8
32. J. Portugali (ed.): *The Construction of Cognitive Maps*. 1996   ISBN 0-7923-3949-5
33. E. Biagini: *Northern Ireland and Beyond*. Social and Geographical Issues. 1996
ISBN 0-7923-4046-9
34. A.K. Dutt (ed.): *Southeast Asia: A Ten Nation Region*. 1996   ISBN 0-7923-4171-6
35. J. Settele, C. Margules, P. Poschlod and K. Henle (eds.): *Species Survival in Fragmented Landscapes*. 1996   ISBN 0-7923-4239-9
36. M. Yoshino, M. Domrös, A. Douguédroit, J. Paszynski and L.D. Nkemdirim (eds.): *Climates and Societies – A Climatological Perspective*. A Contribution on Global Change and Related Problems Prepared by the Commission on Climatology of the International Geographical Union. 1997   ISBN 0-7923-4324-7
37. D. Borri, A. Khakee and C. Lacirignola (eds.): *Evaluating Theory-Practice and Urban-Rural Interplay in Planning*. 1997   ISBN 0-7923-4326-3

The GeoJournal Library

38. J.A.A. Jones, C. Liu, M-K. Woo and H-T. Kung (eds.): *Regional Hydrological Response to Climate Change.* 1996   ISBN 0-7923-4329-8
39. R. Lloyd: *Spatial Cognition.* Geographic Environments. 1997   ISBN 0-7923-4375-1
40. I. Lyons Murphy: *The Danube: A River Basin in Transition.* 1997 ISBN 0-7923-4558-4
41. H.J. Bruins and H. Lithwick (eds.): *The Arid Frontier.* Interactive Management of Environment and Development. 1998   ISBN 0-7923-4227-5
42. G. Lipshitz: *Country on the Move: Migration to and within Israel, 1948–1995.* 1998
        ISBN 0-7923-4850-8
43. S. Musterd, W. Ostendorf and M. Breebaart: *Multi-Ethnic Metropolis: Patterns and Policies.* 1998   ISBN 0-7923-4854-0
44. B.K. Maloney (ed.): *Human Activities and the Tropical Rainforest.* Past, Present and Possible Future. 1998   ISBN 0-7923-4858-3
45. H. van der Wusten (ed.): *The Urban University and its Identity.* Roots, Location, Roles. 1998   ISBN 0-7923-4870-2
46. J. Kalvoda and C.L. Rosenfeld (eds.): *Geomorphological Hazards in High Mountain Areas.* 1998   ISBN 0-7923-4961-X
47. N. Lichfield, A. Barbanente, D. Borri, A. Khakee and A. Prat (eds.): *Evaluation in Planning.* Facing the Challenge of Complexity. 1998   ISBN 0-7923-4870-2
48. A. Buttimer and L. Wallin (eds.): *Nature and Identity in Cross-Cultural Perspective.* 1999   ISBN 0-7923-5651-9
49. A. Vallega: *Fundamentals of Integrated Coastal Management.* 1999
        ISBN 0-7923-5875-9
50. D. Rumley: *The Geopolitics of Australia's Regional Relations.* 1999
        ISBN 0-7923-5916-X
51. H. Stevens: *The Institutional Position of Seaports.* An International Comparison. 1999
        ISBN 0-7923-5979-8
52. H. Lithwick and Y. Gradus (eds.): *Developing Frontier Cities.* Global Perspectives – Regional Contexts. 2000   ISBN 0-7923-6061-3
53. H. Knippenberg and J. Markusse (eds.): *Nationalising and Denationalising European Border Regions, 1800–2000.* Views from Geography and History. 2000
        ISBN 0-7923-6066-4
54. R. Gerber and G.K. Chuan (eds.): *Fieldwork in Geography: Reflections, Perspectives and Actions.* 2000   ISBN 0-7923-6329-9
55. M. Dobry (ed.): *Democratic and Capitalist Transitions in Eastern Europe.* Lessons for the Social Sciences. 2000   ISBN 0-7923-6331-0
56. Y. Murayama: *Japanese Urban System.* 2000   ISBN 0-7923-6600-X
57. D. Zheng, Q. Zhang and S. Wu (eds.): *Mountain Geoecology and Sustainable Development of the Tibetan Plateau.* 2000   ISBN 0-7923-6688-3

The GeoJournal Library

58. A.J. Conacher (ed.): *Land Degradation.* Papers selected from Contributions to the Sixth Meeting of the International Geographical Union's Commission on Land Degradation and Desertification, Perth, Western Australia, 20–28 September 1999. 2001
ISBN 0-7923-6770-7
59. S. Conti and P. Giaccaria: *Local Development and Competitiveness.* 2001
ISBN 0-7923-6829-0
60. P. Miao (ed.): *Public Places in Asia Pacific Cities.* Current Issues and Strategies. 2001
ISBN 0-7923-7083-X
61. N. Maiellaro (ed.): *Towards Sustainable Buiding.* 2001   ISBN 1-4020-0012-X
62. G.S. Dunbar (ed.): *Geography: Discipline, Profession and Subject since 1870.* An International Survey. 2001   ISBN 1-4020-0019-7
63. J. Stillwell and H.J. Scholten (eds.): *Land Use Simulation for Europe.* 2001
ISBN 1-4020-0213-0
64. P. Doyle and M.R. Bennett (eds.): *Fields of Battle.* Terrain in Military History. 2002
ISBN 1-4020-0433-8
65. C.M. Hall and A.M. Williams (eds.): *Tourism and Migration.* New Relationships between Production and Consumption. 2002   ISBN 1-4020-0454-0
66. I.R. Bowler, C.R. Bryant and C. Cocklin (eds.): *The Sustainability of Rural Systems.* Geographical Interpretations. 2002   ISBN 1-4020-0513-X
67. O. Yiftachel, J. Little, D. Hedgcock and I. Alexander (eds.): *The Power of Planning.* Spaces of Control and Transformation. 2001   ISBN Hb; 1-4020-0533-4
ISBN Pb; 1-4020-0534-2
68. K. Hewitt, M.-L. Byrne, M. English and G. Young (eds.): *Landscapes of Transition.* Landform Assemblages and Transformations in Cold Regions. 2002
ISBN 1-4020-0663-2
69. M. Romanos and C. Auffrey (eds.): *Managing Intermediate Size Cities.* Sustainable Development in a Growth Region of Thailand. 2002   ISBN 1-4020-0818-X
70. B. Boots, A. Okabe and R. Thomas (eds.): *Modelling Geographical Systems.* Statistical and Computational Applications. 2003   ISBN 1-4020-0821-X
71. R. Gerber and M. Williams (eds.): *Geography, Culture and Education.* 2002
ISBN 1-4020-0878-3
72. D. Felsenstein, E.W. Schamp and A. Shachar (eds.): *Emerging Nodes in the Global Economy: Frankfurt and Tel Aviv Compared.* 2002   ISBN 1-4020-0924-0
73. R. Gerber (ed.): *International Handbook on Geographical Education.* 2003
ISBN 1-4020-1019-2
74. M. de Jong, K. Lalenis and V. Mamadouh (eds.): *The Theory and Practice of Institutional Transplantation.* Experiences with the Transfer of Policy Institutions. 2002
ISBN 1-4020-1049-4
75. A.K. Dutt, A.G. Noble, G. Venugopal and S. Subbiah (eds.): *Challenges to Asian Urbanization in the 21st Century.* 2003   ISBN 1-4020-1576-3